Climate and Catastrophe
in Cuba and the Atlantic World
in the Age of Revolution

ENVISIONING CUBA

Louis A. Pérez Jr., editor

SHERRY JOHNSON

Climate and Catastrophe

in Cuba and the Atlantic World
in the Age of Revolution

The University of North Carolina Press | Chapel Hill

Publication of this book was supported by the Rachel Carson Center for Environment and Society (Munich).

The paper in this book meets the guidelines for permanence and durability of the Committee on Production Guidelines for Book Longevity of the Council on Library Resources.

The University of North Carolina Press has been a member of the Green Press Initiative since 2003.

Library of Congress Cataloging-in-Publication Data
Johnson, Sherry, 1949–
Climate and catastrophe in Cuba and the Atlantic world in the age of revolution / Sherry Johnson.
p. cm.—(Envisioning Cuba)
Includes bibliographical references and index.
ISBN 978-0-8078-3493-0 (cloth : alk. paper)
1. Cuba—Climate—History—18th century. 2. Climatic extremes—Social aspects—Cuba—History—18th century. 3. Climatic extremes—Political aspects—Cuba—History—18th century. 4. Disasters—Cuba—History—18th century. 5. Social change—Cuba—History—18th century. 6. Cuba—Politics and government—18th century. 7. Cuba—Social conditions—18th century. 8. Cuba—History—To 1810. 9. Caribbean Area—History—To 1810. 10. Latin America—History—To 1830. I. Title.
QC987.C8J64 2011   363.34'9209729109033—dc23   2011033282

Portions of this book were published, in somewhat different form, as "The St. Augustine Hurricane of 1811: Disaster and the Question of Political Unrest on the Florida Frontier," *Florida Historical Quarterly* 84 (Summer 2005): 28–56; "Climate, Community, and Commerce, among Florida, Cuba, and the Atlantic World, 1784–1800," *Florida Historical Quarterly* 80 (Spring 2002): 455–82; "El Niño, Environmental Crisis, and the Emergence of Alternative Markets in the Hispanic Caribbean, 1760s–1770s," *William and Mary Quarterly*, 3rd ser., 62 (July 2005): 365–410. Used by permission.

15 14 13 12 11   5 4 3 2 1

For Mom, Dad, and LPJ

CONTENTS

Figures

Maps

ACKNOWLEDGMENTS

· ·

During the completion of this book, I have incurred many debts, both personal and professional. I am grateful for the funding I received from several institutions, including the Lydia Cabrera Award Committee of the Conference on Latin American History; the Jay I. Kislak Foundation, Inc.; the National Endowment for the Humanities Extending the Reach Research Grant; the Center for Latin American Studies and the University of Florida Libraries at the University of Florida; the Library Company of Philadelphia, Program in Early American Economy and Society; the Rachel Carson Center for Environmental Studies Fellowship, Deutsches Museum, Ludwig-Maximilians-Universität, Munich, Germany; the Spanish Ministerio de Educacción y Ciencia (award HUM2006–00454/HIST); and the Historic St. Augustine Research Institute at Flagler College, Historic St. Augustine Research Foundation. I further acknowledge the generous assistance of the John D. and Catherine T. MacArthur Foundation, the Ford Foundation, and the Christopher Reynolds Foundation via their awards to the Cuban Research Institute at Florida International University, and for the continuing support from the Department of History, the Latin American and Caribbean Center, and the Cuban Research Institute, the College of Arts and Sciences, the Department of Sponsored Research, and the Florida International Foundation/Provost's Office, at Florida International University.

I am also indebted to the various archival repositories and institutions and their professional staff members—all of which made this book possible—including the Archivo General de Indias; the Archivo General de Simancas; the Biblioteca Nacional de España; the Archivo Histórico de la Nación; the Archivo Provincial de Cádiz; the Archivo Nacional de Cuba; the Biblioteca Nacional de Cuba José Martí; the Archbishop of Havana's Archives; the Archive in S.M.I. Catedral; the Archivo Provincial de Pinar del Río; the Archivo Provincial de Santa Clara; the Archivo Municipal de Santiago de Cuba; the Houghton Library, Harvard University; the Massachusetts Historical Society; the Peabody Essex Museum; the Historical Society of Pennsylvania; the Library Company of Philadelphia; the Howard-Tildon Memorial Library, Tulane University; the Library of

Congress; the Diocese of St. Augustine Catholic Center; the P. K. Yonge Library and the Latin American Collection at the University of Florida; the Cuban Collection at the University of Miami; and Special Collections of Green Library at Florida International University.

I am particularly grateful to the many professionals who over the years have encouraged my interest in Cuba. In Spain, José Hernández Palomo, G. Douglas Inglis, Manuel Salvador Vásquez, María Dolores Gonzáles Ripoll Navarro, and Consuelo Naranjo offered helpful advice. Scholars of Cuba who have generously provided guidance from the inception of this project include Allan J. Kuethe, Franklin W. Knight, John R. McNeill, Linda K. Salvucci, Jean Stubbs, and K. Lynn Stoner. This work could not have been completed without the cooperation and friendship of many scholars in Cuba, including María Carmen Bárcia, Jorge Ibarra, Fé Iglesias, Gloria García, Mercedes García, René González, Sergia Martínez, Olga Portuondo, Pedro M. Pruna, and Oscar Zanetti. Christof Mauch and Helmuth Trischler deserve special thanks for creating a "scholarly Shangri-La" at the Rachel Carson Center in Munich, where the final draft of the manuscript was completed, and, as always, every scholar of Cuba is in debt to Louis A. Pérez Jr. To all of these people, I extend my sincerest appreciation.

Former colleagues at the University of Florida to whom I am indebted include Bruce Chappell, James E. Cusick, Keith Manuel, and John Ingram. Special thanks are extended to Latin Americanists David A. Bushnell, Neill Macaulay, Murdo J. MacLeod, Michael E. Moseley, and Robin Lauriault, who over the years have contributed helpful ideas along with uncomfortable questions to help me think through the structure of this book. I owe special debts to César Caviedes, Richard S. Olson, Robert H. Claxton, and Victor Bulmer-Thomas for their solid advice, which helped improve the manuscript immensely. My colleagues at Florida International University, Mark D. Szuchman, N. David Cook, Victor Uribe-Urán, Lisandro Pérez, Uva de Aragón, Damián Fernández, Darden A. Pyron, and Gwyn Davies, have been a continuous source of support. I am privileged to have studied and worked with these friends and scholars. Without the friendship of Bill Waller, Karen Waller, Lynne Guitar, Ron Lewis, Nancy Macaulay, Alexandra Cook, Charlotte A. Cosner, Karen Y. Morrison, Ian Maynard, and Kathy Bauman, life as an academic would have been difficult, if not impossible.

This book is dedicated to my parents, Edgar W. and Marianne Shelly Johnson, *floridanos* from 1933 through the 1950s. Long before satellite images gave early warning of an approaching hurricane, they, like many families in the tropics, instilled in me the resilience to weather many storms. This book is dedicated to their memory.

Climate and Catastrophe
in Cuba and the Atlantic World
in the Age of Revolution

CHAPTER ONE

Cursed by Nature

C LIMATE CHANGE! Global Warming! El Niño and La Niña! These phrases, now part of our daily vocabulary, stir emotions and prompt reactions ranging from fear, to anger, to a feeling of helplessness in the face of impending disaster. For the past several years, the Caribbean, the southeastern United States, and the Gulf Coast have endured repeated hurricane strikes, while the Pacific region has suffered through alternating periods of drought-induced wildfires and torrential downpours. Governments are warned to be prepared for an imminent period of weather-induced environmental crisis caused by a warming cycle in the earth's climate.

Decades of research have made "climate change" household words, but until now the social sciences have rarely utilized scientific discoveries to understand the connections among climate, catastrophe, environmental crisis, and historical change. Drawing inspiration from hard science and contemporary issues, this book will establish that the current phase of climate-induced stress is not unique and that a similar cycle, a fifty-year warm anomaly, occurred during the last five decades of the eighteenth century. In addition, historical climatology demonstrates that in the period under study (1748–1804) barely a year went by when the world did not experience the effects of an El Niño or La Niña cycle, episodes of severe, prolonged drought counterbalanced by hurricane activity in the Atlantic basin. Such scientific facts have little value, however, unless the consequences of environmental stress can be shown to coincide with a known historical narrative.

This book will establish that nexus of science and social science by demonstrating correlations among the late-eighteenth-century climate anomaly, the onset of the El Niño or La Niña cycle, and historical processes. It will argue that—not coincidentally—these phenomena coincided with one of the most critical periods in history, termed the Age of

Revolution.[1] From the mid-eighteenth century through the first decades
of the nineteenth century, the Atlantic world from Boston to Barbados and
beyond underwent political upheavals culminating in the United States'
War of Independence, the French Revolution, and the Haitian Revolution.[2] This book builds upon the foundations laid by hard science and
rests on the scientific data provided by research in historical climatology,
then incorporates the techniques and theories from the field of disaster
studies. It accepts the scientific evidence of prolonged and severe weather
sequences in the latter decades of the eighteenth century, what one of the
leading scholars terms "spasmodic climatic interludes."[3] Borrowing from
multidisciplinary work in economics, sociology, political science, and international relations, it will show how disaster in the Caribbean generated
both positive and negative consequences throughout the Atlantic basin.
The timeline of disaster placed alongside a chronology of political, economic, and social events demonstrates causal relationships between scientific facts and historical processes.[4] This juxtaposition makes clear that
processes and events that traditionally have been attributed to political,
economic, and/or social forces were impacted by, and often caused by,
weather-induced environmental crises.

The Science

The science that underpins this study is based on historical climatology,
particularly studies of fluctuating temperature cycles and climate change.
Beginning with a handful of studies in the 1990s, teams of researchers
all over the globe contributed the results of their individual projects to
an ever-growing body of knowledge about temperature fluctuations occurring over several millennia.[5] These collective efforts have established
beyond a reasonable doubt that the temperature of the planet varies,
sometimes reaching extremes. One such extreme occurred from the mid-
1400s to approximately 1850, during which the earth's climate experienced
cooler-than-normal temperatures, a cycle that is known as the Little Ice
Age.[6] Around 1850, the Little Ice Age began to wane, and the earth entered into a period of warmer temperatures, which the planet continues
to experience to this day. Before that happened, however, the cool cycle
was punctuated by a fifty-year warm anomaly that began around 1750 and
lasted until about 1800.[7] The importance of this warm period in the area

under study was that its effects can be correlated with severe weather events that exhibit the characteristics of El Niño/La Niña sequences.[8]

Until the winter of 1983, when devastating floods hit northern Peru and made the international news, the extreme weather event known as an El Niño/Southern Oscillation (ENSO) was unknown to all but a small group of scientists and geographers. Because the Peruvian flood occurred during the Christmas season, the event was named for the Christ child, El Niño, and the ENSO cycle now is universally recognized by its popular name. The disaster came as no surprise to the handful of climatologists and geographers who for several decades had suspected that the El Niño was a recurring phenomenon.[9] Over time, as climatologists studied global weather systems, they learned that such systems are not just recurring but even are interrelated. In turn, they coined a term, "teleconnections," to explain how a weather phenomenon in one area is mirrored by certain characteristics in another.[10]

For reasons that are still being debated, an El Niño begins with the warming of tropical Pacific waters, which leads to torrential rainfall along the coast of the Americas. Scientists quickly realized that the consequences of an El Niño event are not limited to the Pacific region. While the Pacific Coast suffered from too much rainfall, other tropical zones, including Mexico, the Caribbean, and sub-Saharan Africa, were hit by severe, prolonged drought. Worse still, the end of an El Niño event did not mean the end of environmental stress. Researchers further discovered that El Niño has a malevolent twin sister, La Niña, who accompanies her destructive brother by causing increased hurricane activity in the Caribbean and the Atlantic basin. In some ways, the evil twin is worse, because La Niña impacts tropical regions already stressed from drought.[11]

A primary goal of this study, then, is to establish whether the signature characteristics of the El Niño and La Niña cycles impacted the Caribbean basin from about 1750 through 1804. The existing information, chronological lists of hurricanes for the Caribbean and for the United States compiled more than four decades ago, proved to be incomplete, so documentary repositories throughout the Atlantic world were scoured for evidence to establish that the signature consequences of the destructive twins affected Cuba and its neighboring regions such as Puerto Rico, Louisiana, Florida, and the Caribbean littoral. Yet simply adding to the existing chronology of hurricanes was not enough to indicate the existence of the El Niño sequence. To demonstrate convincingly that the region experienced climate-induced environmental stress, this study also needed to establish the close temporal correlation between drought sequences and hurricane strikes (see Appendix).

Once the wealth of evidence particular to the Caribbean was collected and analyzed, the next goal was to determine if the Caribbean experience could be compared and contrasted to historical weather patterns. Fortunately, the most comprehensive study about the occurrence of historical El Niño and La Niña cycles to date was published contemporaneously with the completion of this book. Authors Jöelle Gergis and Anthony Fowler synthesized data identifying the ENSO cycle worldwide in an article that establishes the frequency and intensity of El Niños and La Niñas for nearly five centuries, from 1525 through 2002. The results of their study add an important comparative aspect as well as confirm the data for the Caribbean. During the fifty-eight years covered in this book (1748–1804), eighteen El Niño cycles were identified; of these, two were very severe events (1770 and 1791) and two were severe events (1799 and 1803). The data for the La Niña, the cycle that impacts the Caribbean by creating conditions favorable for hurricane formation and torrential rainfall, however, are even more significant. Thirty-one La Niña events occurred from 1748 through 1804, and of these, four were very severe and ten were severe.[12] Only eleven years were free of an El Niño or La Niña event, and except for one brief interlude (1774–75), these were nonconsecutive years.[13] Simply put, only once in fifty-eight years did the residents of the Hispanic Caribbean enjoy two consecutive years that were free of the weather hazards of El Niño or La Niña. Their more common experience was to endure a hurricane or similar wet weather every year, punctuated by winters characterized by long periods of severe drought.

Gergis and Fowler's article permits one of the most frequent questions posed by this study to be addressed: Did the latter half of the eighteenth century, in fact, experience unusual weather cycles, or were the droughts and hurricanes simply routine challenges of life? During the colonial period in the Americas (1525–1808), the authors identify three distinct phases. The first, beginning in the middle of the sixteenth century, ran through the early seventeenth century (1520s through 1660s) and was characterized by the "most sustained period of La Niña activity." During the 1650s through 1720, the frequency of the La Niña declined, although there were three instances of very severe La Niña events. Beginning in the 1720s, the frequency of the La Niña sequence again increased, lasting through the period under study in this book. The data show a maximum duration of thirty-six consecutive years of the La Niña between 1738 and 1773, and, significantly, for the starting point of this study, the 1750s, "La Niña dominated 90% of the decade."[14] In addition, Gergis and Fowler's conclusions come with a high degree of confidence. Their evidence (proxy data) came from a variety of scientific and documentary sources, such as dendrochronology, ice core samples, cave samples (for example, stalagmites and stalactites), and lake bed sediments, and were weighted using statistical analyses to confirm the strength of the evidence.[15] This coincidental publication of the results of a decade of scholarship brings hard science to bear upon the documentary data and gives this study an even greater degree of confidence.

The Scholarship

The Atlantic hurricane belt, the zone where residents are in the greatest danger, runs from the outer islands of the West Indies stretching along the northern littoral of South America to the Florida peninsula. Vulnerable populations along the Gulf of Mexico, the Yucatán peninsula, Veracruz, the Texas coast, Louisiana, Mississippi, Alabama, and the vast Florida coastline down to the Florida Keys learn to keep a wary eye on the weather from June through the end of November.[16] The East Coast of the United States is also at risk, and although most hurricanes strike south of Virginia, on rare occasions northern states as far north as New England can feel nature's fury. Hurricane strikes are dramatic events, but these same areas are equally vulnerable to the quietly debilitating consequences of severe, prolonged drought.[17] Given the reality of living in the Atlantic hurricane belt, it is surprising that the natural setting has been underutilized as a theoretical framework to understand the signal events in Atlantic world history.

Any study such as this must begin within the established principles of environmental history, especially humans' dynamic relationship with their environment.[18] Fundamental themes, such as how humans have been dependent upon their natural surroundings and the limitations such surroundings place upon human activity, are relevant to this research.[19] During the period in question, historical attitudes toward nature and the natural world changed over time, as did governmental approaches toward nature and its consequences.[20] Nonetheless, the intellectual inspiration for this book, although related to traditional environmental history, is distinct from it.[21] The logic of the argument is turned around and evaluates how the foremost climate phenomena of the region—hurricanes and drought—brought catastrophe and, in turn, affected humans' economic, political, social, and cultural behavior.

Until Hurricane Andrew devastated South Florida in 1992, disaster received scant attention as a conceptual tool to establish and evaluate historical processes. Andrew brought home to scholars the effect of disaster and its aftermath in a most painful fashion. Disaster as a catalytic event, in and of itself, became the guiding principle in a trend, as researchers began to examine catastrophes, such as the earthquake in Nicaragua in 1772, the Mexico City earthquake in 1985, and Hurricane Mitch in Central America in 1998. What emerged was a growing body of multidisciplinary scholarship that takes all sorts of disasters as its starting point and examines the consequences of such cataclysmic events.[22]

This scholarship provides the many theoretical foundations for this book, such as seminal works in political science that demonstrate that disaster can be a force behind political change but that disasters do not necessarily have to become political. The authorities' behavior in the aftermath of disaster determines whether the population will react in a positive or a negative way, thus making the disaster the trigger that causes a "critical juncture" in political events.[23] The concepts implicit in critical juncture theory rest upon the idea of contingency, that is, acknowledging that many potential paths could be chosen, mostly leading to different outcomes—some positive and some negative. One of the most important themes of this book is that the choices made by royal officials played a fundamental role in the disaster's outcome. This is especially true when dealing with the way in which government authorities provided relief and/or secured provisions in a post-hurricane situation. Another important aspect is the degree to which various branches of government cooperated or competed with each other. Although one of the fundamental tenets of the Enlightenment was to reduce the power of the church, pragmatism and cooperation were the rule during the first four decades of this study. When that policy of cooperation was abandoned in the 1790s, the region was riven by unrest, which, in turn, was exacerbated by continued periods of environmental stress.

A second fundamental tenet, also borrowed from interdisciplinary research, is the domino effect of disaster—that is, how crisis in one area creates a domino or ripple effect in other areas. The concepts are simple. While some areas suffer after a disaster, others reap the benefits of scarcity and shortages. Cuba is the geographic center of this analysis because it was the epicenter of Spanish rule in the Caribbean. The domino effect was evident in Cuba's subordinate colonies, Puerto Rico, Florida, and Louisiana, and even outside the Spanish empire into North America and other nations' islands of the Caribbean.[24] During extended periods of drought and after every hurricane strike, commercial restrictions were temporarily abandoned as foreign traders rushed in to provision the affected areas. The domino effect, thus, lends itself well to the principles of transnationality, an analytical tool that deemphasizes artificially created political boundaries and concentrates on forces (social movements, kinship networks, economic connections) that can cross arbitrarily created lines of demarcation. Obviously so much history is framed in national or imperial terms, but climate and catastrophe do not recognize national boundaries.

Sociology, especially studies of the social chaos after Hurricane Andrew, lends yet additional conceptual tools.[25] Especially useful are studies of post-disaster community self-organizing efforts and the leveling effect of disaster.[26] When an entire community was threatened, social boundaries were, of necessity, set aside as rescue and recovery efforts took priority over the niceties of social ordering. Survivors clinging to the wreckage of their ruined houses cared little for the social status or the color of the arm that reached down to pluck them from the raging current. In other instances, activities such as smuggling that were unquestionably illegal were tolerated and even encouraged when catastrophe threatened. Just as important, the perpetrators were rarely prosecuted for their actions. To the contrary, many were hailed as folk heroes for risking prosecution to provide for the community's desperate needs. Although social boundaries would be restored as life returned to normal, the behavior of particular persons during the emergency never left the community's collective memory.[27] Bravery and decisive positive decisions were celebrated in the form of songs, folktales, and laudatory poetry, while cowardice and impotence were brought to the community's attention in *pasquines* (lampoons) posted in public places—both to be remembered long after the emergency had passed.

Using scientific evidence and interdisciplinary theory to explain historical processes is undeniably appealing, yet the temptation to attribute change over time to disasters and their aftermath must be tempered with common sense. Recently, one of the leading scholars of historic hurricanes pointed out that the Caribbean experienced at least one hurricane every year but that not every hurricane produced permanent change.[28] Indeed, during the historical period in the Americas, there are numerous examples in which an area suffered a severe storm, the population recovered, and life returned to normal as it had been before the storm. Such caution is echoed by members of the scientific community, who express appropriate skepticism about relying upon historical accounts that make claims like "this was the most severe winter in living history." Climatologists are understandably wary of documentary sources that provide data that are not verifiable by scientific measurement.[29] To avoid the pitfall of assigning too much significance to disasters, this study will establish whether or not the hurricane/drought sequence produced "legacies," that is, whether permanent change over time can be attributed wholly or in part to weather-related factors.[30]

In the end, readers will be asked to set aside a recognized, indeed, an instinctive chronology. Atlantic world history is unconsciously predicated on an accepted timeline generated by events in British or American history, but this study presents an alternate chronology based on environmental and weather events. Events such as the Seven Years' War (1756–63), the siege, capture, and occupation of Havana (1762–63), Spain's sequential declarations of free trade beginning in 1765, the American Revolution, the French Revolution, and the Haitian Revolution will be familiar to informed readers of Caribbean history, but the signal events of the Age of Revolution are seen as consequences of environmental crisis rather than as the reasons for historic change (see Appendix).

The Setting

From the time of Cuba's first conquest and original settlement in 1511 through the mid-eighteenth century, Spanish colonial institutions had changed but little.[31] The island fell under the jurisdiction of the viceroyalty of New Spain, and the senior royal official on the island was the governor, who lived in Havana and also held the title of captain general, indicative of his dual administrative/military function. Havana's closest rival in terms of population, Santiago de Cuba, located at the other end of the island, had been important in the early years of contact, but its influence had diminished as that of the western city had risen. In 1607, the island had been divided administratively into two jurisdictions, with the governor of Santiago de Cuba becoming subordinate to the captain general in Havana.[32] Although the Bishopric of Cuba had been established in Santiago in 1522, the bishop resided in Havana and rarely made the arduous journey to his official seat. The primary court of appeal was the Audiencia de Santo Domingo, located on the neighboring island, Hispaniola, to the east.[33] Prior to 1763, the island was largely underpopulated, with the majority of inhabitants concentrated in the two primary cities, Havana and Santiago de Cuba.[34] By the mid-sixteenth century, the indigenous population had virtually disappeared, due to the combined effects of assimilation, conquest, disease, and overwork.[35] Until the mid-nineteenth century, the European-descended population predominated over persons of color, accompanied by an unbalanced sex ratio in both the white and black populations, with men outnumbering women.[36]

Cuba's primary function within the Spanish empire was defense. After a series of raids by French, English, and Dutch interlopers in the 1550s and 1560s, Spain transformed the Caribbean basin by erecting a string of fortifications designed to repel challenges to her dominance.[37] The Spanish navy, the Armada de Barlovento, was assigned to the area to protect trade and treasure routes, and a royal order mandating group sailings (*flota*) was issued in 1561.[38] Because of its strategic location and capacious, protected harbor, Havana became the nexus of Spanish power and was considered to be impregnable.

Defense measures cost significant amounts of money, and the economy of the island rested predominantly upon the *situado*, or military subsidy, sent from Mexico. Even before reforms were implemented in 1764—the famous Bourbon Reforms, which had wide-reaching consequences throughout the Americas—military spending and subsidiary industries fueled the Cuban economy. Enormous sums of money were pumped into Havana's economy at the expense of Mexican taxpayers.[39] Such spending paid for military salaries, construction on fortifications, and indirect expenditures, such as food for the troops. One subsidiary industry was the Real Arsenal, or royal shipyard, located outside Havana's city wall. The city had long been important for careening and provisioning ships in the *flota* and for constructing smaller ships for local commerce, but large-scale ship construction began only in the first decades of the eighteenth century. The shipyard, the "pride of Havana," had increased its importance in the 1740s when the royal monopoly (the Real Compañía de Comercio de la Habana) accepted the financial responsibility to build ships for the royal navy. The island had an abundant supply of hardwoods, which could be cut only with royal permission. Careening of ships provided employment for significant numbers of men, skilled craftsmen and day laborers alike. The revictualing of the fleet provided income for the hinterlands, and while in port, thousands of sailors needed lodging, food, and entertainment.[40]

Agricultural production for export was the second foundation of Cuba's economy. Tobacco was Cuba's primary export and was stringently regulated under Spain's mercantilistic philosophy. A crown monopoly was implemented in 1717, which purchased the crop at controlled prices at terms favorable to the crown and the company.[41] Sugar production was another agricultural enterprise that was focused on the export market.[42] Cattle raising and related industries were also important sources of income. Cattle provided fresh meat for the military garrisons and cities, and what could not be consumed immediately was salted or dried for future use or for export. Hides from slaughtered cattle were an additional commercial product, destined for Spain, to be crafted into shoes, saddles, and tack.[43] Beeswax and honey were important export commodities after hives were brought to the island by Florida refugees in 1763.[44]

One of the fundamental institutions of Cuba's economy and one that plays a central role in this study was the Real Compañía de Comercio de la Havana. This Havana-based monopoly was established in 1740 and was financed by a diverse group of investors, including its primary director, Martín de Aróstegui, several members of the Havana elite, and the Spanish royal family. Operating under mercantilistic principles, the Real Compañía performed several functions. One of its most important responsibilities was to purchase and to ship Cuba's products, primarily tobacco, back to Spain. In addition, it was charged with providing manufactured goods to Cuba. As mentioned above, the Real Compañía also accepted the responsibility of building ships for the Spanish navy, and its other obligations included promoting the immigration of Canary Island colonists and providing food for the *presidios* (military outposts) in Florida.[45]

The Real Compañía also held the exclusive privilege to import slaves into Spanish America, the *asiento*. Until 1740, this privilege rested with Great Britain, which had gained it in 1713 under the terms of the Treaty of Utrecht ending the War of Spanish Succession (1702–14). Because of the British propensity to use the *asiento* as a means to smuggle, Spain rescinded the privilege in 1739 and in retaliation formed its own monopoly company.[46] Such a decision was based upon an economic logic of the previous century. By the time the Real Compañía was fully operational, the rest of the world had already moved to commercial philosophies based upon capitalism and free trade. As early as the 1750s, the Real Compañía proved to be totally inadequate in meeting demand within an ever-changing commercial atmosphere, and in reality, it exacerbated rather than alleviated many of the problems of the island.

The institutional structures of the Spanish empire were often additional obstacles to life in the Caribbean basin. Rivalries among the great European powers, Spain, France, Holland, and Great Britain, characterized the eighteenth century, and such rivalries worsened rather than abated as the century drew to a close. Although the function of Spain's fortified cities in the Caribbean was defense, by the early eighteenth century Spanish policy makers failed to recognize the realities of international politics when making important decisions.[47] With the exception of the viceregal capitals of Mexico City and Lima, nowhere in the Americas was there a large governmental presence. Spain was noted for not having a standing army throughout its 400-year domination of its colonies—the exceptions, of course, were the primary Caribbean cities. There, under ideal circumstances, military governors appointed in Spain ruled in the name of the monarch. Their authority was upheld by contingents of European soldiers, who had no ties to the local communities. This policy reflected metropolitan desires that police functions, especially the never-ending battle against contraband, become the responsibility of royal officials stationed in rural areas.

Yet the isolation of the scattered communities throughout the Hispanic Caribbean meant that such draconian policies separating peninsular officials from the local population were unenforceable, at best. Although most of the men were outsiders, they worked in conjunction with the local *ayuntamientos*, or town councils, out of necessity. The highest-ranking officer was there to enforce the Leyes de Indias (the Laws of the Indies, that is, Spanish America), but local officials had tradition, prerogative, and most important, the realities of daily life on their side. From the time of colonization, the ayuntamientos held considerable power, including the privilege to award land grants and to set prices for the staples of the food supply.[48] The relationship between local populations and outsider officials, thus, ranged from genial cooperation to violent antagonism.[49]

Religion and religious institutions were also central to all aspects of

Spanish rule. At the midpoint of the eighteenth century, the secular philosophies of the Enlightenment were just entering the collective mentality of the Spanish elites, but such progressive ideas rarely penetrated the
thoughts of educated provincial leaders. Even among the most learned,
the overriding belief was that disasters were the will of God. Residents
and royal officials alike held the fatalistic belief that, aside from staying
on the good side of the Almighty, there was nothing one could do to prevent hurricanes.[50] Priests and other religious reinforced such beliefs and
acted as God's intermediaries. At the onset of hurricane season, prayers
and masses were offered for divine mercy, and after the passage of a storm,
Te Deum Masses of thanksgiving were conducted. In times of crisis, priests
and nuns cared for the sick and dying, and priests officiated at the burials
of victims. Faith also had its practical side. In many hamlets, the church
was the only substantial structure in the area, and when threatened, the
population took refuge there from the tempest. Church leaders were always in the vanguard of the relief efforts after a hurricane. Because they
knew their parishioners better than anyone else in the neighborhood, they
were the best prepared to assess the extent of the devastation.

The Scientific Climate

By the mid-eighteenth century, traditional structures and attitudes in
Spain's far-flung empire were experiencing change along with the rest of
the Atlantic world. Even before the beginning of progressive Charles III's
reign (1759–88), royal officials took an active role in observing, recording, and promoting scientific knowledge.[51] From the Caribbean to the
Malvinas (Falkland Islands) to the Pacific Northwest to the Philippines,
the crown sent out investigators to visit strange lands, to encounter exotic
peoples, and to catalog scientific curiosities.[52] These expeditions were not
dedicated to the accumulation of knowledge for knowledge's sake. Instead,
the goal was to advance scientific knowledge in the interest of defense, fulfilling Charles III's desire to neutralize British power and to thwart British
expansionism.[53]

A major beneficiary of the Spanish Enlightenment was the royal navy,
whose navigation manuals and nautical charts became increasingly more
accurate by the end of the eighteenth century.[54] The art of navigation was
tied closely to advances in meteorology as a science, and during the last
half of the eighteenth century, meteorology, although primitive by mod-

ern standards, grew by leaps and bounds. Centuries of keen observation meant that mariners understood when dangerous weather systems were imminent. Cloud movement, opposing tide and wind patterns, and the famous "brick-red sky" all gave warning that danger lay ahead. On land, residents interpreted animal behavior as a sign of bad weather to come. Yet residents still lacked the understanding of hurricane formation and movement. Not until the following century did the true nature and movement of the deadly storms become clear.[55]

Royal bureaucrats were in the vanguard in recording firsthand observations of local conditions, reporting on enemy troop movements, and collecting exotic specimens to be sent to Spain for the monarch's pleasure. Governors, provincial officials, administrators, harbor pilots, customs officials, ship captains, and common seamen all set about collecting data and contributing their observations about the effects of storms on land and sea.[56] Among the most famous of these men were Jorge Juan and Antonio de Ulloa, whose journey to South America in the 1730s became the fundamental text of the Spanish Enlightenment.[57] Although they were the most famous, Juan and Ulloa were but two of many royal officials whose missives to Madrid were responsible for advancing scientific knowledge about hurricanes and formulating royal policy toward a disaster's aftermath. Beginning in the 1750s, scientists in Spain relied on the reports sent by military officials who had served in the Caribbean and had experienced the consequences of hurricanes on a personal level. These unheralded contributors provided firsthand accounts to a growing body of knowledge.[58]

At the forefront of Spanish scientific inquiry was a royal mail system (patterned after the ideas of Benjamin Franklin) established in 1764. It was the mail system more than any other branch that forged ahead with scientific observation of weather phenomena.[59] Pilots and ship captains of the mail system were given specific regulations to which they were required to adhere.[60] By the mid-1770s, harbor pilots in the Spanish Caribbean port cities operated under even stricter rules that compelled them to delay departures if traditional wisdom and weather signs warned of danger.[61] Caribbean ports were closed during the autumnal equinox, and no ship was permitted to leave until the dangerous season had passed.[62] After sustaining considerable losses from a series of storms that struck the Caribbean in the 1780s, in June 1784, Minister of the Indies José de Gálvez sent out another circular order to all captains and pilots detailing additional regulations for royal transports to avoid being caught in storms at sea.[63] By the 1790s, harbor pilots and captains had added the barometer to their arsenal of weapons against nature—the telltale sign of the dropping barometer portended bad weather ahead.[64]

On land, officials in the hamlets and villages wrote increasingly more sophisticated reports about local weather conditions. By the 1770s, on the orders of the captain general, constables (*capitanes del partido*) were required to record details about the nature and characteristics of storm movement and forward that information to their superior officers. Two decades later, in 1791, the constables were further obligated to submit twice-yearly reports about the state of their jurisdictions. They were charged with commenting specifically about the weather, crop conditions, the potential harvest, and the state of "prosperity or misery" of the residents.[65] Individual observers added to the body of scientific knowledge about the weather. One such contemporary observer, Antonio Lavedan, recorded that there would be a calm before the storm, but then the sky would darken much like in a normal afternoon. The telltale warning sign, however, was when the wind came from one direction for a long period of time. If that happened, a hurricane was imminent, and precautions should be taken for survival.[66] Observers were especially interested in drawing comparisons with wind movement in the areas vulnerable to hurricane strikes. Before the circular movement of hurricanes was known, it was thought that the most destructive winds in Cuba came from the north-northeast or from the west, while on neighboring Hispaniola, such winds blew in from the south or the west.[67]

Strategies for Survival

Eighteenth-century residents based their strategies for survival upon risk avoidance and common sense. Most deaths in a storm came from drowning, either from the deadly wall of water along the coast, known today as a storm surge, that obliterated everything in its path; from flooding near the mouths of rivers where the storm surge pushed a wall of seawater upriver; or from heavy rainfall in the interior that caused mudslides that swept away populations with little warning. Near the coast, hurricane-force winds whipped the seas into a froth that was driven inland. Contaminated by saltwater, the wind-whipped mist burned the foliage off trees and ruined stored water supplies in cisterns. Even residents who lived well inland were vulnerable to the deadly threat of rising water. Continuous rainfall eventually saturated the soil to the point at which it could absorb no more water. When that happened, rivers raged out of their banks, and flooding drowned weakened animals and humans and ripped crops out of the soil. When preceded by drought, the raging waters eroded the parched earth. Dead bodies and carcasses contaminated streams and wells, and disease spread quickly. Even minor storms blew the plantains from the trees, giving an abundant supply in the storm's immediate aftermath. But within days, victims were surrounded by the stench of putrefying crops. Conversely, during times of drought, crops withered in the ground, and animals died of thirst and exhaustion.

Avoidance was the logical course of action. From time immemorial, authorities had prohibited building along the coast, and although the prohibition was enacted to minimize smuggling and contact with foreigners, during a hurricane it worked to save lives.[68] After the fall of Havana in 1762, royal officials were resolute that such building restrictions be enforced, and in one instance in February 1771, when war with Great Britain threatened, a general evacuation was ordered.[69] Residents who made their livelihood from maritime activities, such as fishermen and salt rakers, were required to obtain licenses from local authorities before they could put out to sea. On one hand, the regulations were intrusive, but they did serve to provide notice when members of the community had fallen victim to a storm's fury.[70]

Flooding was the greatest danger inland, and building was also prohibited along the river bottomland, which was reserved for tobacco growers by royal decree. Rising waters ruined the tobacco crop, but the infrastructure—houses, drying barns, and mills—would usually be spared if not located on bottomland.[71] Engineers further suggested preventative measures to minimize the threat from flash floods, such as building diversion canals that could be opened when rivers overflowed their banks. The aqueducts would serve the dual purpose of providing irrigation during times of drought.[72] Contemporaries were aware of the importance of preserving the fertile topsoil, "capa fructífera del terreno," and they recommended that farmers build terraces surrounded by hedges.[73] Cuban farmers believed that the darker the soil, the better it was for cultivation, but as the eighteenth century drew to a close, trial and error proved that lighter-colored soils (color pardusco), especially near Güines, were especially suited for tobacco cultivation.[74]

In spite of the recommendations of the most learned scientists of the day, demographic change exacerbated the problem of soil erosion, especially around Havana. Until the 1770s, because of its scant population, Cuba was still densely forested. Forests worked to reduce erosion and limited the damage from flash flooding, but by the 1790s, population increases caused Havana and its hinterlands to double in size.[75] Trees near the capital to the west were felled to make room for agricultural expansion to provision the burgeoning population, and the harvested lumber was sold to the royal shipyard to fuel the shipbuilding industry. By June 1791, urbanization and deforestation west of the capital city meant that a relatively minor storm with unusually heavy rains became a major catastrophe, which, according to some observers, claimed as many as 3,000 lives.[76]

Storm-related deaths also occurred when structures collapsed or when victims were struck by flying debris. The typical architecture of the common people was the traditional wattle-and-daub, thatch-roofed house called *bohíos*, a style inherited from the pre-Columbian inhabitants, the Taino. These classic examples of impermanent architecture were particularly vulnerable to high winds and driving rain.[77] Although of flimsy construction, such houses were quickly and easily rebuilt after storms. City officials, especially in Havana, enacted prohibitions against building with wattle and daub and roofing with thatch. Nonetheless, in 1754, some 470 houses (about 14 percent of the city's total) were poorly constructed. In Santiago de Cuba, 29 percent of dwellings were impermanent; and in the other cities of the island, such as Matanzas, the majority of houses were thatch-roofed, wattle-and-daub huts.[78]

The preferred construction materials, especially in the cities and for those who could afford them, were stone or mortar combined with heavy local woods and tiles. Such construction built well inland was usually able to withstand the worst a storm could deliver. The most substantial of these were built of stone and were roofed with tiles made into the characteristic "U" shape by molding the wet clay over a man's leg and laying the tiles in the sun to dry before firing in a kiln.[79] Yet, although substantial and permanent, during the strongest storms roof tiles lifted from the underlying wood and became deadly missiles, sending razor-sharp shards of tile flying through the air. The only preventative measure was to situate the buildings in such a way as to break the force of the wind.[80]

Flying debris and collapsing structures led to many deaths, and infections caused by injuries increased the death toll. The laconic Spanish reports provide few details, but a contemporary letter from the British colony at the Bay of Honduras (present-day Belize) vividly described the aftermath of the storm surge and river flooding, which was eerily similar to the Cuban catastrophe of 1791: "The distressed inhabitants, without any dry cloathing, or other necessary refreshment, almost exhausted with extreme cold, their bodies every where bruised by the blows they had received from the limbs of trees, logs of mahogany, and other pieces of wrecks floating about in the bush, betook themselves to the erecting a few temporary sheds, and by digging among the rubbish, endeavoured to find some part of their cloathing." The writer also describes what was probably a case of tetanus that led to gangrene: "Captain Edward Davis, who having received a violent cut in the bottom of his foot with a glass bottle, whilst wading through the bush to gain a place of safety, it produced a mortification in his bowels."[81]

Yet, with the exception of the hurricane in June 1791, the number of immediate fatalities from the passage of a storm was surprisingly low, and the survivors were left with a feeling of confidence that since they had endured the effects of at least one deadly hurricane, they could do so again. Out of survival came a sense of capability; knowing what to do meant one could survive a future disaster and cope with its aftermath. The intangible mind-set associated with being a survivor became ingrained in the collective mentality of the population. Time and again, elders related family tales and folklore, recounting the horrifying effects of one or another hurricane only to reiterate the resilience of the community in its ability to survive anything that nature might deliver.[82] Conversely, if populations were subjected to repeated hurricane strikes without respite, a sense of helplessness and hopelessness, "hurricane fatigue," set in. The resulting depression worked against a community's ability to overcome post-disaster challenges.[83]

Medicine and Disaster

Physicians, apothecaries, and other medical practitioners were also in the vanguard of addressing the aftermath of disaster. Among the remedies prescribed to cure the injuries caused by impact injuries and contusions was Cuba's premier crop, tobacco. Ground or chewed tobacco was made into a poultice and placed directly on a wound before it was dressed. This was believed to reduce swelling. Tobacco was also used to cure other skin problems such as ulcers and head lice. Taken as an infusion in water or warm wine flavored with cinnamon or nutmeg, it was a powerful emetic that was sometimes prescribed for colicky babies. Diluted tobacco-infused water was thought to cleanse cloudy eyes.[84]

Impact injuries caused some loss of life, but the underestimated (and understudied) cause of death after the passage of a storm was disease. Contemporary reports only summarized the immediate number of deaths from drowning and injuries, while the death toll from dysenteries and fevers was rarely reported because the casualty figures would not be known until days or weeks later. The innumerable endemic seasonal fevers that contemporary medical observers called *calenturas tercianas* (tertiary fevers) were especially dangerous to European newcomers. Estimates suggest that as many as 30 percent of newly arrived Europeans did not survive the seasoning process of acclimatizing to the tropical disease environment.[85] Physicians were unaware of germ theory, believing that fevers were caused by miasmas or vapors emanating from stagnant water or air. Rather than suspecting unsanitary conditions and contaminated water, as late as 1797 Cuban physicians believed that the passage of a storm blew away the noxious fumes that caused disease.[86] In addition to the dramatic epidemics of yellow fever and smallpox, ordinary fevers often resulted in death from secondary infections and were exacerbated by the debilitation brought on by exhaustion and starvation.[87]

To the European nations' credit, by the mid-eighteenth century, policy makers were aware of the dangers of sending large numbers of troops to the tropics even while they were hardly able to prevent the horrific toll that fevers took on Europeans.[88] As early as the 1740s, metropolitan bureaucrats took seriously the observations rendered by Juan and Ulloa that the theaters of operation that Spanish troops would encounter were "cursed by nature" because of the disease environment.[89] The fall of Havana in 1762 was a catalyst in the scientific and medical realm as well as in the areas identified in the well-studied Bourbon Reforms. Weather-induced seasonal fevers incapacitated the military forces in Cuba and played a pivotal role in the Spanish defeat, and in its wake, royal doctors and officials were charged with enacting a total reform in the way disease was approached. In the following decades, even the most insignificant and isolated outbreak of fever was brought to the attention of the captain general, and measures were implemented to prevent the fever from spreading.[90] Military hospitals were established in Havana and Santiago de Cuba and later in New Orleans, and strict regulations regarding the conditions in the hospital were enacted. Among the most important considerations were to make sure that the bedding in the wards was changed regularly and that beds were equipped with mosquito netting to protect patients from the swarms of biting insects. An additional requirement was to provide a substantial, healthy, and balanced diet.[91] As usual, the Spanish records reveal little, but British commentaries, such as a letter written by one Captain Davidson to a colleague, Captain Garrigues, in Cádiz, tell much about the hospital conditions in Cuba in 1775: "He was obliged to put in [to Havana] by sickness, he conceives himself in gratitude bound to inform the public that during three weeks illness at that port he was treated in the most humane, friendly, and polite manner, that the attention of the physicians as well as the neatness and accommodations of his apartments was every way equal to what he could expect in an English hospital."[92]

The Structure

Once environmental crisis is established beyond a doubt, the fundamental events of Caribbean history are incorporated into the analysis. Readers will be familiar with many chronological landmarks from 1750 through 1804. Imperial issues such as the change in monarchs in 1759 and 1788 play a large part, along with the concomitant events such as the militarization of Cuba, the Spanish acquisition of Louisiana, and Spain's sequential declarations of free trade, beginning in 1765. Social processes, especially an ever-increasing population, put a strain on the imperial provisioning system that could not be remedied through traditional means. Foreign influences, such as North American and British traders from Jamaica engaged in illicit activity and the contraband trade, play a part in developing the argument of this book.

Nonetheless, this study will provide some surprises to informed historians of the Atlantic world. It begins by offering an alternative explanation as to the reasons for the British victory at Havana, which can be attributed to the effects of weather, food shortages, and disease within the Spanish ranks. Chapter 3 demonstrates that continuing shortages in the Hispanic Caribbean contributed to increased liberalization in international trade, known as the first impetus toward "free trade." Chapter 4 examines in detail one of the most active hurricane seasons, in fall 1772, when nine major systems made landfall throughout the Caribbean. It posits the hypothesis that universal scarcity called for desperate measures and forced the Spanish government to trade with North America, which, in turn offered an alternate market for the Patriots' products. This theme, elaborated and expanded upon in chapter 5, examines the war years from 1776 through victory in 1783, the significance of the *Reglamento para el comercio libre* on trade relations between the insurgent colonies and Cuba in 1778, and the subsequent expulsion of North American traders in 1784. Chapter 6 covers the years leading up to the death of Charles III in 1788 and the aftermath during the regime of his successor, Charles IV. By the time of Charles III's death, the relationship between Cuba and the United States was undergoing a gradual breakdown of trade barriers leading to a permanent trade relationship between the two regions. This chapter deals with the unfortunate years of the 1790s when the incompetent Charles IV sat on the Spanish throne. His representatives in Cuba abrogated the economic reforms and the mitigation policies of the previous forty years, bringing the island to the brink of political crisis. War with France in 1793 brought the Cuban military forces in direct conflict with the troops of republican France on the neighboring island of Hispaniola. There the Spanish expeditionary army suffered a series of defeats caused by a fatal combination of incompetence, inclement weather, food shortages, and sickness. Only the arrival of provisions from the young United States and the decision taken by the leader of the Cuban forces to allow these provisions to be unloaded and sent to the frontier saved the regiments from total annihilation.

This book proposes alternate hypotheses based upon the evidence for disaster, disease, and deprivation as the reasons for change in the Atlantic basin by establishing a clear correlation among climate, environmental crisis, and historical processes. Even if the work of climatologists and geographers had not appeared in such a timely fashion, documentary evidence would have left little doubt about the onset of crisis and the gravity of the situation in the Caribbean, the effects of which spread throughout the Atlantic world. By incorporating theory and methods from modern research into the aftermath of disaster, this study makes clear that environmental conditions during the latter half of the eighteenth century were major contributors to making this period a critical juncture in Atlantic world history.

CHAPTER TWO

Be Content with Things at Which Nature Almost Revolted

T HE GOVERNOR OF Cartagena de Indias, Don Ignacio de Sola, was a conscientious bureaucrat. As the ranking official of the South American city that was the departure point for Jorge Juan and Antonio de Ulloa's scientific expedition of 1735–46, Sola knew that he was obligated to inform his superiors in Madrid about natural phenomena and other curiosities.[1] So in spring 1752, he dutifully reported on the extremes of weather and the many misfortunes that had occurred throughout Spanish America over the past year. The governor wrote that unprecedented flooding had caused many casualties in Chile, in the Juan Fernández Islands, in Peru, and in Guatemala. An unusually strong hurricane had hit Jamaica in October 1751, and like many of his contemporaries, Sola confused thunder and lightning with the tremors produced by earthquakes and thought that the two were related.[2] On the front lines of Spain's scientific revolution, he applied rudimentary principles of scientific method and drew a comparison between what happened in the Americas to a rare hurricane that came ashore on the Iberian peninsula near Cádiz.[3]

The unusual phenomena that Sola reported were not unique to Spanish America. Over the summer of 1752, the southeastern coast of North America suffered from a severe drought that ended abruptly when two hurricanes struck the southern British colonies. The first storm hit Charleston, South Carolina, on 19 September, and the second hit the outer banks of North Carolina on 1 October. The Charleston hurricane was so destructive that it became the benchmark to compare the intensity of subsequent storms.[4] The inordinately severe weather was not confined to the western hemisphere. Reports from northern Europe told of a similar catastrophic cycle: "From the year 1751 until 1761, the seasons were cold and wet, not one agreeable summer intervening to enliven the dreary prospect. . . . To the unhealthiness of these years the bad state and dearth of provisions might not a little contribute; the poor being incapable to procure sufficient sustenance were often obliged to be content with things at which nature almost revolted; and even the wealthy could not by all their art and power render wholesome those fruits of the earth which had been damaged by the untoward season."[5]

Such reports of extreme weather conditions are recoverable historical evidence of a climate shift beginning in the mid-eighteenth century, a climate shift that provoked increased temperatures and a fifty-year increase in El Niño/La Niña activity.[6] This chapter establishes the hallmarks of the El Niño/La Niña cycle that ravaged the Caribbean—rapidly alternating periods of drought and deluge—and contextualizes their consequences within political, social, and economic conditions in Cuba. For thirteen years, one period of crisis followed another, often within weeks, and the onset of severe weather caused collateral effects such as food shortages and sickness. The argument shows how Spanish colonial institutions such as the monarchy, local officials, and the royal monopoly both helped and hindered attempts to deal with repeated periods of crisis. The chronological setting begins at the close of the War of Austrian Succession in 1748, continues with the change in the Spanish monarchy and Charles III's ascension to the Spanish throne in 1758, and concludes with his ill-advised entry into the Seven Years' War in 1762. The 1760s were marked by one of the signal political events in Spanish American history: the British siege and capture of Havana in summer 1762. The relatively easy conquest of the formidable city left a deep scar in the historical psyche of Cuba. By studying the ecological and epidemiological conditions at the time of the siege, this chapter will offer an explanation as to why Havana could not hold out against a well-fed and well-reinforced enemy. The cumulative effects of a decade of environmental stress and a winter marked by the effects of the El Niño/La Niña sequence created an environment that facilitated the onset of intermittent, seasonal fevers (*calenturas*). Long the ally of the Spaniards, an array of fevers became their enemy when they incapacitated every Spanish garrison on the island, from the contingents in the primary fortresses in Havana and Santiago de Cuba to the hospitals in and around Havana to the auxiliary outposts along Cuba's southern coast.

The Onset of Crisis: The 1750s

In 1749, just as the Hispanic Caribbean emerged from the effects of another European war that had spilled over to the Americas, the malevolent twins, El Niño/La Niña, brought their unwelcome presence to the region.[7] For the first time in decades, Spain's Caribbean colonies were at peace with their neighbors, and in the summer of that year, the squadron commanded by Lieutenant General Benito Antonio de Espinola en route to Hispaniola sailed without the threat of enemy ships. Yet the perennial dangers of the Atlantic passage remained unchanged, and the fleet encountered a furious storm near Bermuda. Using skills acquired from years of sailing the Atlantic passage, the commander guided his flagship and four other ships in the convoy through the storm with only minor damage. Finally they made safe harbor in the French colony, Martinique, and from there the convoy proceeded to Santo Domingo. Soon the news of Espinola's lucky escape began circulating throughout Caribbean posts. A passenger aboard the flagship, the Marqués de Gandara, wrote to the governor of Santiago de Cuba, Alonso Arcos y Moreno, about the harrowing passage and his safe arrival. Since Espinola could be held liable for any losses incurred because of his errors in judgment, Gandara sought to establish that the commander had done nothing wrong. Most important, the convoy had respected the prohibition not to leave port near the autumnal equinox; nonetheless, they still were caught in the worst summer weather.[8]

The always-perilous ocean passage proved even more dangerous the following year when another mid-Atlantic hurricane struck Spanish shipping. Like the notices of Espinola's fleet, the news was carried to Havana by a vessel in the coastal trade under the command of Captain Ignacio de Anaya. This time the notices were not joyful. In late July, Anaya left Havana for Campeche and Veracruz, where he loaded flour and other provisions. On his return voyage, sailing along the north coast of Cuba, Anaya was within sight of El Morro, the fortress that guarded the entry to Havana bay, but contrary winds and currents prevented him from making port. Anaya fought to maintain his position, but the hurricane winds caught his ship and swept it northward. For seven days and nights, captain and crew battled for their lives, and when the storm abated, they found themselves far north of their destination. They made safe harbor in Virginia, where they encountered what remained of the treasure fleet under the command of Daniel Huonny that had left Havana en route to Spain just a few days before the hurricane swept into the tropical north Atlantic.[9]

Huonny's tale of skill and survival mirrors that of Anaya, except that the storm caught his convoy one day later and 400 miles further north.[10] On 18 August 1750, well before the equinox, Huonny's flagship, *La Galga*, left Havana en route to Spain. The *La Galga* was accompanied by the *Zumeca*, the *Nuestra Señora de los Godos*, the *Soledad*, the *Nymph*, a Havana-based packetboat under the command of one Captain Arison, and a Portuguese warship. Although the *flota* system had been abolished years earlier, it was still safer to travel together on the ocean passage. The fleet navigated the treacherous Bahama Channel without incident, but on 25 August, somewhere off the Florida coast (around 29 degrees north latitude), the wind began to blow from the north with all the force of a major hurricane. Eighteenth-century navigators were not aware of the counterclockwise circulation of storms, but from Huonny's description of the wind direction, it is clear that the convoy was caught in the most dangerous northeastern quadrant. For five days, the men on board struggled to keep the ship from being swamped by the fierce wind and seas. The sailors were able to lower the sails before they were ripped to shreds. They threw the heavy artillery pieces into the sea, and the bilge pumps worked continuously to keep *La Galga* from taking on water. At last, the hurricane's fury abated, and the Gulf Stream caught the crippled ship and carried it northward. On the night of 31 August, Huonny recognized landmarks on shore and realized that he was off the coast of Carolina. Once he had his bearings, he headed his damaged frigate toward Virginia, where he hoped to make landfall. Yet the contrary winds and currents continued to work against the men on board, and on 3 September, another gale blew in from the northeast. Huonny tried to keep his ship away from land, but on 4 September, *La Galga* foundered on a small island about four leagues off the coast of Carolina.[11]

Although he lost his flagship and all of its cargo, Huonny was able to save his crew with the exception of six men. The survivors crafted a makeshift raft out of the wreckage and made it to shore, and from there they proceeded on foot to Virginia. The *Godos* and the Portuguese warship had also been forced aground by the hurricane, but local residents helped float both ships off the sandbar, and a local pilot guided them to port. Both ships were in such poor condition as to be unseaworthy, so the crews and their cargo were carried to Spain on English vessels. The register ship, the *Zumeca*, also foundered in the shallows along the coast but saved all her crew.[12]

The news was not good for the *Soledad*, the *Nymph*, and Captain Arison's packetboat. The *Soledad* came through the storm with no loss of life or cargo but with major structural damage. Once her captain got his bearings, he too headed northward toward the Virginia port. En route, the *Soledad* encountered the *Nymph* aground on a sandbar and unable to free itself. Noting its location, the *Soledad* made it to port, where her captain hired two local ships with full crews to return to salvage the *Nymph*'s cargo, refloat the ship, and escort it to port. After transferring the cargo, the English crew mutinied, took over the salvage ships, and fled into familiar local waters. The *Soledad* gave chase, and one of the pirate ships ran aground and was captured, but the other escaped carrying with it fifty-four chests of silver and several bags of cochineal.[13] The news was even worse for Arison's packetboat—everyone aboard drowned except for two cabin boys.[14]

Even though the hurricanes of 1749 and 1750 did not strike the Caribbean directly, their collateral effects were enormous. In addition to the loss of life, the mainstay of the economy, the *situado* and commodities produced in the Americas valued at 3 million pesos, went to the bottom with *La Galga* or were stolen by the renegade salvage crew from the *Nymph*. The flour on Anaya's ship destined for Havana was also lost, which meant that the entire provisioning system of the Caribbean would have to adjust to threatened scarcity.[15] In a move to avoid drastic food shortages in Santiago de Cuba, Governor Arcos y Moreno sent word to Captain Fernando González, who was in port in Havana loading supplies, to hurry home before the arrival of the equinox to avoid endangering the cargo he carried to the eastern city.[16]

The royal officials in the Spanish Caribbean could not have known that the second mid-Atlantic hurricane in as many years marked the signs of an El Niño/La Niña sequence and that they had more than five decades of extraordinarily bad weather ahead. After suffering from the effects of the 1750 hurricane season, during the winter of 1751–52, drought struck the eastern region of Cuba.[17] By 1752, the characteristic sudden fluctuation of the El Niño/La Niña was evident when two hurricanes brought disaster to the western end of the island in October and November.[18] The cycle continued through 1753, when unusually wet weather plagued the southern coast from Santiago de Cuba to Batabanó.[19] The cycle accelerated and worsened as the decade progressed. The wet winter of 1753–54 was followed by a severe drought (*gran sequía*) in the spring.[20] By the following October, the region was again plagued by fierce storms.[21] As fall passed into winter and then into spring 1755, drought had returned, and this time it extended westward to the neighboring province, Puerto Príncipe (present-day Camagüey), signaling that the sequence was in full effect.[22] Six months later, in August 1755, local officials wrote of flooded *vegas* (tobacco farms) in Oriente,[23] and later in the month came notices that one of the ships involved in the coastal trade had been lost in a storm off Caiman Chico.[24] The winter of 1755–56 brought no respite. The El Niño/La Niña effect heightened as the rainfall continued, first with a severe storm in March 1756,[25] and then culminating in a hurricane in October 1756 that tore through the western and central parts of the island.[26] After three years of fury, the sequence once again returned to severe drought, which had its most catastrophic impact in spring 1758,[27] only to change again from 1759 through 1761, when one hurricane per year disrupted life on the island.[28]

The El Niño/La Niña cycle affected almost every aspect of life. The inclement weather conditions lasted for months at a time and made misery and deprivation characteristics of daily life. Unsanitary conditions and starvation made the onset of disease almost inevitable. In September 1750, Captain General Francisco Antonio Cagigal de la Vega wrote of the prevalence of disease in Havana, which had claimed friends and relatives. He complained that he had suffered from a severe cough and congestion throughout the rainy season.[29] In an unusually candid letter, his counterpart in Santiago, Arcos y Moreno, replied telling of the casualties among officers and troops in his jurisdiction: "The [victims] have reached to the Lieutenant Generals."[30] The perilous ocean passage combined with the shock of seasoning took a heavy toll on Europeans. In 1751, a shipload of Canary Island immigrant families was so sick upon arrival in Santo Domingo that they had to be hospitalized before they could continue to their final destination.[31] For European bureaucrats and military personnel, duty in the Americas was notorious for its danger. Even seasoned veterans were not immune from the ravages of disease. During the extraordinary season in 1751, in the viceroyalty of Santa Fe, Commander José Pizarro, who had served in the Caribbean theater since the 1730s, was sick from the day of his arrival, proving without doubt the unsalubriousness of that position.[32] Commanders knew that the chain of command could be vulnerable if sickness among the officers reached critical levels. In 1753, the viceroy of New Spain, Juan Francisco Güemes y Horcasitas, Conde de Revillagigedo, who had previously served as captain general of Cuba, sought to mitigate the threat if there were a gap in the chain of command brought about by disease. To "avoid impertinent confrontations," the viceroy mandated that if the governor were absent, incapacitated, or dead, the reins of government would fall to the sergeant major of the garrison.[33]

UNTIL RECENTLY, the psychological consequences of sequential crises have attracted scant attention outside the medical profession. Hurricane strikes on Florida in summer 2004 and the more recent tragedies along the Gulf Coast in 2005 and 2008 have created renewed interest in studies that seek to determine the effects of "hurricane fatigue" on affected populations.[34] Such a framework explains the melancholy and sadness that pervades the correspondence of Cuban officials, especially that of Arcos y Moreno, who described the situation in Santiago de Cuba in 1751 as "a festival of cadavers."[35] The governor of Oriente wrote of the death of his fellow officer, Gaspar Tabares, and of the sympathy he felt for the victim's widow and children in such a remote post. Following the dictates set down in the Laws of the Indies, the family had followed Tabares to his remote assignment, and now it would be a huge undertaking to arrange for their return to Spain. The governor wrote of how much he regretted the widow's plight and that of other women caught in similar circumstances. Arcos also worried about his wife, Doña Teresa, who suffered from the effects of recent childbirth complicated by the deleterious effects of the rainy season.[36] The governor acknowledged his responsibility to do his duty, yet his despair and resignation were evident in his observation: "For me this death and others is a vivid reminder of the anticipation of seeing myself in the same situation."[37]

In addition to coping with death, fear, and loss of loved ones, a serious consequence was the loss of the residents' primary sources of income. All Caribbean cities stood to suffer because of the loss of the 3 million pesos on *La Galga*. Fortunately, another warship, *La Begoña*, left Mexico just a few weeks after the departure of the ill-fated fleet of 1750. Its captain, Joseph Duque, avoided the storm, and in mid-September 1750 he arrived safely in Havana with a portion of the *situado*. Joseph de Montero, the officer charged with guarding the precious funds, delivered them personally to Cagigal.[38] Upon learning of *La Begoña*'s good fortune, Arcos y Moreno observed that "the troops will be very happy to receive their salaries, but it would be very welcome to receive a fresh shipment of flour."[39]

On most occasions, however, the outcome was not positive. When the *situado* did not arrive, the salaries of the army, the navy, and the bureaucracy were not paid, setting off a cascading trickle-down effect that extended to all ranks of society. Captains such as Anaya and González who were employed by the royal administration were asked repeatedly to forgo payment for their services. Local provisioners who raised cattle and hogs for urban consumption were still obligated to drive their livestock to the towns' slaughterhouses, but they, too, had to wait for payment until the garrison was solvent. Other residents—tavern keepers, innkeepers, merchants, artisans, petty traders, farmers, and prostitutes—who relied upon the military presence all suffered a reduction in their income when the military and the bureaucracy were not paid.

The second-most-important source of income was from the tobacco harvest, which suffered from rapidly alternating periods of drought and excessive rainfall. In a normal year, the far-eastern end of the island could be expected to produce nearly 450,000 bundles (*manojos*) for shipment to Spain. Oriente province alone produced nearly 150,000 bundles. But in March 1752, Arcos y Moreno reported that many of the *vegas* were not planted during the past planting season because of the lack of rainfall. The drought meant that the Real Compañía could anticipate receiving no more than 93,000 bundles that year.[40] Production recovered the following season and almost returned to normal levels, but over the winter of 1753–54, wet weather ruined the harvested tobacco, which waited on the wharves of Santiago de Cuba to be shipped to the monopoly's offices in Havana.[41] The following summer, in 1754, just as the El Niño/La Niña cycle accelerated, a new governor, Lorenzo de Madariaga, arrived to take command of the eastern province. Almost immediately, he wrote that the drought had destroyed that year's harvest.[42] The drought continued through the winter and into spring 1755, bringing desperation to the officials in the eastern city.[43] When the cycle suddenly reversed, in August 1755, the purchasing agent of the Real Compañía in Santiago worried about his ability to fulfill the quota because the streams and rivers had inundated the tobacco *vegas* and destroyed the crop in the ground.[44]

As the cycle accelerated, the malevolent effects of the El Niño/La Niña sequence spread westward, where they affected another of Cuba's primary industries, cattle raising. Drought first appeared in 1754 in Puerto Príncipe, and the governor, Martín Estéban de Aróstegui, the brother of the director of the Real Compañía, faced a similar crisis to the one that plagued his colleagues throughout the island. Two years later, in 1756, a strong hurricane struck the central and western parts of the island, laying waste to ranches, tobacco *vegas*, small farms, and haciendas.[45] Puerto Príncipe had barely recovered in 1758 when Aróstegui informed Madariaga that the province would not be able to provide its quota of cattle for the military garrisons because drought had reduced the herds to critical levels.[46]

In addition to reducing the quantity of the primary products for export, the ever-changing inclement weather also disrupted the other mainstays in the food supply. The hinterland's function was to provision the populated cities, and petty farmers outside Havana and Santiago de Cuba grew domestic subsistence crops such as yuca, other root crops such as boniato and *ñame*, plantains, squashes, and a variety of tropical fruits. On market days, they offered their crops for sale, along with chickens and other domesticated animals. As mentioned previously, large ranches maintained herds of cattle and swine, which, in addition to being one of the island's primary exports, were shipped to the cities to provide fresh meat for the military garrison and the town residents. Many urban households, especially around Havana, owned garden plots on the outskirts of the city to provide the necessities of life.[47]

Luxury provisions that so pleased European palates had to be imported, and the responsibility for providing such items fell to the Real Compañía. Cured ham, wines, olives, and olive oil were brought from Spain, but flour was a problem. Under mercantilistic principles of imperial self-sufficiency, flour was supposed to be exported to Cuba from Mexico, but more often than not, the Mexican wheat crop was insufficient to meet the Cuban demand.[48] Early in its existence, the Real Compañía negotiated the right to purchase flour from British merchants in Jamaica, who, in turn, received their flour from the North American colonies, primarily Pennsylvania. The Real Compañía was allowed to import foreign flour into Cuba according to the ratio of one barrel of flour per slave. The concession was ostensibly granted to reduce the drain that the slaves would put on local food supplies, but it was obvious that the flour rarely went to feed the slaves—rather it was sold to local bakers, who baked it into white bread for the European tables. Sometimes, the system worked as intended. In 1752, for example, a frigate arriving from Jamaica brought 100 barrels of legal flour along with its slaves.[49] The following year, 143 barrels of flour accompanied the 150 male and female slaves that came from the British island.[50] Thus, even under the best of circumstances, Cuba was not well supplied with provisions. Even so, few Spanish bureaucrats approved of purchasing flour from the British. Going outside the imperial system violated the fundamental premises of mercantilism, and it also meant that the preferred chain of supply from Veracruz was undermined. Worse still, precious Spanish silver went to foreigners instead of remaining within the Spanish imperial economy.[51] Throughout the 1750s, attempts were made to quantify and regulate the foreign goods that came into the island, and crown officials were required to submit annual reports on the quantity of such products that came through their port.[52]

The overlapping responsibilities of each branch of government frequently collided. Royal governors were required to enforce imperial regulations, including defending the monopoly granted to the Real Compañía. On the other hand, local officials in the ayuntamiento, who referred to themselves as "fathers of the republic," were responsible for maintaining a sufficient supply of food for the public. The commissioners were in charge of the slaughterhouse, where they set prices and collected taxes on each head of cattle and hog brought into town for slaughter. Their privileges included policing the public markets to make certain that weights and measures were accurate, and they also set the price that was charged for bread based upon the price of flour imported into town.[53]

When crisis struck, all normal functions were disrupted. In the face of impending disaster, such as began in 1751, royal and local officials worked together to solve problems. Of primary concern was replacing destroyed food supplies. Governors recognized the urgent needs of local victims, but they also knew that commerce with foreign colonies was absolutely forbidden. The only way to justify issuing such special licenses was if the initiative came from local leaders, and the ayuntamiento members were happy to oblige. After meeting in emergency session, the council members drafted a formal request asking that a trusted member of local society be permitted to sail to other ports in search of food. More often than not, both the governor and the citizens were in agreement that an emergency existed, and the request was just a formality. Local governors routinely authorized voyages to nearby colonies—the practice was commonplace but also illegal—and governors could only issue licenses to sail on a one-time basis.

Such licenses permitted local captains to sail to other Spanish cities, but common sense and urgency dictated that they try to obtain provisions in the closest ports. On one such occasion, as the first effects of El Niño/ La Niña began to be felt in 1749, Captain Pedro Jiménes from Santiago requested and received permission to sail to Bayamo in an attempt to purchase meat.[54] A frequent and favorite destination was the northern coast of South America, especially Cartagena and Caracas, but frequently the inclement weather conditions that plagued Cuba also affected many parts of the Spanish mainland. In April 1748, Captain Antonio Ramírez sailed to Cartagena in search of corn, only to return empty-handed. He did carry an apologetic letter from Ignacio de Sola to Arcos y Moreno telling the Cuban governor that there was no corn in his province.[55] As the crisis deepened in 1751, Arcos y Moreno again called upon his comrades, Sola and Lieutenant General Felipe Ricardo in Caracas, only to learn that the severe drought in Cuba had also ruined the harvests on the mainland. Sola could manage to scrape together twenty-five *fanegas* (a measure of approximately twenty-five pounds) of corn that was only fit to feed Santiago's horses.[56] As other vessels returned with the disappointing news that no food could be obtained in any of Spain's Caribbean ports, Cuban captains were allowed to go outside the Spanish system and trade with foreign ports belonging to friendly nations. They routinely contacted their allies in the French colonies, especially Saint Domingue.[57] As early as the 1740s, an implicit reciprocity was in effect between Saint Domingue and Santiago de Cuba, and in April 1751, Arcos y Moreno sent a ship to Guarico (Cap Français, present-day Cap Haitien) to purchase provisions and to inquire about enemy troop movements. But again the captain of this expedition returned home without success.[58]

By June 1751, every legal avenue for obtaining emergency foodstuffs had been exhausted, and in desperation local authorities conspired to put a measure in motion to provide for the needs of the population that, at best, skirted the bounds of legality. Indeed, it is unclear which government entity—the governor or the ayuntamiento or the Real Compañía itself—was responsible for creating the conditions under which an unnamed English brigantine under contract to the monopoly carrying 102 slaves sailed into Santiago harbor in July. The slaves were particularly unwelcome because they added additional mouths to feed to the already burdened city, but the brigantine carried 524 barrels of greatly needed flour. The cargo blatantly violated the established flour-to-slave ratio, and the governor, in a show of compliance, sequestered both the slaves and the flour. Clearly, this was an unacceptable solution for a variety of reasons. The flour was withheld from public consumption and was in danger of spoiling in government warehouses; meanwhile, the government was forced to feed, clothe, and house the slaves while local officials argued over the best course to pursue.[59] The governor and the ayuntamiento entered into a debate over the pros and cons of releasing the flour for public consumption versus following the letter of the law and destroying the excess cargo. When such evasions of the letter of the law occurred, local officials took great pains to legitimize their actions, and their rhetoric, skillfully crafted to exhibit their knowledge of royal regulations, outlined the consequences all would face if they were accused of contraband. They acknowledged that the cargo of flour was clearly a "case of excess" that should not be permitted and that under normal circumstances they would never violate royal regulations. At the same time, however, they stressed that Oriente faced a dire situation, that all legal avenues had been exhausted, and that they had no choice but to release the illegal cargo.[60] To justify their actions, in the following February, in 1752, Arcos y Moreno reported to the captain general that the contraband flour had saved the population from certain starvation and that it was no longer necessary to procure provisions from Caracas.[61]

The desperate circumstances did not abate during the winter of 1753, and by early summer 1754, another local captain, Antonio de la Fuente, the master of the sloop *Nuestra Señora de Loreto*, received permission to sail to the Río de la Hacha or other Spanish colonies in search of food. The wording of de la Fuente's commission clarified his mission and also protected the captain from possible charges of being involved in illicit commerce: "If you can find no food anywhere in the Spanish Caribbean, you have permission to go to foreign colonies, noting carefully every port that you enter. You are specifically prohibited from carrying contraband in any form, and if caught, the items will be confiscated and you will be subjected to the severest of penalties."[62]

Sailing under such circumstances was always a risky business. Sometimes a captain received authorization in one jurisdiction only to run afoul of royal officials in other cities who refused to recognize the legality of the license. The personal and professional rivalries inherent in the Spanish imperial system contributed to the animosity among royal and local officials. Smaller towns and villas such as Bayamo and Guanabacoa bitterly resented their subordination and obligation to the larger cities, Santiago de Cuba in the case of Bayamo and Havana in the case of Guanabacoa.[63]

In addition, routine rotations of royal officials further complicated matters. One such rotation occurred in 1754, when Oriente's governor, Arcos y Moreno, was replaced by Lorenzo de Madariaga while at the same time the governor of Puerto Príncipe, Luís de Unzaga y Amézaga, was replaced by Martín Estéban de Aróstegui.[64] Both new appointees were forced to cope with the drought that had spread throughout the eastern end of the island from Puerto Príncipe to Oriente.

Ignacio de Anaya, the valiant coastal captain who had guided his ship and crew safely through the hurricane in 1750, was caught up in a controversy generated by imperial dictates, local needs, and bureaucratic shuffle. In 1754, before being rotated out of his position, Arcos y Moreno had granted Anaya a blanket permission similar to that received by de la Fuente to sail "from one port to the other" in search of provisions. Anaya, a skillful navigator who knew the Caribbean well, was a logical choice, and before long he returned with fifty barrels of foreign flour. Meanwhile, Arcos y Moreno had been replaced by Madariaga, and although Santiago de Cuba desperately needed the flour, the new governor questioned the validity of the crisis that had prompted his predecessor to grant the blanket authorization to sail to any port. Furthermore, regulations demanded that he send Anaya to the captain general in Havana for a disposition of his case.[65]

As Anaya's fate hung in the balance, evidence of the crisis continued to mount. In November 1754, reports of desperation in Puerto Príncipe came from an almost unimpeachable source: the new lieutenant governor, Colonel Martín Estéban de Aróstegui.[66] Since his brother was the head of the Real Compañía, Aróstegui had a vested interest in maintaining the monopoly's privileges, yet as he took over his position, he immediately faced scarcity in food supplies. Aróstegui did not have the authority to send ships out to find food because Puerto Príncipe was subordinate to Santiago de Cuba, and so in March 1755 he wrote to Madariaga, asking for provisions and for permission to dispatch a boat to nearby cities.[67] He stressed that supplies for the garrison were dwindling rapidly and that the subsistence crops that were grown locally had failed because of the drought. Santiago, indeed, did have some flour—fifty fresh barrels that had recently arrived on Anaya's boat—but by allowing the flour to be distributed, Madariaga would implicitly acknowledge that the voyage had been justified. The governor responded that flour was also scarce in Santiago de Cuba and that it was being consumed very rapidly since there was

nothing else to eat because the drought had destroyed the other subsistence crops. Nonetheless, he agreed to share two barrels of the flour with
the smaller town, but he refused to grant permission to seek provisions
from other ports.[68]

By allowing the flour to be released for public consumption, the decision about whether to prosecute the case against Anaya became a fore-
gone conclusion. Instead of arresting the captain, Madariaga allowed
him to sail to Havana on his own recognizance and to present himself to
Captain General Cagigal with a letter exonerating him of all culpability.
The letter's wording was specific: "Please forgive the captain of the boat
that brings this letter but because of the severe drought that destroyed our
immediate supplies and harvest, he has been forced to sail from port to
port in search of foodstuffs."[69] Shortly thereafter, the tribunal in Havana
returned a verdict of acquittal, and Anaya returned to Santiago de Cuba
and to his duties as a captain in service to the crown.[70]

At the same time that Anaya was being exonerated in Havana, Madariaga was faced with another arrival under even more irregular circumstances. A local captain, Carlos Basabe, arrived in Santiago de Cuba on a
French boat via the Dutch colony, St. Eustatius, with no documentation
except a copy of a royal order that he argued authorized his voyage. Basabe
was also sent to Havana to have his case adjudicated, and Madariaga explained the charges to the captain general: Basabe had traveled to a foreign
nation without a passport, he returned to Santiago de Cuba in a foreign
ship, and the royal order seemed not to cover what he had done.[71] The
judicial proceedings concerning the outcome of this case have not been
located, but subsequent events suggest that he, too, was exonerated, possibly because of the intervention of his influential cousin, Francisco Xavier
de Palacios.[72] By the following year, Basabe was again employed in the
Cuban coastal trade bringing Mexican flour to Oriente.[73]

Cuban royal officials' actions underscore the complicated issues of the
illicit trade. Royal officials were obligated to pursue possible cases of contraband lest they, too, run afoul of royal laws. Failure to respond would
mean they would be accused of complicity, and being found complicit
in the crime of contraband meant the loss of one's position, the forfeit
of one's entire estate, and a long prison sentence. Conversely, positive
performance meant that career officials could expect to advance in royal
service.[74] Two of Cuba's captains general, Juan Francisco de Güemes y
Horcasitas (Conde de Revillagigedo) and Francisco Antonio de Cagigal

were rewarded with the most prestigious position in the New World, the viceroyalty of Mexico, after serving in Cuba for many years. The governor of Santiago de Cuba, Arcos y Moreno, went on to serve as the president of Guatemala, where he remained until his death in 1760.[75] Luis de Unzaga y Amézaga, the lieutenant governor of Puerto Príncipe, had a long and distinguished career, ultimately rising to the position of captain general of Cuba in 1782.[76]

Nonetheless, the task facing royal officials in the Caribbean was herculean. The inaccessible southeastern coastline between Trinidad on the south and Baracoa on the north was ideal territory for the symbiotic trade that developed among the Cuban hamlets and foreign smugglers of all nations. Even while Spain and Britain were at war, English smugglers freely conducted their trade along the southern coast near Bayamo.[77] In August 1752, Governor Arcos y Moreno warned the founder of Holguín, Joseph Antonio de Silva, to be on guard against French contrabandists on the north coast.[78] Dutch smugglers operating out of their free ports in St. Eustatius and Curaçao were also regular visitors to Cuba's contraband coastline.[79]

In part, the problem was exacerbated because the crown maintained the monopoly privileges granted to the Real Compañía, whose usefulness was long past. The company alienated the population by charging high prices for imported goods and paying artificially low prices for tobacco and hides. Local residents resented the Real Compañía's monopoly and the enforced scarcity that it maintained. As a consequence, tobacco farmers and cattle ranchers found dealing with smugglers to be better than fulfilling their obligations to send their beeves to Havana and Santiago de Cuba or to sell their tobacco to the Real Compañía for ridiculously low prices.[80]

The resentment against the Real Compañía was compounded by rivalries among the cities and towns. All villages and hamlets in the east were subordinate to Santiago de Cuba and, as mentioned previously, were required to provide cattle and other livestock to the public markets. One of the most acrimonious battles was between the ayuntamiento of Santiago de Cuba and the city fathers of Bayamo. Santiago de Cuba's leaders complained repeatedly that the *bayameses* preferred to smuggle with the English islands rather than to fulfill their responsibilities and obligations as Spanish citizens.[81] Individual villages were only too happy to inform on the contraband activities of another, if only in the hope of diverting royal attention away from themselves and onto rivals. At the most personal level, smugglers informed against their rivals, unpopular governors, monopoly factors, or heads of the military detachments stationed throughout the island. These men became the victims of local malicious gossip intended to ruin their reputations and impair their ability to stop the contraband trade.

Such obstacles contributed to royal officials' absolute inability to eliminate the illicit trade. If the reward for remaining on the right side of the law was not clear for law-abiding civilians such as Anaya and Basabe, the alternative was far worse. For every captain who was exonerated, dozens were arrested for complicity, and for these men, punishment was swift and severe. As early as 1743, a royal order declared that anyone convicted of smuggling would be subject to the death penalty, although there is no evidence that such an extreme punishment was inflicted during this time.[82] Instead, the most common punishments were confiscation of the criminal's and his family's wealth and exile to another colony. The destination was often the fortress of Apalachee in Florida. In 1750 alone, approximately forty men from villages along the southeastern coast of the island were sentenced to exile to the Florida fort. For example, in August, a sentence of exile was imposed on Francisco Macedo, who was well known in Santiago de Cuba and the surrounding areas for his scandalous life. In pronouncing sentence, Arcos y Moreno concluded that "we are certain that we have rid our city of a dangerous wastrel."[83] The southern coastal town of Trinidad was notorious for its complicity in the illicit trade, and criminal behavior pervaded all ranks of the town's society.[84] In March 1750, Arcos y Moreno wrote to Cagigal that Pedro José de Acosta, one of the villa's leading citizens, had been given a sentence of exile.[85] A list of condemned criminals sentenced a month later included notable Phelipe Fontayne, in addition to his commoner comrades Pedro Barrancas, Matias Suárez, Pedro Milan, Manuel Pérez, Eugenio Arcila, and Juan Manuel Pérez.[86] The captain general unfailingly approved severe sentences, such as the punishment imposed on Estéban Castellanos and Manuel del Pozo, who were sentenced to six years hard labor in Florida in 1754.[87]

The depth of local involvement in illicit commerce is well illustrated in the notorious case of notable Pedro Carrión. In summer 1752, a small sailboat captained by Juan Díaz de Paz was making its way to Trinidad from Jamaica. Díaz de Paz had received a passport from the Spanish consul in Kingston to travel to Cuba because his wife wanted to convert to Catholicism. As the boat approached the Cuban coastline, Diáz de Paz and his crew observed illicit activity on Cabo Cruz involving some of the area's leading citizens. According to Diáz de Paz, Juan Garvey, Joseph Saravía, Pedro Carrión, and Baltasar Pérez were openly unloading bales and chests onto the beach, unconcerned that their smuggling operation had been discovered. Díaz de Paz continued his voyage to Trinidad, where he reported the illicit activities to the royal authorities. Immediately thereafter he filed a claim for his share of the confiscated goods.[88] Royal officials leaped into action and arrested the four men and their three accomplices, Joseph Rodríguez Matanzas, French citizen Luis Guiral, and Juan de Velasco, and all were thrown into the dungeons in Santiago de Cuba's fortress. In addition, all their property was confiscated.[89] Yet the accused smugglers did not remain behind bars for very long. In September, an angry Arcos y Moreno sent a circular letter to the local officials of towns and hamlets of Oriente province announcing that Carrión, Juan Velasco, Luis Guiral, and Joseph Rodríguez Matanzas had escaped and demanding that all honest citizens should be on the lookout to apprehend the fugitives.[90] Carrión was still at large in November, and local gossip circulated that he had escaped to the settlement at Río de la Hacha or Cartagena de Indias.[91]

A few months later, two more condemned smugglers, Juan Joseph Sanchi and Ignacio Pereira, staged a similar daring escape. In late 1752, the men were arrested and held in jail in Santiago de Cuba for complicity in contraband, and in January 1753, Cagigal de la Vega approved their sentences of exile to Florida. Thereafter, the details of the story differ only in the audacity of subsequent events. One version has it that the two criminals escaped from prison while they waited to learn whether their sentences of exile had been approved by the captain general in Havana. The other version of the story—by far the most romantic—related that both convicted criminals had already been transferred onto Ignacio de Anaya's boat awaiting transportation to the fortress in Apalachee and years of hard labor when fate or local sympathizers intervened (just in the nick of time) and facilitated their escape. In any case, the outcome was the same: both men disappeared into the Sierra Maestra, the rugged mountain chain that surrounds Santiago de Cuba to the north.[92]

Such incidents undermined Bourbon officials' attempts to curb local autonomy, and after 1755, a hardening in the attitude of royal officials is discernible. In April of that year, a royal order expressing His Majesty's extreme displeasure with the liberties that had been taken up until then was sent to all of the governors around the Caribbean basin. Thenceforth, they were forbidden from using the precedent of previous voyages to justify contact with foreign colonies. The royal order also made clear that the governor would be held responsible for containing illicit activity, admonishing that "you are the official who knows what happens in our ports."[93]

More stringent laws came with a greater degree of enforcement in an increased military presence. As justification for its actions, the crown needed to look no further than the case of Second Lieutenant Francisco de Veranes, who came from one of Santiago de Cuba's leading families. In 1750, in response to reports that four Dutch and English ships were engaged in contraband activity near Manzanillo, young Veranes led a patrol along the coast to apprehend the perpetrators and their local accomplices. The smugglers eluded the patrol, and Veranes was unable to apprehend or even to identify the guilty parties. When he submitted his report to the captain general, the superior officer assumed the worst of the young man; he was accused of complicity in the contraband trade and was arrested along with several other men believed to be his conspirators.[94] The governor of Santiago de Cuba, Arcos y Moreno, came to Veranes's defense, writing letters to the captain general and to the Council of the Indies, asserting: "I will never be persuaded that Don Francisco de Veranes could commit such a grievous error."[95] The captain general ultimately was forced to release the lieutenant, but he was able to use the case to argue for greater metropolitan involvement in local affairs. After the Veranes case, Cagigal argued successfully for the wisdom in appointing a lieutenant governor to Bayamo to act as the king's eyes and ears in that notoriously wicked town.[96] In addition, in 1752, Puerto Príncipe was brought under the jurisdiction of Santiago de Cuba. Beginning in mid-decade, smaller cities were assigned contingents of soldiers led by a mid-level officer (usually a captain) who was assigned the title of lieutenant governor and put in charge of administering justice. Detachments of soldiers were sent to suspicious areas, where they conducted regular patrols and received commendations for the number of apprehensions they made. New laws went into effect about how to deal with *vagos y vagabundos* (vagrants and vagabonds). A new level of scrutiny was extended to the ranks of the enforcers, especially to soldiers who were accused of looking the other way while illicit activity flourished in areas where they were supposed to be on patrol. Soldiers sent from New Spain were thought to be particularly perverse, and the most recalcitrant among them were sent to Apalachee, where they could perform "a special service to the monarch."[97]

By mid-decade, the stricter measures began to yield results, and with more troops available to go on patrol, more apprehensions were effected. In June 1755, Madariaga sent several smugglers to Havana who had been captured as a result of the heightened vigilance and increased patrols.[98] In December 1755, one such patrol, headed by Francisco José de Ortíz, was on duty in a rugged area to the east of Santiago de Cuba known as Punto Verracos when he discovered a cave containing unguarded containers of flour and clothing. Ortíz sent Baltázar Mejía back to Santiago de Cuba to get a small boat, and the illicit goods were transferred to town.[99] Patrols also began on the northern coastline within the jurisdiction of Puerto Príncipe, where in September 1757 Juan Fernández Parra, on routine patrol, discovered a beached small craft. The officer burned the suspect boat and then headed inland to all of the suspicious places where contraband activity was commonplace. There to his surprise he encountered a representative of the Real Compañía, Francisco de Roxas Torreblanca, who professed his innocence, asserting that he, too, was engaged in efforts to eliminate the contraband activity in the area. Just two days later, Fernández intercepted a sloop with contraband near the entrance to the Bay of Tánamo.[100]

As enforcement measures tightened, smugglers switched their tactics, resulting in a noticeable change in the goods carried in the contraband trade. Before the increased measures, consumer goods, especially clothing, were the primary commodities; now smugglers chose to deal in provisions, particularly flour. Ortíz's seizure in 1755 involved several barrels of flour, and Fernández's seizure two years later in Puerto Príncipe involved salted meat and forty-one barrels of flour, which "appeared to be English" in origin.[101] Smugglers and their accomplices wrote openly about how the changes in Spanish metropolitan policy affected their activities. In 1755, for example, the intendant of the neighboring French colony, Saint Domingue, ostensibly an ally of Spain, complained that "the Spanish trade was almost dead, except in provisions."[102] Yet in spite of the Spanish royal administration's vigorous attempts at eradication in the 1750s, contraband was never eliminated entirely. A powerful invisible enemy, the weather, worked against them. Shortages in provisions continued to be a problem, and smuggling, the time-honored method of providing for a community's needs, became even more vital for survival.

Siege and Disease

While the environmental crisis in the Caribbean continued, in 1756 war resumed between Europe's two great powers, France and Great Britain. In spite of the family connection between France and Spain, for the first few years of the war, Spain's monarch, Ferdinand VI, and his pro-British ministers kept their country out of the conflict.[103] Ferdinand's death in 1758 brought his half brother, Charles III, the king of the Two Sicilies, to the Spanish throne. Charles did not share the pro-British sentiment of his late brother, and his enmity toward Britain was exacerbated by a series of grievances that he had held for years prior to ascending to the throne of Spain. For the first two years of his reign, diplomats on both sides sought to negotiate a resolution to their mutual grievances, but by fall 1761, the negotiations broke down, and it became clear that Spain would enter the war against Britain on the side of France.[104]

Studies of Spanish participation in the Seven Years' War center on the American consequences of the conflict, particularly the cataclysmic capture and occupation of Havana by the British from August 1762 through July 1763. Such scholarship is unanimous that the fall of Havana represented a watershed in Cuban history.[105] Cuban writers bemoan their loss, while contemporary English accounts present the jubilant point of view of the victors.[106] By the early twentieth century, Cuban writers, influenced by trends in economic history, focused on the economic consequences of the ten-month occupation. The dominant theme of such studies is that the brief interlude introduced capitalism into the backward Spanish imperial economy. Scholars believe that because of their insatiable desire for African slaves, many Cuban creoles willingly collaborated with the British merchants. Some historians take the argument one step further and maintain that the British occupation provided the impetus that brought sugar cultivation to prominence on the island.[107] Other, more contemporary studies, see the fall of Havana as a watershed not so much due to the event itself but because it became the catalyst for extending military, economic, administrative, cultural, and social reform measures—the famous Bourbon Reforms—to Spanish America.[108] For the most part, such studies are concerned with contextualizing what happened in Cuba within the empire or within the process of imperial reform.[109] These historians recognized that Cuba was the "laboratory for reform," and they concentrated their attention on the reforms relevant specifically to the island.[110]

Detailed examinations of events in Cuba share the characteristic of attempting to explain why Havana fell. Several interpretations argue that the city was unprepared and that it was surprised and humiliated when the British captured it so easily. The surrender is attributed to an antiquated military structure and a lack of troops and supplies and to the idea that Spain's strategy was fundamentally flawed. One study argues that "Spain was humiliated at Havana primarily because the crown had failed to adjust its colonial defenses to the changing military realities in America following the War of Jenkins' Ear."[111] In recent years, the impact of disease, especially yellow fever, has been offered as an explanation for the events in 1762, noting with irony that fever had become part of Spain's defensive strategy and that such a strategy broke down when the killer epidemic did not appear until two weeks after Havana surrendered.[112] A body of revisionist work takes issue with the literature that claims the Spanish army was unprepared. The first study argues that reform measures began in Havana as early as 1761, that the military and its leaders were adequately prepared, and that Spain was, indeed, capable of defending Cuba.[113] The second study moves the analysis away from Havana, to Santiago de Cuba, examines the garrisons and troop strength in the east, and emphasizes that military strategists believed that a British attack would be in Oriente rather than against Havana.[114] Both studies maintain that Havana was not properly warned of the imminent invasion and/or that the capital was not reinforced in time.

Although it is certain that the capture of the supposedly impregnable city was caused by a combination of events, the environmental and epidemiological circumstances during the previous decade weighed heavily in the outcome. For twelve years prior to the war, alternating periods of drought and hurricanes led to food shortages and sickness. Nowhere was this combination more evident than in Oriente province, where fever, long the Spaniards' ally, became their adversary. Simultaneous with the assault on the capital, the major garrisons in eastern Cuba were incapacitated, reducing their effectiveness in sounding an alarm, and the troops that could have been used for reinforcements could not make it to Havana until it was too late to repel the siege. Concomitantly, the capital city under siege experienced similar adverse conditions and was unable to hold out against a well-reinforced and well-provisioned adversary.

By fall 1761, the Spanish crown knew that it could not avoid going to war against Great Britain, and by late that year the notification went out to vulnerable ports in the Caribbean.[115] Preparations began in earnest in Havana and in the fortresses and watchtowers that ringed the capital city.[116] Among the preparations were efforts to obtain sufficient provisions to maintain the city if attacked. Local officials were warned to take precautions in getting provisions from the French colonies and to proceed very cautiously if having to deal with the English.[117] In May, two ships arrived from Campeche bearing provisions, and the local treasury board allowed the captains to unload their cargo without paying the customary duties.[118]

At the other end of the island, the defenses of Santiago de Cuba, which mirrored those of Havana, also prepared for war. The city was defended by a fortress also known as El Morro that faced southward toward the mouth of the bay. A smaller fort, La Estrella, stood on the opposite shore from El Morro and trained its menacing guns on the entrance, ready to catch any interlopers in a withering crossfire. Like Havana, Santiago de Cuba was ringed by a defensive perimeter of small coastal watchtowers, which housed detachments of soldiers and artillerymen. Stretching along the coastline to the west was the most remote post, Guaycabón. Closer in, the outpost at Cabañas guarded the coastline and the west side of the entry to Santiago Bay from a distance of about three miles. Its counterpart on the eastern shore of the bay was Aguadores, a substantial outpost. The remaining garrisons, Juraguá and Juraguacito, were located along the coastline to the east.[119]

For Cuba's eastern residents, it was an inauspicious time to go to war. For the previous decade, misery, chronic food shortages, and disease had been part of everyday life, and conditions did not improve as they faced their old enemy. The governor of Santiago de Cuba, Madariaga, like his counterpart in Havana, Juan de Prado, began preparations to establish a provisioning system that could mitigate the effects of a siege. Early in his tenure, he had sought to improve Oriente's road system in order to bring subsistence crops to the city. Madariaga ordered additional training drills for veteran troops and militia alike to improve their readiness. An increased number of patrols ordered for the five outposts along the coast would guard against a surprise attack and also serve to reduce locals' ability to carry on their illicit trade.[120]

Yet as late as October 1760, the effectiveness of Santiago de Cuba's auxiliary garrisons remained questionable, when a strong storm struck Juraguá and damaged the barracks and the powder magazine.[121] The military official of the garrison, José Joaquín Cisneros, immediately sent his men on patrol, and they came across a small boat wrecked to the east of Punto Verracos, at the same place where Francisco Ortíz had discovered several barrels of flour in 1755. Cisneros sent a second lieutenant back to investigate, but the young officer was captured and made a prisoner by the smugglers. When the lieutenant failed to return, Cisneros sent out another party of soldiers and the young officer was rescued, but (as always) the perpetrators fled into the rugged countryside. Few were surprised when the suspects were identified as several local men and a pilot who was a native of the province (*criollo de Cuba*).[122]

To the west of Santiago de Cuba, sickness and misery afflicted the town of Bayamo. In February 1761, Bayamo's ranking official, Francisco Tamayo, reported that his town had been stricken by an unidentified fever that had spread to all of the city's population. He personally had been sick for many days and had been unable to perform his routine duties, including leading the regular patrols that the governor had implemented.[123] The already-debilitated population suffered another blow when a strong hurricane struck the area in October 1761. Tamayo wrote that the furious storm had brought rising floodwaters, drowned people and animals, and caused considerable damage to the hospital that was under construction.[124] Immediately, Madariaga organized relief supplies to be sent to the town, but the barrels of flour and wine that the governor ordered did not arrive until two months later, in November 1761. Even then, the lieutenant governor, Juan Joaquín de Landa, suspected that the captain of the boat that brought the flour and wine had skimmed off a percentage of the cargo.[125]

In February 1762, a Spanish fleet, including its flagship, the *Galicia*, under the command of Manuel Benito Erasun, brought the news that war had been declared between Spain and Great Britain, and defense preparations began in earnest.[126] The *Galicia* also brought the second battalions of the regiments of Aragón and España to Santiago de Cuba. The original strategy was to send a portion of these men on to Havana via Batabanó on the south coast of the island, but upon arrival most soldiers were so sick that they had to be disembarked to recuperate. In one moment, nearly 2,400 additional troops swelled the population of Santiago, compounding the stress placed on the city.[127] Conscious of the demands that so many more soldiers would place on the town, Erasun reassured Madariaga that the *Galicia* carried several barrels of flour to provision his troops, but when local officials broke open the barrels, they discovered that the flour was already spoiled. Hoping to salvage at least part of the precious commodity, Erasun ordered that the spoiled flour be made up into hardtack.[128] The situation eased as some soldiers began to recover and were sent on to Havana; nonetheless, many of the troops remained in the hospital in Santiago de Cuba for the remainder of the summer. The *Galicia* also carried 2,000 new rifles, which were unpacked, cleaned, oiled, and tested, ready at a moment's notice should they be needed.[129]

Upon learning of the declaration of war, a *junta de guerra* (war council) led by Madariaga and made up of senior officers was convened in the governor's house. One of the participants was the commandant of El Morro, Miguel de Muesas, who would draw upon his experience when he became governor of Santiago de Cuba in 1767 and governor and captain general of Puerto Rico in 1769. Muesas formulated a comprehensive strategy for defending the region. The garrisons at Cabañas, Aguadores, Juraguá, and Juraguacito were identified as the most likely places for a surprise attack to be launched. These outposts were ordered to be on immediate alert for a suspicious number of sails, which would signal an enemy approach. If such an offensive were launched, the artillery at each outpost would be vital in repelling the attack, so the commanders of the outposts began repairs to make sure the ramps and carriages that positioned the cannon to optimize their accuracy (*hormigón*) were in good condition. The garrisons were ordered to resist any invasion for as long as possible and then, if unable to defend their position, to retreat to El Morro, impeding the enemy's advance all along the way. The military planners also formulated a strategy for how they would respond in the event of an attack on El Morro and the city, drawing upon previous attempts to capture the city in 1741 and the defensive measures put into place by Governor Francisco Antonio de Cagigal.[130]

One of the most important components of the defense plan was a strategy to gather sufficient provisions "to withstand a siege of five or six months." Muesas was concerned because the existing scarcity of foodstuffs made it difficult for him to accumulate sufficient reserves, but he was confident that "at the first sign of enemy sails," he could call upon the neighboring villages for provisions. Beef cattle and other livestock would be driven to town and held in a pasture between the common well and a field battery, where they would be safe under the watchful eyes that would be in El Morro. The defensive plan also sought to deprive the enemy of its own provisions upon invasion, and Muesas knew that any invading force would suffer greatly, since there was no potable water within two leagues of the city.[131]

Yet in spite of the best-laid plans, the situation in Oriente went from bad to worse when the hurricane season arrived early that year. In June, Muesas received the news that severe storms had dismasted a French cruiser and forced it to seek shelter in Santiago harbor.[132] Summer brought no respite from the bad weather. Military subordinates all along the coast complained that the continuous rains disrupted their ability to perform their duties, such as going out on patrol or working on the fortifications. A second lieutenant from Juraguá regretfully informed the governor that he "could do nothing because of the rains."[133] The captain of Aguadores, José Perés, reported in July that the rain had saturated all of the rope used for fuses for the cannon, and by September he was expressing concerns that the continuous rainfall had put the maintenance projects seriously behind schedule.[134]

The crisis deepened as subsistence crops began to suffer from the unending rainfall. One by one, reports of hunger among the troops made their way to Madariaga in Santiago de Cuba. By late June, El Morro's second in command, Hilario Remírez de Esteños, son of the sergeant major of Havana, warned of an impending shortage of casabe.[135] Casabe is bread made from yuca, a root crop that is grown throughout the year. Yuca is one of the first subsistence crops to be ruined when flooding occurs because it is not harvested until it is ready to be consumed.[136] By July, the commander of Aguadores, Péres, also wrote that he had no casabe for his men.[137] The head of the post in Juraguá, Rafael Antonio de Sierra, had a sensible solution to the problem that faced his unit. He requested permission to send soldiers outside his district to buy cattle.[138] By October, Juraguá still faced starvation. By then, the outpost's correspondence was being maintained by Joseph Plácido Fernández, possibly because Sierra had fallen victim to the fevers. Summarizing the cumulative effects of weather, Fernández wrote of the "great scarcity" of provisions. The rain had ruined the crops and further damaged the poor roads, thus preventing supplies from reaching his post. Over the summer, Juraguá had received only two shipments of casabe, and he pleaded with Madariaga to allow him to send soldier Bernadino Ricardo to Holguín for food.[139]

The main garrison in El Morro was especially hard-hit. Not only did the commissary have to feed its own troops, but it also had to deal with the additional personnel that had arrived in February. From the original warning sounded by Remírez de Esteños in June, the situation deteriorated rapidly. By 1 September, the supply of casabe had run out and all that was left were the stored supplies.[140] The officers in charge of the garrison were faced with the dilemma of whether and when to break into the carefully hoarded supplies intended to withstand a lengthy siege. The responsibility fell upon the shoulders of Miguel de Muesas. On 4 September, Lieutenant Sebastián Julián Troconis reported to his commanders on the provisions that had been collected "for the use of the present war and for resupply of other posts." He informed the junta that there were 200 barrels of flour, 45 barrels of biscuit or hardtack, 197 barrels of salted pork, 50 barrels of rice, and 145 barrels of corn in the warehouse in his charge. The pasture near the fortress held 500 head of beef cattle and 70 head of other livestock such as swine.[141]

The decision facing Muesas was not easy, since he had no way of knowing whether another British squadron would appear on the horizon ready to attack, as had happened in the past. Adding to his worries, for the next ten days he received a steady stream of complaints from junior officers that their men were starving.[142] By 10 September, one company's food had completely run out, so Muesas gave permission for three officers and seventy men to disembark onto land, where they received one-third barrel of hardtack.[143] The following day, 11 September, reports arrived from another company of soldiers who had not eaten for two days, and Muesas ordered that one barrel of salted meat and another barrel of biscuit be released from the food reserves to feed them.[144] And so, as early as June, just as the British sailed toward Havana, conditions in the east were already set that contributed to or exacerbated the outbreak of fever that began by early July and that ravaged all of the garrisons in Oriente until the fall.

The fever appeared earliest in the four auxiliary garrisons that guarded against a surprise enemy attack. The first outbreak was reported in one of the posts located at the mouth of the bay, Cabañas, on 16 July, when commander Pedro Valiente wrote to the governor that he was suffering from a "*calentura terciana*."[145] Almost simultaneously, the corresponding garrison on the eastern littoral, Aguadores, reported a similar outbreak. The well-fortified outpost was particularly hard-hit. The commander, Luis Trufa, complained that he had been struck by a severe headache accompanied by a high fever.[146] The next report from Aguadores was submitted by Trufa's subordinate, Joseph Péres, who described the symptoms of the *calenturas* that afflicted his superior and the rest of the soldiers under his command: severe headache, fever, and chills. Among the victims was one soldier who had the usual headache and fever, but his condition was worsened by recurrent vomiting.[147] In an unusual move, afflicted officer Bernardo Ramírez asked to be relieved of his position because he was too ill to perform his duties.[148] Almost simultaneously, the illness struck Juraguá. On 23 July, Juraguá's commanding officer, Francisco Casals, wrote that the fortification held only one artilleryman and seven helpers, and of the twenty-five regular soldiers, two were already sick.[149] Thus, by early July, Santiago de Cuba's defensive perimeter was compromised, and garrisons intended to sound the alarm in case of attack had been incapacitated by the debilitating disease.

Just as the rash of hunger and fevers descended upon the garrisons in Oriente, the British squadron appeared in front of Havana. The surprised commanders in Havana's El Morro had no warning of the enemy's approach. The British plan to attack Havana was well executed and audacious. Instead of taking the usual course from Jamaica around Cuba's western point and approaching Havana from the west, the British had made the daring move of sailing through the Windward Passage between eastern Cuba and the French colony, Saint Domingue. The fleet had navigated the notoriously dangerous Bahama Channel with great skill and appeared off of Havana on 6 June.[150] Thus, in spite of Santiago de Cuba's vigilance and given the environmental conditions and the state of health in the outlying posts, the British fleet had sailed undetected past the incapacitated watchtowers on Cuba's southeastern coast.

Once aware of the danger, the governor of Havana, Juan de Prado, and the *junta de guerra* met to establish an emergency plan to defend the fort and the town. Ships were brought into the city, and a chain was raised across the mouth of the bay to prevent enemy ships from entering.[151] Havana's troops not housed in El Morro were placed under the command of Juan Ignacio de Madariaga, captain of the warship *El Tigre* and brother of Santiago de Cuba's governor.[152] The junta made Juan Ignacio de Madariaga the commanding general of the troops in the rest of the island, and he was charged with establishing a defensive perimeter around Havana to block the British advance in every way possible. He was also charged with impeding the sustenance of the enemy by taking comestibles from the villages around the city.[153] At the same time, an urgent call went out to the lieutenant governors throughout the island to send reinforcements, and Puerto Príncipe's governor, Martín Estéban de Aróstegui, replied that he would send infantrymen and members of the mounted cavalry as soon as possible.[154] By 23 June, the enemy's intention to conquer the city was clear, and the junta approved a plan drafted by Juan Ignacio de Madariaga to ask Santiago de Cuba to send men, munitions, and provisions. The plan was to bring the reinforcements via Jagua on the south coast. These reinforcements would join with the forces outside the city, and Juan Ignacio de Madariaga would lead a counterattack against the invaders.[155]

By the time the urgent request for help arrived in Santiago de Cuba, the military structure was already suffering from the cumulative effects of hunger and disease. Lorenzo de Madariaga convened a junta in Santiago's El Morro, and with the counsel of his fellow officers, it began to organize forces for a counterattack on the British.[156] For his part, Muesas juggled the need to provide provisions for the expedition with retaining enough food to stave off hunger in his debilitated troops.[157] Five companies from the regiments of Aragón, Havana, and Edinburgh, made up of several commanding officers, 283 men, and a chaplain, were embarked onto the warship El Arogante, and it left Santiago sailing along the southern coast toward the outpost at Jagua. Notices were sent to the villages surrounding Santiago de Cuba announcing the situation in Havana. All men capable of bearing arms were to present themselves for duty to their local authorities, and from those groups, selected civilian militiamen would be sent to Havana. Within days, men from all ranks of life rushed to volunteer for duty in the infantry and cavalry to defend the most important city in the Spanish Caribbean.[158]

But, ironically, as Oriente's men began to gather to take part in the planned expedition to lift the siege of Havana, they carried the fever from the outlying garrisons to Santiago de Cuba, where it afflicted the command structure in El Morro. Concomitant with the call to arms, Juan Álvarez, the commanding officer of the garrison at Cabañas, replied that he was unable to respond because of the chills and fever "that they [the doctors] have told me are calenturas tercianas."[159] In September, Álvarez was still in bed, frustrated because of his continued inability to "take up arms against the enemy."[160] The situation remained serious in Juraguá. Antonio Marín, of the regiment of Aragón, fell victim to the fevers, and he wrote that the medicine that the doctors gave him did not relieve his misery.[161] Marín eventually recovered, and as soon as he was well enough, he was transferred to Aguadores to relieve Bernardo Ramírez and boost the number of officers at that key post at the mouth of the bay.[162]

But the most serious situation was that the main fortress in Santiago de Cuba, El Morro, began to suffer from the effects of the disease. The first definitive outbreak occurred in early September, along with reports that the supply of casabe had run out. By 11 September, Muesas acknowledged the grim news. His subordinate officers were continually asking for food for the troops, one soldier was already in agony, and an officer suffered from the effects of a terrible fever.[163] Muesas would discover that the *calenturas* afflicted almost every unit in his command, and sickness raged among Santiago de Cuba's troops well into the cooler autumn months.[164] By October, young Hilario Remírez de Esteños also fell ill and requested to be relieved of his post because he was too ill to perform his duties adequately.[165] Meanwhile, fevers continued to plague the outlying garrisons. In Juraguá, Commander Manuel Hernández offered a grave prognosis for Second Lieutenant Pablo Gansia; the severe fever had sent the young soldier into a delirium.[166] In mid-November, the commanders of remote posts nursed their sick soldiers and coped with a reduction in the ranks in other commands such as Bayamo.[167] As late as December, Francisco de Torralbo summed up the situation: "There is not one available militiaman, all of the regular soldiers are sick, and I have only five artillerymen who are fit for duty."[168]

Meanwhile, the situation grew more desperate in and around Havana. In that city's besieged Castillo del Morro, the defenders valiantly tried to hold on while militia units from Guanabacoa and mounted dragoons resisted British maneuvers to encircle the city. By 25 June, the enemy had positioned units around the town, from the hills near Guanabacoa to Jesús del Monte, and the Spanish cavalry had to retreat to San Juan.[169] Juan Ignacio Madariaga constantly changed his position to avoid British patrols while he awaited reinforcements from Santiago de Cuba and other towns.[170] On 14 July, Prado requested an additional 800 men from Santiago de Cuba, and he was encouraged by the rumor that the first reinforcements had arrived in the harbor at Jagua.[171] More encouraging news arrived a few days later when Lorenzo de Madariaga replied that he was sending from 1,000 to 1,500 of the new rifles held in readiness in Santiago de Cuba, and he reassured Havana's commanders that Spain's allies throughout the Caribbean knew of the urgency of the situation.[172]

Soon the defenders around Havana began to experience the same adversity that plagued their counterparts in Santiago de Cuba. Early in the siege, the enemy began shelling the fortresses, and the destruction was considerable within a circumference of three to four leagues. Prado ordered women and children to leave the city, but the evacuation was hampered by the onset of continuous summer rainfall. The evacuees walked barefoot through knee-deep mud, and coaches and wagons were abandoned along the roadside, stuck in the morass created by the unending rain.[173] One contemporary account observed that "even the hardest heart would be touched by the sight of the poor, the sainted nuns and the delicate women walking through the thick underbrush with the roads full of mud from the rains."[174] By 20 June, any woman who remained in Havana was ordered to leave. If she resisted, she was forcibly removed.[175]

In July, food shortages set in, afflicting the men in Havana's El Morro, the remaining civilians in town, and the troops under Juan Ignacio de Madariaga's authority in the countryside, and soon thereafter the debilitating seasonal fevers appeared in the military and civilian ranks alike. To replace the rapidly diminishing regular forces, all able-bodied men were summoned to Havana, and citizens too aged or infirm to participate in defending the city were tasked with collecting plantains and other fruit. Sugar plantations were ordered to contribute part of their slave gangs for similar tasks.[176] By 24 July, Prado worried that "we are beginning to suffer from a shortage of flour to make hardtack and we also are not getting an adequate supply of casabe." He urged Juan Ignacio de Madariaga to send "shipments of every possible commodity," especially the vitally important bread made from yuca.[177] The residents in Villa Clara (known then as Cuatro Villas) responded to the call, milling wheat and collecting plantains, boniato, casabe, and cornmeal. Still, the situation worsened, and by late July, officials were obliged to reduce the soldiers' ration because of the lack of flour, a problem for which they offered a novel solution.[178] They proposed that the soldiers be paid the equivalent of a half ration in cash so they could try to purchase rations on their own.[179] In spite of the siege, a few *pulperias* (taverns or small stores) remained open, even though their hours of operation had been limited to the morning. Military officials recognized their utility and used them as conduits of information, posting notices on their doors that all men must arm themselves in any way they could.[180]

With the enemy on their doorstep, not surprisingly, royal officials took a no-nonsense approach to the impending crisis in the food supply. When civilians were ordered out of the city, military officers were motivated by a desire to reduce the drain on the available food reserves as much as for the women's and children's safety.[181] Several of the city's leading citizens were arrested for hoarding, and seven black men were whipped for stealing vegetables.[182] Providing food to the enemy was a capital offense, and perpetrators received no quarter. Instances of alleged treason prompted an eyewitness to write: "Almost every day men, white and black, are hung for being criminals and others for being found to have provided the enemy with vegetables, meat and other comestibles, and some who were in the jail have had their throats cut."[183] Such extreme justice also extended into the countryside, where a mulatto man was caught gathering provisions. When he was searched, he was discovered to be in possession of a passport from the British officers, and like the others, he was hanged on the spot from the nearest tree.[184] Community outrage was so great that the sacristan of the cathedral had to leave the city during the dead of night to bury those who had been executed for selling food to the enemy.[185]

When the bombardment began, hospitals in the city were pressed into action to treat the wounded. The hospital of San Juan de Dios, the oldest in the city, treated white males; the convent of San Agustín was set up to treat black and mulatto men; and the convents of Santa Clara and Belén were women's facilities.[186] Boticario Blas de Fuentes was normally assigned to San Juan de Dios, but when the authorities decided to evacuate the city, de Fuentes was ordered to the countryside to serve in the field hospitals.[187] There he treated the infirm, dispensed medicines, and supervised the bleedings, which were routine practice for treating fevers. As the number of casualties rose, the few doctors in Havana were stretched to their limit, and de Fuentes moved continuously from the country to the city to treat the sick in El Morro and in the corresponding fortress on the opposite side of the bay, La Punta.[188] Working alongside de Fuentes was Leandro de Tagle, a brother in the order of San Juan de Dios, who also attended to the infirm in the hospital in town. Tagle's responsibilities began where medicine left off: caring for the spiritual needs of the wounded and sick and administering the last rites to the multitudes of dying men. By the last days of the siege, there were so many casualties that Tagle worked around the clock, sometimes getting only an hour of rest before being called out to administer to another dying person.[189]

Soon doctors and priests alike were overwhelmed by the numbers of men who needed their care.[190] Food shortages hurt many patients' recoveries, and by mid-July commanders in the countryside received orders to collect chickens and eggs to send to the hospitals for the sick.[191] With the onset of seasonal fevers, physicians and other caregivers were even more powerless to nurse their patients back to health. On 14 July, Prado frantically requested additional troops from Santiago de Cuba, explaining that the remaining defenders in Havana were succumbing to the twin dangers of hunger and disease.[192] As the situation got steadily worse in the overcrowded hospital of San Juan de Dios, almost all of the patients were sick with *calenturas*.[193] There, the commandant of the navy, Lorenzo de Montalvo, estimated that San Juan de Dios was caring for nearly 1,200 sick men—1,080 soldiers and 95 civilians.[194]

The only hope for the defenders lay in the arrival of reinforcements from outlying towns and from Santiago de Cuba. As soon as he received notification of the enemy's assault, Martín Estéban de Aróstegui began preparations to send a relief contingent to Havana from Puerto Príncipe.[195] The governor and military officers in Oriente also dispatched reinforcements to Havana, but one of the great mysteries in Cuban history is why the additional troops never arrived. The column marching from Puerto Príncipe was but one day away when Prado was forced to surrender (thus becoming one more contentious issue between Cuban historiography and Spanish historiography). The fate of the other contingent from Santiago has never been established.

The service record of Juan de Lleonart, from a distinguished Catalán family, recorded years later, reveals what happened to the soldiers from Oriente.[196] Lleonart was one of five captains, each of whom led a full complement of seventy men aboard the *Arogante* when it sailed from Santiago de Cuba in July.[197] The *Arogante* arrived in the bay of Jagua on 24 July. It is not clear whether the contingent left the ship; but without question, sometime during the time that they were in port, the seasonal fevers struck the relief column. In 1788, recounting his many years of service, Lleonart recalled that "he had come from Santiago de Cuba in 1762 to relieve the siege of Havana but he was forced to retreat from Jagua because of his illnesses."[198] Lleonart's assertion was verified by Lieutenant Colonel Isidro de Limonta, who wrote that Santiago de Cuba sent nearly 500 troops to reinforce those in the capital.[199] Meanwhile, as the southern contingent was forced to retreat, the northern relief column from Puerto Príncipe desperately struggled to get to the capital. Like their counterparts, though, they would have faced the same food shortages and hunger and would have been forced to march over impassable roads much like the ones the evacuees took when fleeing the city. In addition, the column was likely exposed to contagions that brought on the fevers as they passed through Matanzas, which would have slowed their progress considerably.

After the siege was lifted, the lieutenant governor of Matanzas, Simón José Rodríguez, told of a variety of *calenturas* that had affected the garrison and residents of the city under his command.[200]

Meanwhile, behind the British lines, spies kept the officers informed of the deteriorating situation in Havana, and at the end of July reinforcements arrived from North America.[201] Along with fresh troops, the besiegers received a shipment of flour from Philadelphia to alleviate their hunger and the fatigue that accompanied it.[202] Around 24 July, the British attacks began to escalate; on 30 July the walls of El Morro were breached, and the fortress surrendered. Thereafter, the full weight of the British assault concentrated on La Punta, which was quickly surrounded by British troops. Left with no alternative, Prado sent an offer of surrender to the British commander, George Keppel, Lord Albemarle, and a message to his subordinates: "In just a few hours our shining light will be extinguished by an irreparable disaster."[203] On 11 August 1762, the sergeant major of the Spanish garrison, Antonio Remírez de Esteños, delivered Prado's offer of capitulation to Albemarle. Three days later, the British were formally in possession of Havana, where they remained for approximately ten months, until 30 June 1763.

Two weeks after Albemarle took possession of Spain's "Key to the New World," yellow fever descended upon the city. Historians have long noted the irony in the arrival of yellow fever two weeks too late to prevent the surrender of Havana, while completely overlooking just how close the Spanish and creole defenders were to defeating the British forces. In early July, Albemarle worried about his ability to continue the siege, telling his colleagues that the increasing sickness among his troops, the intense heat of the weather, and the approaching rainy season were circumstances that "prevent[ed his] being too sanguine as to ... future success."[204] By 17 July, he wrote to the British secretary of state, worrying that reinforcements and provisions from North America had not arrived.[205] In retrospect, Albemarle acknowledged again that the army was "severely sickly," and if the North Americans had not arrived when they did, he would have been "forced to do something desperate."[206]

The mysterious fever or combination of fevers generated by deteriorating environmental conditions that afflicted Cuba's military from one end of the island to the other have been overshadowed by the more virulent outbreak that decimated the victorious British army. Yet while the British forces suffered greatly from the tropical environment, the grinding, debilitating, and cumulative effects of discomfort, hunger, and disease took just as heavy a toll on the defenders, limited the ability of each city's hinterlands to provision the garrisons, and impeded reinforcements from Santiago de Cuba and Puerto Príncipe from lifting the siege.

Because of the imprecise state of medicine in the eighteenth century, it is almost impossible to identify the disease or series of diseases that afflicted the population with any degree of certainty. By process of elimination, Asiatic cholera may be ruled out because true cholera did not arrive in Western Europe until the 1820s and in Cuba until 1833.[207] The disease was also not yellow fever. In every case, the correspondents were experienced officers and military doctors who knew yellow fever when they saw it, and of the dozens of letters from the officers in the auxiliary garrisons, not one uses the term "*vómito negro.*" Without fail, every letter used the term "*calenturas*" or "*calentura terciana.*" Eyewitnesses and victims describe the classic symptoms: chills, fever, severe headache, and vomiting, and many victims lapsed into a delirium.

Because the area suffered from continuous rainfall, the cause could have been any of the filth-borne and waterborne diseases, such as typhoid fever, bacillary dysentery, or amoebic dysentery. Typhoid fever is a very good candidate, especially since it strikes when populations are most vulnerable due to malnutrition.[208] The various forms of dysentery are also potential pathogens.[209] Fevers could also have been the mosquito-borne diseases such as malaria or dengue fever, but malaria is less credible because one veteran complained that the medicines that the doctor prescribed did not work.[210] If the affliction had been malaria, the victim should have responded to the administration of quinine, since the drug was normally part of the eighteenth-century physician's pharmacopea.[211] On the other hand, fevers such as dengue or "blackwater" fever are very good candidates, especially since dengue is "explosive," that is, "a very high proportion of susceptible people are attacked within a short time."[212] Military units are particularly susceptible, and according to one expert, "an epidemic of dengue can render a military command unfit for duty."[213]

Given the imprecision in diagnosing with certainty prior to the discovery of bacteria, any explanation cannot go beyond the speculation of attributing the outbreak to the generic remittent or bilious fevers that were common in the tropics. It is also highly likely that *calenturas tercianas* could have been a combination of two or more diseases that struck simultaneously or sequentially. The fevers afflicted both newly arrived troops who would not yet have been seasoned and native creole soldiers; victims ranged from the peninsular lieutenant of the regiment of Aragón to the Cuban-born son of the sergeant major of Havana. Once afflicted, the victims had little chance for a speedy recovery, for a variety of reasons. For the better part of the summer, the entire island suffered from nearly continuous rainfall. In addition to the early hurricane in Oriente in June, reports throughout the summer described rains that impeded progress on one or another project and that turned the roads into impassable morasses. Continuous rainfall would have created and perpetuated a vicious cycle of disease, food shortages, and unsanitary conditions, each contributing, in its turn, to the conditions that caused the epidemic in the first place.

Misery, disease, and food shortages prevented an effective defense of Havana, and after August 1762, British troops were in control of the city and its hinterland. The British occupation is as controversial as the capture was. Nationalist Cuban historians interpret the period as one of great courage, disgraceful collaboration, and terrible cowardice. Albemarle permitted the town council (ayuntamiento) of Havana to meet and to serve as a mediator between local citizens and the occupying forces. Consequently, the ayuntamiento maintained its position of authority within local society. Albemarle also needed the cooperation of members of the creole elite. Men such as Sebastián de Peñalver and Gonzalo Recio de Oquendo have been tarred as traitors because they cooperated with the British, while on the other hand, the bishop of Havana, Pedro Morel de Santa Cruz, is lauded as a national hero because of his obstructionist behavior. The bishop's measures to thwart the British at every turn led to his exile to St. Augustine for the duration of the war. With Prado's surrender, soldiers in the regular Spanish army were evacuated to Spain, and many civilians fled to the countryside, where they remained until Havana was returned to Spanish rule in June 1763. The few civilians who returned to Havana suffered the misery of scarcity and the humility of occupation.[214]

One of the most contentious issues in Cuban historiography has been advanced by economic historians, who have studied the consequences of imposing free trade on the Spanish mercantile system. The general theme of such studies is that the unrestricted ability to introduce slaves into the island gave an impetus to Cuba's nascent plantation economy.[215] Such scholarly interpretations may now be scrutinized in the light of new evidence that suggests that the Cuban economy was not capable of absorbing a large number of slaves.[216] At the time of the British invasion, tobacco *vegas* and petty agricultural enterprises (*estancias*) dominated the countryside.[217] Early into the occupation, British slave merchants recognized that trade with Havana was less lucrative than they had expected. As early as October 1762, Philadelphia merchant Richard Waln complained to his colleague in the slaving entrepôt of Barbados: "Those who have shipped goods from Barbados or Jamaica to the Havanna in expectation of great markets are much disappointed."[218]

Yet although Liverpool and London slave traders may have come away disappointed, North America provisions shippers reaped the benefits of food shortages, since the crisis did not abate with the temporary change in imperial control. Only days after the capitulation, the head of the Spanish navy in Havana, Lorenzo de Montalvo, asked Albemarle for 10,000 pesos to provide necessary rations for the sick in the hospital.[219] Havana's town council also turned to the British commander to address the urgent provisioning needs of the population. At the top of the list was the need to provide food to the area's residents. The farms around town could provide eighty head of cattle daily and could also contribute salted meat, but those domestically produced provisions only supplied a portion of the population's needs. Thus, Havana's town council requested that the British governor bring as much flour as possible from North America.

Faced with a surly and starving population, the solution to the British commanders' problem was readily at hand, and they turned to Philadelphia, the "breadbasket of the colonies."[220] Philadelphia's merchant ships were already off the coast of Cuba since they were involved in bringing provisions to the besieging army. It was only logical that they would continue the trade in flour and biscuit as long as necessary, but common sense dictated that they would not return to their homeport empty.[221] Even though Havana had been vanquished militarily, the island could offer its "*frutos del pais*," primarily rum, sugar, and molasses, in return for supplying the vital commodity.[222]

The route to market in Pennsylvania followed a pattern established for products that were captured as prizes of war from enemy ships on the high seas en route to Europe. Such prizes of war, especially those from the French sugar-producing islands, were taken to the nearest British port with an Admiralty court, in this case, New Providence (present-day Nassau) in the Bahamas. There the enemy ship was condemned and the captain was awarded the ship and cargo as a prize.[223] North American ships leaving

Havana followed the same route and sailed to New Providence, where the captain declared his cargo and received a receipt that differentiated between "prize sugar" and "foreign sugar." Upon arrival in Philadelphia, the consignee presented the receipt to customs officials and paid the duty on importing foreign products into British colonies. This was the route followed by the sloop *Abigal*, which, even before the outcome of the siege was known, brought three casks of rum into the port of Philadelphia totaling 154 gallons of liquor.[224] After the British took control of Havana, ship's master James Wilson successfully delivered his cargo of seventy-one hogsheads of sugar aboard the *Discrete* to consignees James Foulke, Conyngham and Company, and Usher and Mitchell on 22 November.[225] Arrivals with foreign sugar were differentiated from ships carrying prize sugar, such as the cargo aboard the *Tyger*, which arrived on 4 December. Designated as "French prize sugar," the *Tyger*'s cargo was awarded to its owner, Thomas Clifford, and subsequently sold to the leading merchant houses in Philadelphia.[226]

Private vessels carrying Cuban products simply unloaded their cargoes, reloaded Philadelphia's flour and biscuit, and sailed back to the island. Two such ships were the sloop *Lovely Peggy*, owned by Conyngham and Company, which arrived on 5 October, and the brig *Albemarle*, which docked in Philadelphia on 19 November. Both had stopped over in New Providence to obtain the legal paperwork to bring in foreign sugar.[227] The frenzy of activity surrounding the *Albemarle* can only be imagined, for in just six weeks the ship was reloaded and reprovisioned, and by 3 January it cleared the port for its return voyage.[228] Likewise, the schooner *Industry*, laden with nine hogsheads, three tierces (*tercios*, a measure calculated at approximately 200 pounds), and four barrels of sugar stopped in the Bahamas, but its turnaround took several months.[229] The *Industry* did not sail back to Cuba until March of the following year. Meanwhile, merchants loaded the sloop *Adventure*, and it cleared port bound for Havana on 13 December.[230] On its way outbound, it probably crossed paths with its fellow Philadelphia ship, the *Marquis de Granby*, inbound from New Providence and laden with 4,800 gallons of foreign molasses, which arrived on 17 December. Its owners, Samuel Purviance Jr. and John McMichael, were assessed a duty of six pence sterling per gallon, and they promised to pay the duty on the molasses within one month.[231]

In spite of Philadelphia's notoriously bad weather, especially when the Delaware and Schuylkill Rivers froze solid and closed the port, the city's merchants dispatched one ship per month from Philadelphia to Havana throughout the winter.[232] With the arrival of spring, commercial traffic between the two cities increased. In March, five ships laden with flour cleared for Havana; in April, five more cleared; and two more private ships cleared in May.[233] In nine short months, Philadelphia's most prominent merchant houses benefited enormously from the provisions trade, a lesson that would not be forgotten in the coming years.

After the fall of Havana, the Spanish army was sent back to Spain and a lengthy court of inquiry was convened to assign blame and impose the appropriate punishment. There was no shortage of blame to go around, providing fodder for historians, who have sought to ascribe culpability to almost every aspect of the siege. Prado and the men in his war council defended themselves by charging that the fall of Havana was caused by a lack of proper troops, the lack of materiel and provisions, and the failure to notify him of the enemy's approach. He was particularly critical of the fighting capability of the local militias, a charge hotly contested by Cuban historians. Cubans counter with charges of cowardice and incompetence, asserting that if Prado had held out for one more day, reinforcements could have saved the city. Indeed, militia units from Matanzas were only a day away from Guanabacoa, but the professional soldiers sent from Santiago de Cuba never arrived. The defenders in Havana did not know that their reinforcements were forced to turn back because of the debilitating effects of the fevers.

Although the effects of misery, disease, and food shortages do not appear in the proceedings against the men responsible for the fall of Havana, the punishments handed down by the tribunal in Spain are revealing. Only the most senior officers responsible for the capitulation received any punishment, and even then, their punishments were mild in comparison to what they could have received from a vengeful court. Other officers involved in the military actions were not punished at all; instead almost every officer not directly involved in the surrender of Havana went on to a distinguished military career in Caribbean service.[234]

Yet the lessons from the fall of Havana were learned well, and the mistakes of 1762 would not be repeated during the reign of Charles III. Charles III's famous militarization projects for Cuba and elsewhere have come to be recognized as keystones in a radical military reform program in Spanish America. Less well studied are the connections among disease, food shortages, and imperial policy, which also commanded the attention of Charles's reformers. After 1763, a thorough overhaul of the commercial system included serious attention to the problems in maintaining an adequate and reliable food supply, which, in turn, became a key tenet of the gradual movement toward free trade (*comercio libre*). Yet in spite of their best efforts, the malevolent El Niño and La Niña sequences worked against Bourbon reformers at every turn. The consequences of these climatic phenomena will be explored in subsequent chapters.

CHAPTER THREE

It Appeared as If the World Were Ending

T HE END OF THE Seven Years' War in Europe and in the Americas brought momentous political and territorial changes. Great Britain emerged as the winner, while her primary rival, France, was vanquished. Spain was dragged into the war because of her commitment to her French relatives and suffered a major defeat when Havana fell in 1762. The Treaty of Paris ended the war in 1763 and resulted in territorial realignments in North America. France was forced to give Canada to Britain, and Spain relinquished the Floridas to secure the return of Havana. In compensation, France ceded Louisiana to Spain.[1] By 1764, the once-extensive French empire in the Americas was reduced to the sugar-producing islands in the Caribbean and French Guyana, while France's Bourbon cousin, Spain, had "added another desert to her empire."[2]

Among the residents in the ceded colonies, the unwelcome change in sovereignty brought confusion and created resentment, and neither Britain nor Spain dealt efficiently with their newly acquired territories. Britain assumed control over Canada and, to prevent political and ethnic rivalries from developing between the French colonists and British North Americans, closed the frontier across the Appalachian mountains. In Florida, almost all of the Spanish residents evacuated to Cuba and were gradually replaced by British colonists.[3] In Louisiana, Spain inherited a surly population of Frenchmen and attempted to placate them through economic and political concessions. While Great Britain celebrated its victory, a far different atmosphere pervaded the court of the Spanish monarch, Charles III. The fall of Havana generated a debate that would lead to sweeping reforms in virtually every area of imperial policy, the well-studied Bourbon Reforms.[4]

The immense body of literature about the administrative, political, fiscal, social, and cultural dimension of the Bourbon Reforms needs not be revisited at length, except to note that virtually every study of Cuba depicts the return to Spanish rule as celebratory. The officials chosen to implement the royal wishes—Ambrosio Funes de Villalpando, the Conde de Ricla; Field Marshal Alejandro O'Reilly; and engineers Silvestre de Abarca and Agustín Crame (or Cramer)—had a clear mandate and unlimited authority to carry out their monarch's wishes.

Implementing these changes and especially ameliorating the food shortages in Cuba would prove to be a far more difficult task. For the first three years, 1763 through 1765, weather cooperated with the European powers, but by 1765, the return of the El Niño/La Niña sequence was evident throughout the Caribbean. Havana province experienced intermittent drought; Oriente province suffered through repeated drought/deluge sequences. Elsewhere, the interlocking provisioning system was strained by hurricanes in Puerto Rico, and Louisiana received the brunt of two storms. In many areas, the aftermath of disaster brought political unrest, particularly in the form of slave uprisings on other Caribbean islands. Of all the war-weary European nations with colonies in the Americas, France and Spain were the least prepared to deal with the escalating environmental crisis; the colonial subjects were at a loss to explain what was happening to them.

Trial and Error: 1763–1766

On 30 June 1763, the Conde de Ricla received Havana from the British commander, and the following three years became a period of trial and error. The Bourbon reformers scored some notable successes in overhauling Cuba's troop structure and beginning massive fortification projects around Havana. Fiscal reforms began with the arrival of an intendant, Miguel de Altarriba, in February 1765, and a new mail system under the auspices of an experienced administrator, José Antonio Armona, began the same year.[5] Later Cuban officials would follow the lead of mainland officials and expel the religious order, the Jesuits, from the island in 1767. Yet the problem of adequately provisioning the population of Cuba would present a perplexing paradox. First and foremost, the goal was to fortify the city to prevent such a disaster from ever happening again, but the massive militarization projects required a large increase in population. Ricla and O'Reilly brought 2,675 Spanish army troops, and an equal number of navy personnel arrived with the commander of the navy, Juan Bautisa Bonet.[6] In subsequent decades, thousands of Spanish troops were assigned to Cuba, many of whom ultimately stayed after their enlistments expired.[7] Extensive construction projects required workers, a problem that was addressed in three ways. The first was to use slave labor. After 1765, a sizable number of slaves could not be absorbed by the private market, and they wound up working on Havana's fortifications.[8] The second solution was to sentence criminals from all parts of the empire to hard labor in Cuba. Such men also became part of the work gangs assigned the task of augmenting and reinforcing the original fort in the city, La Fuerza, repairing the city wall, and refortifying the watchtowers of La Chorrera, San Lázaro, and Cojímar.[9] In some cases, European craftsmen were employed in Havana's rebuilding, but as head engineer Luis Huet complained, such skilled workers were scarce. His solution was to allow soldiers to moonlight on their off-duty hours, being paid extra for their labors. In every case, the influx of new arrivals contributed to population increase. Most of the newcomers were Europeans who refused to eat casabe, the bread made from the Caribbean staple carbohydrate, yuca, and demanded that they be provisioned with white bread (*pan de harina*). Unlike lower-status workers elsewhere, even the free colored militia members and king's slaves had access to the royal bakery, which provided much of the bread for the city.[10]

As royal advisers debated the interlocking problems, it became clear that militarization was incompatible with the existing provisioning system. More troops meant a greater demand on the provisioning capabilities of Cuba in particular, and on the Spanish imperial economy in general. By 1765, royal officials realized that in order to feed the increased number of soldiers and sailors, they could not continue to do business as usual. Bureaucrats recognized that they had to take steps to provide consumer goods, foodstuffs, and slaves, while at the same time be able to market Cuba's products in an effective manner. Such demand required a change in the existing economic organization, especially in how goods circulated throughout the empire. Their solution was a complex and interconnected plan to bring provisions to the island while excluding foreigners from the Spanish imperial economy.

The first facet of their plan was to retain the existing regulations that required Mexico to provide flour to Havana and her satellite cities. For decades previous, the interior Mexican provinces had been obligated to send part of their crops to the Caribbean colonies, and in spite of the challenges of the 1750s, royal bureaucrats saw no reason to alter this aspect of the system. The Mexican economy functioned well on several levels, and any alteration met stiff resistance from the *consulados* (merchant guilds) of Veracruz and Mexico City.[11] In their minds, the problem with Mexican production was not supply but rather making the delivery system more efficient. At the same time, consumer goods such as cloth, hardware, and china, along with luxury provisions such as wine, olives, olive oil, cured ham, and additional flour, continued to come from Spain. A free trade declaration in October 1765 permitted registered ships to carry such products and provisions from nine cities in Spain to Cuba.[12] Although labeled "free trade," the Spanish government did not envision this system as extending beyond the Spanish empire, and such trade was still restricted to Spanish ships manned by Spanish crews sailing among Spanish ports.[13]

A second component of reform was to address the issue of domestic food production. If the fall of Havana had demonstrated anything, it was the inability of Cuba's small farms to provision the military garrison adequately. Field Marshal O'Reilly submitted a detailed opinion on how to increase Cuba's food production. One of the suggestions was to create agricultural communities made up of retiring or separating soldiers. These veterans would perform a dual function by also forming the basis of Cuban rural militia companies. The proposal suggested that commercial regulations be relaxed because restricted commerce only played into the hands of civilians who willingly purchased from British smugglers because of existing shortages of necessary goods. While promoting greater freedom of trade within the Spanish imperial system, O'Reilly called for strict enforcement against contraband activities and vigorous punishment of violators of the new decrees.[14] He also called for measures to increase the white population by encouraging European immigration. Foreigners, particularly Germans, were welcome, but they must be Catholic. O'Reilly believed that Cuba's civilian population would increase by 600 families per year.[15] The success of O'Reilly's plan can be seen in the number of towns that sprang up between 1763 and 1800, and Havana's hinterland expanded rapidly as a consequence of the explosive population increase.[16]

The final measure enacted by the Spanish crown was the least popular with the residents of the Hispanic Caribbean. In spite of a chorus of complaints, the monarchy authorized the reestablishment of a monopoly company for Cuba. Since 1740, the original Real Compañía had controlled the commerce of the island, but with Charles III's ascension to the Spanish throne in 1759, its power was curtailed.[17] In 1763, the old Real Compañía was replaced by a new monopoly chartered in Cádiz, the Compañía Gaditana de Negros. The Compañía Gaditana retained the prerogative to provide slaves to Spanish America, and so it usually was referred to by its original name, the Asiento de Negros, or simply the Asiento.[18] Allowing the Asiento to continue its existence was a strategic compromise because of its ability to fulfill the demand for slaves.[19] The monopoly's directors remained directly allied with transatlantic slave traders via factors in Cádiz, and the Spanish crown chose to rely on it as it had in the past.[20] Initially, the Asiento planned to sail directly to Africa, purchase slaves, and take them to an entrepôt created in Puerto Rico. The operation in Puerto Rico served as a quarantine and seasoning facility, and from there the slaves were distributed throughout the Caribbean basin. Almost immediately, this plan proved to be unworkable, and just one year after its creation, the Asiento turned to British slavers in Jamaica, Barbados, and other British colonies.[21]

Given the recent trends in scholarship and the unfortunate name retained by the new company, studies have focused on the Asiento's slaving activities, while its importance in providing provisions to the Caribbean has been downplayed.[22] The Asiento is invariably seen as a negative—almost medieval—influence, but fundamentally it was a broker between foreign providers and Hispanic consumers for the two items that the Spanish empire could not supply. By granting the monopoly permission to deal with foreigners, the crown intended to remove foreign ships from Cuba's ports, which would reduce smuggling and minimize the opportunity to engage in espionage. The Asiento retained the privilege to import flour along with the slaves, theoretically to feed the slaves so as to avoid impacting local food supplies, but the Caribbean governors dispelled that notion when they complained to the Spanish governing body for the American colonies, the Ministry of the Indies, that the Asiento was selling the flour to their residents.[23] The quantity of flour was directly proportional to the number of slaves, that is, for every slave the Asiento brought in, it was permitted to bring in one barrel of flour. Before and after the decree, the proportion was routinely violated.[24] Yet, then as now, the law of supply and demand determined whether the commercial enterprise would be a success, and the reorganized Compañía Gaditana was a resounding failure. Key to its failure was—in its own estimation—its obligation to import what Cuba did not need—slaves—while it was limited in its ability to bring in the much-desired flour produced in North America.

Confronting Problems in Cuba

Under these circumstances, the Conde de Ricla came to Cuba with orders to implement reforms. The new governor and the men in his retinue harbored a great deal of resentment against the British empire in general, against British merchants who were exploiting Cuba's riches, and against Cuban creoles who were accused of collaborating with the enemy.[25] From the moment he arrived in Havana, Ricla was determined to make life as difficult as possible for the remaining British subjects in the city. Although the terms of the peace treaty signed in February 1763 promised that British merchants would be allowed to remove their goods, Ricla refused to let them leave with their merchandise. To no avail, the heads of merchant houses in London petitioned the British secretary of state, the Earl of Halifax, who proposed to send several ships in ballast to retrieve their property.[26] By November 1763, all of the British officials had left the city, and Albemarle had left for Jamaica, leaving the remaining merchants to fend for themselves.[27]

At the same time, the island became more and more desperate for provisions. Just days after Ricla's arrival, Lorenzo de Montalvo, now the intendant of the Spanish navy, who had been one of the intermediaries between the British authorities and the Cuban population, wrote to the Marqués de Cruillas, the viceroy of Mexico, asking for flour, ham, and vegetables, along with 300,000 pesos.[28] A month later, in August 1763, when his first request went unanswered, Montalvo repeated the need for flour and other comestibles to provision the increasing number of troops.[29] The situation was equally difficult in Santiago de Cuba, which had not had the benefit of North American flour during the occupation. In February 1764, the new governor of Oriente, Juan Manuel Cagigal, the Marqués de Casa Cagigal and the son of the man who had been captain general in the 1750s, appealed to the captain general for permission to send a ship to Jamaica because of a shortage of foodstuffs in the city of Santiago. In a lengthy letter to the captain general, Cagigal wrote about the misery and deprivation that plagued the eastern end of the island.[30] Three months later, Montalvo's and Cagigal's petitions were reinforced by the intendant of Havana, Altarriba, who also urged that Spanish ships or ships sailing under Spanish colors be allowed to travel to Jamaica and New York to purchase provisions on an emergency basis.[31] After another dire summer, proponents of direct contact with British provisioners received an unexpected show of support from a surprising quarter, the administrator of the Real Compañía, Martin José de Alegría, who had been sent to Havana to report on the viability of retaining the monopoly in its existing form.[32] Alegría had rendered his conclusion to Ricla and the Council of the Indies: the company would remain an unprofitable venture because of the need to transport provisions from Spain and the ever-present threat of war that made such imports prohibitively expensive. His letter reinforced Cagigal's argument that Santiago's captains desperately needed permission to travel outside of the Spanish system to find enough provisions for the city's population to survive.[33]

The conflict over food for Cuba is characteristic of the multiplicity of interests that competed for influence and power on many levels. On an imperial level, the conflict was between Britain and Spain, although at times France, Denmark, and Holland also could influence policy. Within the British empire, a strong antagonism grew between metropolitan officials in London and provisioners in the thirteen colonies, primarily Philadelphia. Sorting out the combatants in the Spanish empire is a bit more complicated. On one hand, rivalries existed between Cuba and Mexico, in which Cuba was the perennial recipient of the largesse of the mainland. In return for their obligation to supply the Caribbean, Mexican merchants demanded exclusivity on many products that they argued should be provided from within the empire. On the other hand, the monopoly company in Cádiz set up to bring desirable goods to Cuba countered that since they were taking the financial risks, they should be rewarded with preferential treatment. Within Spain itself, yet another rivalry existed between old factions in the traditional gateway to the Americas, Cádiz, and newcomers in the northern cities of Bilbao, El Ferrol, and La Coruña.[34] Finally, all had to contend with the fiscal watchdog for each region, the intendant, who was charged with increasing tax revenues, restricting frivolous spending, and guarding His Majesty's financial interests in every way possible.

Political rivalries within Cuba complicated matters further. The new administrator of the mail service, Armona, who reported directly to the minister of the Indies, possessed sweeping powers of his own, including the power to limit what was carried on vessels contracted to carry correspondence back and forth from Spain. His authority over boats carrying mail increased the potential for conflict with the commander of the naval forces in port. Yet the most significant conflict on the island was between old Cuban families who could trace their lineage to the earliest settlers of the island and newcomers such as Ricla and O'Reilly who came with the obligation to reform Cuba's defensive posture within the empire. Almost immediately a bitter personal and professional rivalry grew up between these two groups, which only deepened in the subsequent decades. Each faction had its supporters at court, and the fortunes of each rose and fell, depending upon which faction held power in Spain.[35] Under such conditions, reaching a consensus was virtually impossible, and the residents of all of the islands in the Caribbean were caught in the middle.

While bureaucrats and functionaries argued over how to feed the island, local residents resorted to their time-honored reliance on contraband. Not waiting for permission to trade with nearby foreign colonies, the residents of eastern Cuba took matters into their own hands and set out in search of food for their communities. During the month of August 1763 alone, two cases of contraband involving creole vessels were sent to the Council of the Indies for adjudication; in each case, local ships were caught leaving the island with mules and horses to trade with foreign colonies for food.[36] Another seizure involved the capture of an English sloop while it was cruising off the island waiting for the coast to be clear.[37] Oriente's residents also sought to secure provisions by smuggling excess provisions in legally documented cargoes. In July 1764, customs officers seized a large quantity of illegal goods on the frigate *Nuestra Señora del Rosario* when it sailed into Trinidad from Cartagena de Indias.[38] Three months later, in October 1764, 67 *tercios* (13,400 pounds) of illegal flour were discovered aboard the packetboat *Santisima Trinidad*, and royal officials were also successful in preventing a British ship, *La Bretagna*, from unloading its illegal cargo in Oriente province.[39] The rise in contraband activity prompted another pessimistic letter from the administrator of the Asiento in Havana, Alegría, who wrote to the beleaguered governor in Santiago de Cuba, offering words of condolence but little hope for improvement. According to the Asiento's own representative, the company's prices were exorbitant, and so it would be impossible to contain or eliminate the contraband trade.[40]

Royal administrators' attempts to deal with Havana's food crisis were undermined by the presence of a handful of British merchants, who remained in the city in spite of the officials' efforts to remove them.[41] With hunger widespread, the *ingleses* were an unwelcome reminder to the authorities and population alike that flour and other comestibles could be obtained in nearby foreign ports if only they could go outside the Spanish commercial system. The British merchants, therefore, posed a challenge to metropolitan bureaucrats, who sought to implement freer trade on their own terms. Spain faced the dilemma of determining how to prevent foreign traders from capitalizing on the empire's enormous purchasing power while still obtaining what only they could supply, but the government did not have a viable solution to the crisis at hand. After allowing the British merchants eighteen months to settle their affairs on the island, a royal decree ordered them out of Havana in 1765.[42] Representatives of the ten remaining English merchant houses were told to inventory their remaining stock in preparation for their departure.[43] Meanwhile, the men and their families were held under house arrest.[44]

Environmental Crisis Returns

While royal administrators grappled with these problems, in 1765 another El Niño sequence descended on the Caribbean. The first area to feel the effects was Oriente, where drought returned in early 1765. In April, Cagigal wrote a top secret missive to the minister of the Indies, the Marqués de Grimaldi, reporting that there was a severe food shortage in Santiago de Cuba, and that even if had been enough food, there was a shortage of workers for the harvest.[45] The island-wide drought also ruined crops around Havana, forcing Ricla to face the grim reality of the situation.[46] Citing dire necessity, he gave his approval for a voyage to New York to purchase food and other provisions. The contractor, Francisco Salvatore, proposed to sail his packetboat, La María, to the northern city to purchase butter, lard, cheese, wine, cider, linseed oil, candle wax, and various medicines. The packetboat never left Havana because Altarriba inexplicably revoked the concession and the voyage was canceled.[47] Shortly thereafter, Ricla was recalled to Spain, and the inspector of the army, Pascual de Cisneros, succeeded him as interim captain general pending the arrival of the new appointee, Antonio María Bucareli. Cisneros fared little better with the intendant when the two clashed over who should have authority to oversee the unloading of 691 badly needed barrels of flour that arrived on an English frigate under contract to the Royal Company.[48] Four months later, the administrator of the royal mail system, José Antonio Armona, joined the fight when he protested the exemption from a tariff concession that was granted to Salvatore for another shipload of provisions that he had managed to find for the city.[49]

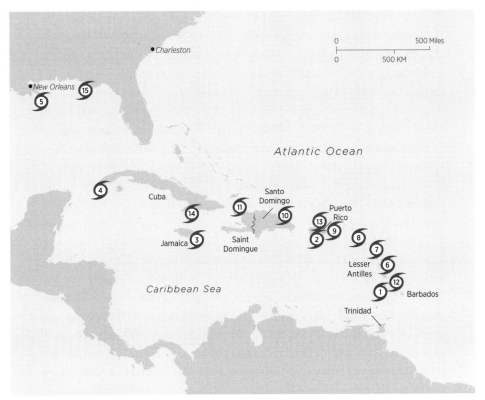

Map 3.1 Hurricane strikes in the Caribbean basin, from August to October 1766. Sources: see appendix 2.

Amid such political and economic chaos, in summer 1766, La Niña re-turned to plague the Caribbean basin. Every major area from the Lesser Antilles to the western Gulf of Mexico suffered repeated and severe dam-age from mid-August onward.[50] The heavily populated French island colonies were particularly hard-hit, but the British, Danish, and Dutch is-lands also sustained serious damage. Martinique was the first island to feel nature's wrath when it endured an intense early storm on 13–14 August— 440 people died in the storm, 580 people were wounded, and 80 ships were damaged or destroyed.[51] This powerful hurricane went on to strike Jamaica, where equally devastating consequences were reported.[52] A month later, in September, the northern islands of the Lesser Antilles were in the line of fire when a slow-moving hurricane caused extensive flooding on Montserrat as it lingered three days over the island. Subsequent storms later in the month devastated Saint Domingue, Tortuga, St. Christopher,

St. Eustatius, and Jamaica.[53] Skirting the north coast of Hispaniola, the September storm made its last landfall near Tortuga, where it destroyed the salt pans of that island. Finally, on 6–7 October, the southern islands, which had escaped destruction early in the season, became the victims of the late-season system.[54]

The storms that devastated the rest of the Caribbean also caused extensive damage in the Spanish colonies. Puerto Rico was particularly hard-hit by the August system that ravaged Martinique.[55] The slow-moving September hurricane that destroyed Montserrat also ruined the majority of the subsistence crops in the eastern end of Puerto Rico.[56] Barely a month later, the October storm brought additional destruction; Puerto Rico was in the path of the deadly system that hit Guadeloupe with a twenty-five-foot storm surge.[57] This hurricane continued westward, where it came ashore in eastern Cuba on 10 October, striking the mountains of Oriente province and inflicting significant damage on the village of Santiago del Prado.[58] Reports from the lieutenant governor, Bartolomé de Morales, described how the hurricane blew in from the southeast and lingered over the mountains for twenty-four hours, dumping copious amounts of rainfall. The mountainous terrain could not absorb the runoff, causing the river to rise out of its banks, sending waves of floodwaters cascading down the mountainside into populated areas. By some miracle, the raging waters were diverted into two currents before they reached Santiago del Prado, thus saving the village from total destruction.[59] Yet the problems caused by the hurricane did not end with the storm's passage. Two weeks later, a sergeant of the grenadiers took to his bed with fever. He fought off the disease and survived, but the population of royal slaves did not fare as well. In a possible attempt to avoid spreading the fevers, slaves were separated into groups according to their ages, and out of humanitarian concerns, the oldest male slaves were exempted from paying religious fees for their burial in the event that they fell victim to the fevers. As the fever raged through the town, Morales reported that "removing the cadavers will be difficult, but [he would be able] to accomplish the task without help." He closed his missive to Cagigal with the familiar prayers for the governor's health and safety.[60]

An earthquake had devastated the area just a few months previous, and as the disasters continued to hit the region, they overwhelmed local resources. Neighboring towns were ordered to contribute food, but conditions were no better elsewhere. As late as April of the following year,

Puerto Príncipe "suffered from no small measure of scarcity," wrote lieutenant governor Juan de Lleonart, but he assured Cagigal that he would mobilize the residents to begin providing *tasajo* (dried beef) to send to Santiago de Cuba.[61] Commercial interests were not exempt from the obligation to contribute to the welfare of society, and the Asiento was ordered to use funds in its coffers to aid the victims.[62]

Although escaping a direct hit, Havana and the western end of the island also endured the effects of poor weather and scarcity, giving the new captain general, Antonio María Bucareli, his first experience in coping with disaster and exercising his authority to deal with an emergency in his jurisdiction. Just days before receiving news of the tragedy in Oriente, he had received a copy of a new royal order reiterating the regulations under which governors could operate in emergency conditions. One of the restrictions was that no governor could grant permission to sail to foreign colonies.[63] At the first reports of misery, Bucareli responded to Cagigal's urgent plea for help, promising that as soon as the rains of "this cruel season" let up, he would send some relief.[64] Weeks later, Bucareli informed Cagigal that he had 201 barrels of flour on hand, and that although he needed at least five barrels for daily consumption for Havana, he would send 300 pounds of flour on the first ship.[65]

Conditions throughout the Spanish Caribbean continued to deteriorate when two severe storms struck the western Gulf of Mexico, undermining Spanish efforts to bring order to its newly acquired colony, Louisiana. The first storm, in early September, grazed the northern Gulf Coast and finally made landfall on the Texas coast. It missed the populated areas along the Mississippi River, but it delayed construction on fortifications planned for the entrance to the river at Balisa.[66] The second hurricane, on 22 October, however, did serious damage when five Spanish convoy ships laden with the *situado* (subsidy) were blown off course and sank at the mouth of Mobile Bay.[67]

The collateral effects of two hurricanes drained the limited funds in Louisiana's treasury and contributed to the political instability that threatened Spanish rule.[68] The man chosen as governor of Louisiana, Antonio de Ulloa, was among the most capable and talented men in Charles III's group of military officers.[69] Intelligent, educated, and loyal, and with decades of experience in the Americas, these men were, nonetheless, unprepared to deal with metropolitan intransigence, new geopolitical realities, and the environmental crisis that they faced. Nowhere did this unfortunate combination of circumstances come together in more vivid fashion than in Louisiana from 1766 through 1768. In December 1766, immediately after the hurricanes sank the ships carrying the money to sustain his colony, Ulloa wrote to Havana pleading for replacement funds. In an effort to relieve the suffering of Louisiana's residents, Ulloa reported that emergency provisions could be obtained from the British colonies to the north, but that these purveyors of flour and other comestibles demanded payment in a timely fashion.[70] For months thereafter, in January, in February, and in March 1767, Ulloa's pleas went unanswered. In February, the governor authorized Baltazar Toutant Beauregard, the captain of La Campeleon, to sail to the French colony, Saint Domingue, with survivors from one of the shipwrecks. En route, he was charged with stopping in Havana to present letters to Bucareli requesting immediate relief.[71] Hope soared when the mailboat, the Postillión de Mexico, arrived in Louisiana on 15 March, only to be dashed once again. The Postillión carried no funds for the struggling colony, and Ulloa waited another three months, until 60,000 pesos finally arrived.[72]

Louisiana was not alone in its distress. By November 1766, most Caribbean islands faced immediate starvation because of the destruction of the food supply on hand. Crops already harvested and in storage were ruined when roofs gave way under the weight of water and the wind. Crops in the ground rotted in standing water or were washed away. Coastal flooding destroyed salt works that could have been used to preserve what little food could be found. The disaster for Martinique and the French colonies was compounded because France had lost its agricultural colonies, and its normal supply routes had been disrupted when it ceded Canada to Great Britain and Louisiana to Spain.[73] The French colonies faced not just an immediate and local crisis of subsistence but also an imperial crisis of supply and delivery. The situation on Martinique and Montserrat was exacerbated because most of the vessels in the ports were destroyed. Martinique lost eighty vessels in the August storm, the September storm destroyed every ship in port in Montserrat, and thirteen additional ships were lost at St. Christopher (St. Kitts).[74]

The situation went from bad to worse. The unusually active hurricane season of 1766 was accompanied by an epidemic of insurrection that swept the Caribbean basin.[75] One of the earliest uprisings occurred in Savannah la Mar, Jamaica, in early October when slaves on several plantations went on a rampage and killed several white settlers. Almost simultaneously, a series of squalls hit Kingston and damaged many vessels in port. The unlucky coincidence complicated the post-disaster situation by hampering the government's ability to put down the rebellion.[76] The relatively minor uprising in Jamaica paled in comparison to a full-scale rebellion on Grenada, one of the hardest-hit islands in the October storm. There, in late December, 600 to 700 escaped slaves established maroon communities in the mountains from which they raided farms and plantations in the lowlands.[77] In Louisiana, Ulloa faced serious resistance in New Orleans, where French residents resented the cession of their territory to Spain and forced him to delay his entry into the city. He and his retinue took up residence on the barrier island of Balisa, some thirty miles downriver from New Orleans, while negotiations were conducted with the recalcitrant French citizens on settling their grievances and surrendering the city to Spanish rule. In the middle of such uncertainty and discomfort, a small contingent of soldiers mutinied against the intolerable living conditions on what amounted to little more than a sandbar in the middle of the Mississippi River.[78]

The timing of all of the revolts throughout the Caribbean basin suggests that they were launched when the conspirators felt that they had the greatest chance for success. Indeed, such a conclusion conforms to rebellion theories that gained popularity in the 1980s and early 1990s.[79] More important, however, the timing of the rebellions is consistent with a recent trend in post-disaster theory that demonstrates a strong correlation between the destabilizing effect of disaster and the propensity for political unrest.[80] In the aftermath of a disaster such as a hurricane, all state systems are strained, and people who see themselves as oppressed tend to take advantage of a vacuum in authority to change their oppressed status.

It is at this point that the authorities' behavior at every level from the metropolis to the individual Caribbean islands becomes significant. As could be expected, metropolitan nations sent substantial numbers of troops to quell the rebellions in their respective colonies. British authorities transferred several army regiments to Jamaica, and in short order local residents optimistically wrote that the leaders of the insurrection had been captured and hanged.[81] In Grenada, the rebel leaders were hunted down and executed in particularly gruesome fashion.[82] In Louisiana, Ulloa's loyal soldiers held firm and thwarted the mutiny of twelve conspirators, who were sent to Havana for two years of hard labor on that city's fortifications.[83]

Yet if the harsh military response represented the stick, it was accompanied by the carrot in the form of economic concessions granted to most Caribbean islands by their metropolitan governments.[84] Britain was among the earliest nations to enact a Free Ports Act for its colonies, in 1766.[85] In the aftermath of the hurricane of 1766, local officials on Martinique acted of their own accord and allowed foreign flour and biscuit (hardtack) to be imported into the colony.[86] After the mutiny in Louisiana, Ulloa granted economic concessions to the local residents. Among these concessions was permission to purchase flour from British settlements upstream from New Orleans in the Illinois territory.[87]

Predictably the local governors' actions in Martinique and Louisiana brought a storm of protest from metropolitan merchants, who complained that the measures were little more than a subterfuge for contraband. Yet both Martinique's and Louisiana's governors' actions were approved at the highest level of royal authority. Martinique's initiative was upheld by the royal government in Paris on 30 March 1767.[88] Two weeks later, on 14 April, the Spanish metropolitan government rescinded the stringent royal order issued the previous year and officially granted discretionary powers to its Caribbean captains general that allowed them to purchase foodstuffs from foreign sources in an emergency.[89] The decree stands in sharp contrast to previous policy, in place since the mid-1750s, under which officials were severely reprimanded when they granted permission to local captains to seek provisions on neighboring islands in times of hardship.[90] Indeed, just days before the new order was issued, Bucareli had prohibited Cagigal from granting a license to sail to Jamaica, telling the eastern governor to seek provisions in Bayamo, where there always was a good supply.[91]

The connection between the liberalization in Caribbean trade and the ecological crisis generated by the hurricanes of 1766 has, until now, gone unnoticed. Most political and economic interpretations see the movement toward free trade as the spread of classic laissez-faire economic policy. In Britain, reform measures are usually hailed as a challenge to the West India Lobby, sugar planters from the British Caribbean islands, who controlled economic policy until the mid-eighteenth century.[92] With regard to the Spanish American colonies, most scholars, subject to a healthy dose of Black Legend pessimism, see the measures as long overdue steps along a continuum that led to Spain's gradual and unwilling acceptance of Enlightenment economic philosophy. They also ascribe any movement toward liberalization to Spain's inability to control smuggling among her possessions and the British colonies. In Louisiana, in particular, Ulloa's actions are seen as a futile attempt to stem the tide of provisions from a vigorous Yankee economy pushing westward, along with a belief in the manifest superiority of free trade. Even the most optimistic interpretations see the privilege as part of a generalized movement toward the inevitable reform of Spain's antiquated economic system.[93]

Yet when analyzed as a consequence of an environmental crisis, the concessions of 1767 can be seen as not merely one discrete moment of imperial weakness but as a rational and astute response to an emergency. Royal officials were clearly aware that in disaster's aftermath a strong correlation existed between the way they responded and the degree of discontent in their subjects.[94] Given the proximity of rebellions in Jamaica and Grenada, they recognized that times of crisis were not the time to enact punitive measures; instead, the commonsense reaction would be to implement a plan to alleviate the impact of the disaster on the civilian population. The royal order of 1767 removed all of the Caribbean basin captains general from the restrictions on trading in foodstuffs, and in doing so, it gave them extraordinary power, albeit in a local context. When disaster threatened, these men would have unprecedented autonomy in setting aside both metropolitan and local regulations that would hamper recovery efforts. Nonetheless, although the concessions were a step toward liberalization, they cannot be compared to the free port fever that swept other nations' Caribbean colonies in the 1760s.[95] The British, the French, and the Dutch opened their ports to other nations' ships for their own political reasons. Spain shrank from opening her harbors because of the foreigners' propensity to smuggle, or, worse still, to engage in espionage.[96] Instead the decree placed the decision in royal governors' hands. They were the ones who could decide when crisis in their area was so severe that Spanish ships could go to the free ports and purchase provisions. Foreign ships were still unwelcome in Spanish harbors.

The liberalization decree also marked a fundamental shift in the definition of contraband. From the inception of European rivalries in the Americas, contraband to the Spanish colonies flourished under a multitude of disguises. Illicit goods included almost everything that was not provided through legal channels, but the concessions of April 1767 signaled an important change.[97] Durable goods and slaves remained on the list of articles that could only be imported in limited quantities by approved carriers; but food, especially flour, became a negotiable item that under certain conditions could be imported into Cuba and her satellite cities outside the monopoly's control and at the captain general's discretion.

Throughout the Atlantic world, the change in what was now legal in the Hispanic Caribbean came with repercussions for the sources of supply. For some suppliers, the concession was irrelevant or inconsequential. Great Britain produced manufactured goods that remained on the prohibited list, so suppliers of textiles, cutlery, and china still needed to sell their goods to factors in Cádiz or Bilbao, who, in turn, sold them to Spanish companies for shipment to the Americas. For the large transatlantic firms engaged in slave trading, the concession was also irrelevant, since the slave trade to Spain's Caribbean cities remained in the Asiento's hands. The situation for North American producers was different. They did not produce, indeed they were prohibited from producing, manufactured products. On the other hand, their primary products were exactly what the Caribbean needed in times of crisis: provisions such as flour, Indian corn, and rice. Not surprising, when the news became public that both the French and the Spanish markets in the Caribbean were now open—albeit on a limited basis—the British colonies along the eastern seaboard rejoiced. The temporary concessions granted to cope with post-hurricane shortages were first announced in newspapers in Charleston and New York in October 1766, and by January 1767 the notices appeared in Philadelphia.[98] Up to this time, the most lucrative foreign markets in the Caribbean had been the free ports such as St. Eustatius.[99] With direct access to the French cities, North American provisions now could make their way to the Spanish colonies via Martinique, Guadeloupe, or Saint Domingue or continue to filter in quasi-legally down the Mississippi River to New Orleans. Moreover, North American provisions that sustained Jamaica and the Bahamas (and were smuggled into Cuba) now could become legal in crisis situations. Significantly, though, Spain's concession was intended to be a short-term expedient—temporary and local in scope. Throughout the Caribbean basin, it was understood that invoking the privilege was only to be done in a grave emergency. Spain remained opposed to opening its colonies to foreign trade, and necessary provisions were still supposed to come from Mexico or Spain or be imported by the Asiento.

If local residents and governors used the privilege to trade with foreign colonies at will, the monopoly company's representatives in each city were ever ready to complain to their directors that their privileges were being violated. After the disastrous summer of 1766 and the subsequent decree of 1767, the Asiento's monopoly on the foreign flour trade was undermined further. The Asiento, like its predecessor, was wildly unpopular in the cities that it was charged with serving. Nowhere was the rivalry among local political authorities, the residents, and the Asiento more visible than in Puerto Rico. From fall 1766 through spring 1767, British factor John Kennedy became a victim of the enmity between political and economic forces. Kennedy, a man with long experience trading with the Caribbean, was among the Philadelphia merchants who traded with Cuba during the British occupation of 1762–63. His ship, the *Lark*, was one of the last vessels to leave Havana in April 1763 in anticipation of the return of Spanish sovereignty.[100] After the Asiento's debacle in its attempt to trade directly with Africa, the directors signed a contract with a slave trading enterprise organized in Cádiz and owned by Englishman John Brickdale, and in 1766, Brickdale sent Kennedy to Puerto Rico to oversee his operation. In September, the factor arrived in San Juan harbor, but the governor, Marcos de Vergara, refused to allow Kennedy to leave the Asiento's ship until a formal visa was granted.[101]

While the factor awaited the governor's decision, the first hurricane struck the island, followed three weeks later by the second, even more devastating storm. The Asiento's boat was sent to other ports in search of provisions (presumably with Kennedy on board), while the governor, taking no chances, wrote to Madrid outlining the circumstances of the dispute.[102] In February of the following year, the Ministry of the Indies replied that the monarch had upheld the governor's refusal to allow the Briton to land.[103] After surviving two hurricanes and the winter aboard ship, Kennedy was stricken with yellow fever, and based upon humanitarian concerns, at last he was permitted to disembark to be treated for his illness.[104]

The cavalier treatment of the man who was so vital to their business success, combined with the threat to their exclusive privilege, greatly angered the directors of the Asiento. During the winter of 1766–67, they complained bitterly and often to the Ministry of the Indies on how the governors of the Caribbean continually usurped their privileges.[105] The thrust of their complaints was the familiar charge that commerce with foreign ports would encourage contraband and espionage.[106] For their part, the Caribbean governors countered that it was the company that violated imperial laws by sending ships with crews of French and English sailors, in clear violation of the decree of 1765.[107] The governors further criticized the Asiento, charging the company with everything from price gouging to negligence. One of the most serious complaints came from Cumaná (present-day Venezuela), where the town council claimed that the Asiento was responsible for a smallpox epidemic because it had brought in slaves who had not passed through the required quarantine measures.[108]

Bureaucrats at court could overlook the incessant bickering between Asiento factors and Caribbean governors because after the liberalization decree in April 1767 the Spanish Caribbean enjoyed an uncharacteristically plentiful supply of food. They may even have believed that they had finally solved the provisioning dilemma, lulled into a false sense of security. In January 1768, the head notary of the treasury in Havana, Pedro Antonio de Florencia, compiled his annual report for the previous year. He noted with satisfaction that eight boats had arrived in port with 1,017 *tercios* of "good quality" flour, according to the evaluation of Havana's most competent bakers. As required, they had shipped a portion on to the other cities on the island. In March 1767, one such shipment of 150 *tercios* had been examined by José Robles and Vicente de Quintana, residents and licensed bakers in Havana, who swore that the flour was fresh and well packed for the journey. The shipment was transferred overland to the southern port, Batabanó, where it was loaded onto the goleta the *San Francisco de Asis*, under the command of captain Pedro Fernández, for its journey to Santiago de Cuba.[109] Six months later, on 21 August, the sloop *La Lipe,* commanded by Juan de Framategui, left for Oriente loaded with 200 *tercios* of flour, which had just arrived from Mexico. Like the cargo that went out in March, *La Lipe*'s shipment had been certified as fresh and well packaged.[110] A grateful administrator of the city council of Santiago de Cuba, Juan Garvey, conveyed thanks to officials in Havana in late October 1767, reassuring the capital that "the shortages [in Santiago de Cuba] had been eliminated."[111] To metropolitan and Cuban bureaucrats, it probably appeared that the difficult times had passed.

The coming of the new year and events in Puerto Rico, however, rekindled the ongoing conflict between the monopoly and local governors and revealed the inadequacies of the existing system. The conflict began when two ships from Philadelphia arrived in San Juan harbor—the sloop *Hibernia*, captained and partially owned by James McCarthy, and the frigate *Charming Susanna*, with Thomas Connor at the helm. Both ships had sailed from the northern port in November 1767, declaring their destination to be Jamaica, but when they arrived in the Puerto Rican capital on

23 December, both captains claimed that they sailed under the auspices of the Asiento.[112] Together, they carried a total of 1,439 barrels of flour contracted to the monopoly's new factor, Vicente de Zavaleta, under the terms of a contract that had been negotiated with a commercial firm in Cádiz, Arturo Moylan and Company.[113] The contract was executed in Cádiz, but Moylan and Company was a branch of another company in London run by James Moylan and his brothers, who had been in business since 1722.[114] The skeptical governor of Puerto Rico, Vergara, whose victory over the monopoly in preventing John Kennedy from disembarking on Spanish soil was still fresh in his mind, seized the cargo as contraband, declared both ships as forfeit for smuggling, and detained both captains and their crews, awaiting a resolution from the Council of the Indies.[115]

While the captains and crews languished in Spanish custody, the governor and the factor sought guidance from their superiors in Spain, each justifying his decision. For his part, Vergara was simply performing one of the most important tasks of his position, fighting the illicit trade. Zavaleta's superiors in Cádiz, the Asiento's directors who had arranged the shipments from Philadelphia, were equally eloquent that in order to provide fresh flour in equal quantities as the number of slaves, they had to contract with the British colonies, ideally Philadelphia or New York. The decision from the Council of the Indies arrived in early 1768. Charles III approved of Vergara's actions and upheld his decision to confiscate the cargo and impound the vessels. Zavaleta, as the representative of the Asiento, was held responsible for any costs incurred in Puerto Rico. The monarch further authorized payment for the flour. Royal officials were ordered to take good care of the precious commodity and to store it in the royal warehouses to make sure that it did not spoil, pending a decision on its distribution. The ships and their crews were released and allowed to return to their home ports, and by late March, Connor, McCarthy, and their men had returned safely to Philadelphia.[116]

Meanwhile, in Havana, similar circumstances reinforced the difficulties of the existing commercial system. In February 1768, the mailboat *Quiroz* was sailing its regular route between Veracruz and Havana with correspondence and passengers for Spain. Once in Havana, its officers tried to buy four barrels of flour that had arrived in November of the previous year and been stored since that time in government warehouses. When royal officials broke open the barrels, they discovered that the flour was spoiled, and they were forced to declare 200 barrels unfit for consumption. At the same time, North American captain Joseph Carbo sailed his packetboat into Havana harbor "in distress," asking for firewood and water to continue his journey. He just happened to have fine, fresh flour aboard his ship. The junta was faced with the problem of whether to buy Carbo's flour to feed Havana's residents or to send him on his way. After a discussion in which the participants debated the "royal interests that every minister must observe," the officials concluded that the "urgent need" in Havana warranted their violating the rules. Carbo was allowed to sell his flour to the royal warehouses, and Altarriba himself authorized the payment of 5,500 pesos.[117]

By February 1768, both major ports in the Hispanic Caribbean were well stocked with precious North American flour, so there was virtually no market for a cargo of Mexican flour that arrived in Havana in March aboard the Spanish register ship the *Nuestra Señora de Dolores*. The *Dolores*'s captain, Antonio Aday, was given permission to proceed on to Santiago de Cuba with his cargo of 150 *tercios* of flour.[118] There he was greeted by a firestorm of controversy. After leaving Batabanó the previous August, *La Lipe* had arrived in Santiago de Cuba with its cargo of flour. By October, the flour had been off-loaded and officials were making good progress in distributing the vital commodity to the residents. One recipient was Julián Marín, who took possession of 260 pounds for the navy.[119] On 8 January 1768, the town council authorized that any excess flour be stored in the royal warehouse at government expense.[120] Thereafter, events in Santiago de Cuba took the familiar downward turn; when royal officials opened the barrels, they discovered that the flour was rancid. The administrator for the royal treasury asked three reputable bakers to render their opinion, and all three agreed that the flour was in the last stages of spoiling. Unlike Havana, however, Santiago de Cuba's residents would not take the impending food shortage quietly. Violent protests broke out in the streets, with angry citizens voicing their anger over the poor quality of flour that always made its way to Oriente. Only the arrival of Carbo with 200 barrels of flour shipped from Havana saved the public from severe hunger and the political situation from deteriorating into full-scale rebellion.[121]

Then the usual recriminations began. City fathers could not ignore the grumbling by the public over its justified refusal to accept inferior flour. The town council met in emergency session, hearing conflicting and heated testimony from locals, who complained that Havana was always the first recipient of fresh flour. The townspeople railed against royal officials, who, in turn, sought to defend the imperial restrictions.[122] A man with long experience in the intracoastal trade, Francisco de Arrate, master and captain of the packetboat *San Juan and San Guillermo*, outlined the complicated procedures involved in the flour trade. Even though he only sailed between the western and eastern ports of Cuba, he still was required to secure permission to sell his cargo in his port of arrival. He maintained that the delay in receiving permission caused the spoilage, and not coincidentally, that the majority of the spoiled flour came from Veracruz.[123] Arrate received support from a powerful quarter when administrator Garvey added his endorsement to the complaint.[124] Cagigal forwarded Santiago de Cuba's protest on to Havana. Altarriba, the consummate bureaucrat, responded angrily that "there was no doubt that the flour that went out on *La Lipe* was in good condition, and the only delay was in the normal time usual for the voyage."[125] Tellingly though, in a separate, unofficial letter, Altarriba confided: "When the residents have access to foreign flour, they despise ours."[126] The final insult was that the cost of the 200 barrels of flour was deducted from Santiago de Cuba's portion of the *situado*.[127]

Conditions were equally precarious in Louisiana. Ulloa had received 60,000 pesos in June 1767, which were entirely spent by August simply to pay the debts in arrears. That month, he dismally reported that he had stopped paying the salaries of the French officials who remained in New Orleans to help with the transition to Spanish rule. The head of the Spanish commissary had stopped payments for flour, other provisions, and ammunition brought downriver by French and English traders. Ulloa warned of the "general and continuous clamor" of the residents of Louisiana who depended on the royal treasury.[128] By the turn of the new year, the last of the funds in the treasury were gone, leaving locals and administrators without the wherewithal to provide for daily rations. Ulloa sent a grim warning to Havana: "The results will be a complete and painful disaster, because as I have already informed Your Lordship, there are no resources over here upon which to recur in an emergency."[129] By the following February, in 1768, the price of victuals had increased, not because of scarcity but rather because the provisioners had lost considerable sums of money by extending credit to locals only to be told that there were no funds to pay the invoices. One Captain Moore from New York, "one of the principal providers of flour and salt meats of this colony," waited more than six months for payment.[130] In July 1768, Ulloa granted him a license to sail to Havana, accompanied by a lieutenant in the Spanish navy, Andrés de

Valderrama, to present his voucher directly to treasury officials.[131] By August 1768, Louisiana's reputation for insolvency had spread up and down the eastern seaboard, from New England to Pensacola and upriver to Illinois. Merchants were warned to "stop making trips down here as there is no money to do any trading."[132]

The Hurricane of Santa Teresa, 1768

Such were the conditions in the Spanish Caribbean on the eve of hurricane season 1768. On 15 October, a horrific storm struck western Cuba.[133] Named for Santa Teresa for the day on which it struck the island, the system came ashore on the south coast near Batabanó and exited via the north coast near Havana, ravaging the western half of the island and cutting a path of destruction approximately 150 miles (50 leagues) wide on either side of the storm's center (roughly from present-day Pinar del Río to Matanzas).[134] The south side of the island, ringed by shallow coastal marshes, sustained a direct hit, and the deadly storm surge swamped fourteen of the ships waiting to be loaded with tobacco at the wharf at Batabanó.[135] Four more boats were beached in the shallows, and another vessel was carried nearly three miles inland.[136] In Havana harbor, fifty-five ships lay at anchor when the storm hit. The winds blew in from the east, then from the north, then from the west, accompanied by a fifteen-foot storm surge. As the winds came in from the east, the ships in port broke loose from their moorings and were thrown up against the city wall, one against the other. Then the wind shifted direction, and vessels that had so far escaped damage were scattered about the eastern and southern shores near Regla and the mouth of the Río Luyanó. Only two frigates, the *Juno* and the *Flecha*, escaped harm, because they had new anchor cables and because they were moored in the middle of the bay. Most other ships were seriously damaged, and several, including the packetboat *Despacho*, were destroyed.[137] His Majesty's brigantine, the *San Juan*, rode out three terrible days at sea between Cuba and Jamaica at the mercy of the storm. Afterward, the ship limped into port in Jagua on the southern coast to be repaired.[138] The countryside was particularly hard-hit, and in the storm's aftermath, reports of the devastation began pouring into the captain general's residence from far and wide.[139]

A close examination of the recovery efforts made in the hurricane's aftermath gives powerful evidence of the enlightened and responsible attitude that pervaded the ranks of Charles III's officials. Until his reign, it was everybody's—and nobody's—responsibility to cope with disaster, and any relief usually came from the church. Yet the much-heralded tension between church and state, so characteristic of the Bourbon Reform era, could be put aside in the face of catastrophe. In the aftermath of the storm, all acknowledged that the hurricane was an act of God and gave thanksgiving for safe deliverance. Charles III's officials immediately began working to lessen the effects on the civilian population.[140] The obligation shifted to the government, and the responsibility began at the top. An experienced, well-informed, Enlightened despot even before he ascended to the Spanish throne, Charles III was at the forefront of modern economic ideas.[141] The hurricane of 1768 marked an important watershed, and thereafter the monarch began a judicious policy of disaster mitigation that continued throughout the next twenty years of his reign.

Once aware of the magnitude of the crisis, Charles III's surrogate on the island, Bucareli, took decisive action that placed the burden for recovery on all sectors of society, both private and public. As soon as the immediate danger had passed, he ordered military patrols to ride out from the capital and from remote detachments to declare martial law and to prevent looting.[142] In town, martial law was declared, a curfew was imposed, taverns and small shops were ordered to close at dusk, private homes were asked to put torches on the fronts of their houses, and groups of soldiers patrolled the streets.[143] Bucareli personally led the efforts by riding his horse and visiting *estancias* and *vegas* and offering assistance to the suffering residents.[144] The commandant of the Castillo del Morro, José de la Cuesta, organized groups of soldiers to comb the harbor in rowboats and canoes to search for drowning victims of the storm surge.[145] The few small boats that escaped harm were pressed into service for the patrols and to ferry soldiers, provisions, and medicines to the numerous beached vessels that ringed the shores of the bay.[146] The commandant of the navy and the seamen onboard the grounded ships worked tirelessly to save the newest warship, the sixty-gun *Santiago*, another warship, the *America*, and other vessels that were damaged in the storm.[147] The island's property-owning citizens were responsible for reporting the damage to their *estancias, ingenios,* and *vegas,* and these reports were sent on to Havana by the local constables (*capitanes del partido*).

Given the ferocity and the extent of the storm of 1768, it is perhaps surprising that there were not more casualties—37 deaths and 117 injuries.[148] To begin with, the population outside of Havana was sparse, with the majority of rural residents clustered in scattered hamlets in the interior.[149] Another factor was population distribution. Although Havana's population was substantial and concentrated near the city, regulations prohibited building within a specified distance, about 4,500 feet or three-quarters of a mile, from the coast.[150] Havana and its barrios outside the city walls contained more than 40,000 people, yet when the fifteen-foot storm surge struck the north coast, it did not cause the immense loss of life of the flash floods of subsequent hurricanes because the population lived well inland.[151] In addition, the fertile bottomland of the rivers, which was vulnerable to flooding, was also off-limits since it was reserved for tobacco cultivation.[152] Finally, the destructive and irreversible process of deforestation caused by population increase and economic expansion had not yet begun in earnest.[153] The authority to cut lumber rested with the royal navy, and this exclusive privilege was maintained until 1789.[154] With the exception of the hurricane in Oriente in 1766, which struck a mountainous region that already had runoff flooding because of its terrain, the massive mudslides caused by deforestation and subsequent soil erosion were still twenty-five years in Cuba's future. Most casualties from drowning were reported on the shallow and swampy south coast. When the storm surge hit Batabanó, Captain Antonio de la Guardia was aboard his boat in harbor, and he and a sailor from his crew were never seen again and their bodies were never recovered.[155] Later, search parties discovered six badly decomposed bodies caught in the mangrove roots along the swampy southern coast, victims who were never identified.[156]

The majority of the victims died from injuries sustained by flying debris or were caught when the structures in which they had taken shelter collapsed upon them. Even so, the casualty figures are modest considering that the storm destroyed over 5,500 houses in Havana province.[157] In the village of Jesús del Monte, where most of the structures were wattle and daub, several residents lost their lives when houses and outbuildings were torn apart by the wind. After visiting all of the farms in his district, the *capitán del partido* returned to his own devastated plantation, where a slave remained in critical condition and a ten-month-old baby had perished.[158] On the *sitio* of widow Antonia de Solis Puñales, a slave died when he was pinned under a fallen palm tree, and a vagrant was killed by flying debris.[159] Guanabacoa suffered five people killed and forty-five wounded as a result of the storm,[160] and in the barrios to the west, six were killed, including three men and three women.[161] One graphic report came from the administrator of the Jesuits' property, who told of a pregnant slave, Catarina Mollombe, who died during the storm. The plantation's surgeon managed to save her unborn child, delivered by cesarean section after her death.[162]

Unlike in rural areas, in Havana most houses and other buildings were constructed of wood or stone, and the city walls offered a degree of protection from the wind. Nonetheless, many structures, especially the church towers, sustained considerable damage. The hurricane destroyed the parish church when the bells, clock, and clock tower came crashing through the roof of the building to the ground.[163] Eight more of Havana's churches suffered damage to a lesser degree, as did the jail, the city council's building, and a portion of the city wall.[164] Substantial buildings also fared better in the barrios outside of the city, where only 43 mortar-and-tile structures were completely ruined, as compared to 218 wattle-and-daub structures.[165] Detailed reports from the commanders of the fortifications around the city told a similar tale. Doors were blown in and roofs blew off powder magazines, exposing the precious gunpowder to the rain. Near the Puerta de la Tenaza, a portion of the city wall fell to the ground. The oldest fortress in the city, La Fuerza, lost all four doors and the bridges that spanned the moat encircling the building. All of the windows in the lodgings of the regiment of Lisbon were blown in, and walls fell in the apartments of officers Benito Saavedra and Francisco Bello. Even the *prición de distinguidos* (the prison for distinguished criminals) was deemed unsafe because of the loss of several windows; no mention was made as to whether the damage allowed the offenders to escape.[166]

The outlying rural areas reported nearly total devastation. Immediately after the storm, militia captain José de San Martín rode out to inspect the damage in the countryside surrounding Guanabacoa and reported that in his district the parish church, the hospital, two convents, and 1,405 houses were partially or totally destroyed.[167] Simultaneously, his brothers-in-arms, captains Miguel Peréz Barroso and Francisco Rodríguez and lieutenants Tomás Alvarez, Antonio Montiel, and Manuel de Beralles, took to their horses to search the countryside for victims. Juan Carrasco left his position as supervisor of a sugar mill to help in the search and rescue efforts, and even aging Andrés Pulgarón of the disbanded militia company joined in the efforts.[168] Three of the villa's clergymen submitted a poignant report to the Council of the Indies. They wrote: "Not only were houses and crops ruined but the secondary consequences were that many families were now homeless. Many more were sick and wounded from the wounds and blows that they had received during the hurricane, and [this misery] was followed by a shortage of food because the rain has continued to fall and nobody could go out to the fields to see if anything could be salvaged."[169]

Reports were equally dismal from other towns and villages, especially to the east and south of Havana. The parish priest of San Miguel del Padron spoke of the "*horrendo temporal*" (horrific storm) that destroyed the parish church, damaged the militia barracks and over 160 *estancias* in the district, and forced the villagers to construct makeshift shacks (*chozas*) to protect from the elements.[170] The captain of Regla, Francisco Blandino, brought the victims into his own house, which had come through the storm unscathed.[171] A long account written by the overseer of the former Jesuit plantation, Río Blanco, located to the east of Havana, exemplified some of the information that officials desired about the storm's trajectory. The hurricane began at one o'clock in the afternoon and raged until seven o'clock at night. After enduring six hours of continuous wind, rain, and flying debris, "it appeared as if the world were ending."[172]

Once survivors were rescued, government, ecclesiastical, and private mitigation efforts swung into full gear. The responsibility for dealing with the impending famine fell upon the local constables, and their first task was to evaluate how much of the crop was destroyed. The most important food of the common people was the plantain, *el pan de los pobres*, the poor man's bread, and all accounts indicated that the crop was totally destroyed.[173] In addition, major subsistence crops such as yuca, corn, and rice were ruined by flooding, even in areas that did not sustain a direct hit, and crops that had been harvested were ruined when torrential rains breached even the most secure roofs.[174] Another important commodity was salt—vital for food preservation. Most salt pans were located in coastal areas or on adjacent shallow islands, and torrential rains and coastal flooding ruined both the source of supply and the harvested salt that awaited transportation to the cities.[175] Livestock were particularly vulnerable to flooding and flying debris, and the carcasses of dead animals posed a significant health hazard in the hurricane's aftermath. In almost every storm, regardless of intensity, collateral damage was done to the tobacco crop, which provided income for most of Cuba's farmers.[176]

Since reports from every village reiterated the news of total devastation, as soon as the storm had passed, Bucareli immediately sent any ships that were still seaworthy to Veracruz, Campeche, and Cumaná to secure provisions and meat.[177] Notices also went out to governors of the areas of Cuba that were not affected to contribute a portion of their normal crop to help the recovery.[178] Compliance was not voluntary, and persons who refused were punished through fines and sentences of forced labor.[179] One of the traditional enclaves of resistance was the Isle of Pines off the southern coast, a virtual fiefdom controlled by the Duarte family since the earliest years of settlement. Bucareli was not pleased when provisions were slow in coming from the ranches on the small island. A scathing letter from the captain general to the island's constable, Nicolás Duarte, prompted the man to action, and in short order boatloads of salted meat arrived in Batabanó for shipment to Havana.[180] Another cause for censure was speculation and/or price gouging. Immediately after the storm had passed, Bucareli declared that prices would be held at the same levels as before the hurricane.[181]

Soon supplies began to arrive from unaffected parts of the island.[182] By mid-November, the first of such shipments arrived from the village of Alvarez, accompanied by a note from the district's captain, Rosendo López Silvero, who reported that the hurricane was not as damaging in his district and so his residents had banded together to send what foodstuff they had to Havana.[183] Food was also sent from the eastern end of the island, and salt came from the salt pans of New Granada (present-day Venezuela).[184] Treasury officials diverted provisions intended for the garrison to provide for the slaves on the Jesuit plantations.[185] In the neighborhoods surrounding the city, militia members, who usually were prohibited from engaging in commercial activity, were allowed to sell plantains (at pre-storm prices) to the hungry people.[186] Palm thatch for roofs was often in short supply after the storm because the high winds had stripped the palm trees bare. To remedy the problem of providing shelter for thousands of homeless victims, hamlets to the west of Havana that had escaped the brunt of the storm provided the thatch for rebuilding.[187] As the recovery continued, the question of how to pay for the repairs was raised. The captain general decreed a tax of four pesos per *caballería* (33 acres) of land and three per house, which was divided equally among the residents. Ecclesiastical properties and members of the militia and regular forces were exempt because they were already serving in the recovery efforts.[188]

The obvious gravity of the situation would have permitted the captain general to implement emergency measures, but the cautious bureaucrat stopped short of opening the island to foreign trade, at least officially. But for some areas, such as the remote settlements on the south coast of the island where Havana's ability to enforce existing laws was tenuous at best, emergency supplies arrived through networks created by illicit commerce. As in the past, when disaster struck, opportunists from neighboring areas rushed into the affected areas with provisions for the starving victims. Just days after the storm, a military patrol along Cuba's south coast came across a boatload of Jamaican smugglers providing "emergency foodstuffs" to the Spanish residents in the settlements along the San Juan River. When questioned, the smugglers admitted that they had come because they had heard of the suffering and starvation due to Cuba's crops being ruined, and it appears that they were allowed to leave unmolested, although empty-handed.[189] Smuggler Carlos Yons was not as fortunate. He was caught red-handed when his small boat was shipwrecked near the San Juan River, with five "stolen" male slaves and one female slave on board.[190] Another boat discovered in the bay of Jagua was undoubtedly engaged in illicit commerce with local residents, but it managed to escape from the military patrol.[191] Yet government permissiveness only went so far. One Mr. Cabot, the captain of a French merchant ship, sailed into the port of St. Marys on the north coast, having lost his mainmast and all of his provisions. He requested safe harbor in order to repair his damaged boat, but he came to the attention of the authorities because he requested permission to pay for the repairs through the sale of 800 pieces of china he carried as part of his cargo. His request was denied because he had a previous charge of smuggling under similar circumstances pending before the judicial system.[192]

The ensuing crisis paralyzed the western end of the island for weeks, and the widespread reports of misery and destruction were also accompanied by disturbing notices of "disorders." As had happened on other Caribbean islands after the hurricanes of 1766, slaves, forced laborers, and military deserters took advantage of the post-storm confusion and fled. One of the most vulnerable plantations was Río Blanco, which had been confiscated from the Jesuits.[193] Bucareli's decisive leadership was responsible for averting a potential revolt on the Jesuit plantation. Immediately following the storm, several slaves fled into the woods, and the captain general sent out parties of soldiers to capture the fugitives.[194] Smugglers provided an escape route off the island. As for Carlos Yons, there is no way to determine whether the slaves he had on board when apprehended were payment for foodstuff or escapees seeking their freedom. The village of Jagua on the south coast, hard-hit by the storm, provided an ideal escape route for four military deserters who fled to Jamaica aboard an English ship. The number of *negros y guachinangos fugitivos* was also a concern for Jagua's captain, Andrés Brito Betancourt.[195] Some fugitives made their way to the barrios outside of Havana where *capitanes del partido* noted with anger the number of "vagabonds who have infested our village."[196] Like his counterparts, the captain of San Miguel del Padrón took preventative measures and sent militia members out on patrol to "prevent any disorders" that might erupt in the post-storm confusion.[197]

Within a few weeks, western Cuba was well on the way to recovering from the hurricane of Santa Teresa, but the event would linger in the memories of countless survivors. Through decisive leadership, Bucareli engendered a spirit of cooperation and voluntarism throughout the island. The southern coast near Batabanó was greatly affected, but by 9 November, Juan José Galán proudly reported that the workers had managed to construct a provisional warehouse and that they had saved 2,000 *tercios* of tobacco.[198] The eastern province of the island had been especially generous in providing meat and other provisions, and by December Bucareli wrote to the captain of Cuatro Villas (present-day Las Villas), Jose Antonio de Azorena: "Everything has been done to my satisfaction."[199]

Largely because of the decisions made by Bucareli, the mitigation efforts after the hurricane of 1768 are almost a textbook model of the way to respond to a potential catastrophe. In the storm's aftermath, several reports were sent back to Spain to be examined and debated by the Council of the Indies and Charles III. In every instance, Bucareli's peers—even his rivals—were effusive in their praise of his conduct during the crisis. The head of the royal navy, José Antonio de la Colina, referred to Bucareli as "our beloved governor," who while all of the troops and civilians were in a state of consternation, exhorted those in positions of authority to go out on patrol to preserve the public order. It was Colina who reported to Madrid that Bucareli and the ecclesiastics had provided for the victims from their own funds.[200] The administrator of the mail system, José Antonio Armona, offered an even-more-precise description of the captain general's leadership. As the fury of the hurricane waned, Bucareli paced impatiently at the portal of his house in order to be able to act immediately once the danger had passed.[201] Yet the most extraordinary expression of appreciation was a sixty-four-stanza poem dedicated to the captain general and his efforts to bring recovery to Havana as "a magnificent and excellent leader."[202] In December 1768, the captain general received the approval of the minister of the Indies, Julián de Arriaga, who praised the governor's efforts in coping with the disaster. In short order, Bucareli would receive an even-greater recognition from Charles III. In 1770, he was awarded the most coveted position in all of Spanish America, that of viceroy of Mexico.

Just as it seemed that nothing could destroy the feeling of accomplishment, in December 1768, a French merchant frigate sailed into Havana bay carrying unexpected visitors and unwelcome news: aboard was the governor of Louisiana, Ulloa, his retinue, and several Spanish soldiers who had been forced out of Louisiana by a mob of rebellious French inhabitants on 29 October. A stunned Armona conveyed the news to the Council of the Indies[203]—the disaster that Ulloa had predicted just months previous had come to pass. Now royal officials in Cuba had to deal with the recovery efforts on the island and the rebellion in Louisiana. Ulloa's untimely arrival compounded the problems because the majority of officers and soldiers were occupied with the recovery efforts. Neither Ulloa nor Bucareli could respond to the rebellion, and the insurrectionists controlled Louisiana over the winter of 1768–69.[204] The following spring, Spain's most celebrated general, Alejandro O'Reilly, whose authoritarian demeanor was well known, returned to Cuba.[205] He stopped in Havana long enough to put together a military expedition of more than 2,000 soldiers to crush the rebellion in New Orleans.[206]

The New Orleans rebellion is universally and invariably analyzed in a political context and as a purely local event. Revisited in the context of other rebellions and acts of resistance, the rebellion follows the pattern of countless cases of post-disaster insurrection.[207] The aftermath of the Louisiana revolt also conforms to the classic metropolitan response. In the case of outright rebellion, particularly slave rebellion, the response is swift, unequivocal, and brutal. Spanish reaction to the rebellion is notable for its brutality, yet in spite of his heavy-handed methods in dealing with the local population, O'Reilly faced the same problems as Ulloa, with one noteworthy exception: his authority was reinforced by more than 2,300 Spanish soldiers. On the surface, at least, public order was maintained, especially after O'Reilly executed the ringleaders of the rebellion.

Yet the fundamental problems of Louisiana did not disappear with the arrival of a large army of occupation, and indeed, the situation worsened.

Over the winter of 1769–70, severe food shortages in New Orleans compelled O'Reilly to grant permission to merchants, mostly foreign ones, to import flour and other provisions.[208] During that winter, Louisiana spent 70,000 pesos, nearly half of its 160,000 peso budget, for flour purchased from the "English."[209] The flood of flour into New Orleans continued unabated, and by the early 1770s, royal officials were not even making an attempt to control the flow.[210] Henceforth, Louisiana's flour trade would be dominated by North Americans and, to the Asiento's dismay, would set an example for the remainder of the Spanish islands in the Caribbean.

CHAPTER FOUR

The Violence Done to Our Interests

A T 9:00 A.M. ON 10 OCTOBER 1773, during the height of hurricane season, a meeting was convened onboard the *fragata de correos* (mail frigate) *El Quirós*. The participants contrasted sharply, from the grizzled, veteran captains of the coastal and international trade, to the tattered group of harbor pilots that guided ships into port, to the elegant, well-heeled appearance of the captain general, the Marqués de la Torre, and the chief administrator of the mail system, José Antonio de Armona. Despite their differences in status, the men had come together on the orders of the monarch with one specific purpose: to determine whether two of His Majesty's ships, the *Grimaldi* and the *Quirós*, should be allowed to leave port and sail through the dangerous Old Bahama Channel to Spain. From the captain's quarters, the group could look north past the formidable fortress of Havana, El Morro, to the straits of Florida. What the men observed was cause for alarm. Swirling winds caused the flag above the fortress to whip wildly. Darkening storm clouds were gathering on the horizon, which lowered ominously as they watched. As the experienced captains debated the visible weather signs, other members of the meeting warned of the danger inherent in the conjunction of the moon and the autumnal equinox. Throughout the Americas, a full harvest moon in the month of October was universally known to mariners as the most dangerous time of the year.[1] The men were reminded of tragedies of the recent past: the loss of the packetboat *Colón* in November 1771 and the horrific storm of Santa Teresa on 15 October 1768, which "still made people tremble."[2] Finally, the junta agreed that the mail ships should remain in port a few days more. Satisfied that their work was done, the group adjourned with the knowledge that their decision probably saved the ships and their crews from the fate of other mariners who had braved the Atlantic during the autumnal equinox.[3]

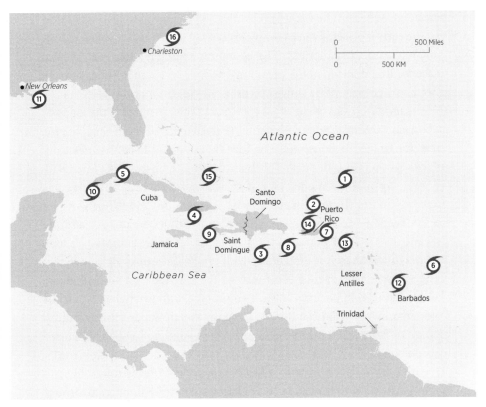

Map 4.1 Hurricane strikes in the Caribbean basin, from June to September 1772. Sources: see appendix 2.

The caution evident in their deliberations came about because of the events of the recent past. For five years, the residents of Spain's Caribbean islands had suffered through a period of environmental crisis. Following the disaster of 1768 and clearly evident by 1770, the signature hazards of the El Niño/La Niña cycle were widespread throughout the Caribbean basin. Severe drought initially disrupted the smooth functioning of the imperial commercial system, especially in Mexico and other areas of the mainland. Four years previous, the hurricane of Santa Teresa had initiated changes in the way the crown responded to disaster. That experience would prove invaluable, and by the 1770s, royal officials had experience, legislation, and precedent at their disposal to face future catastrophes. Yet no one could have foreseen the extraordinary hurricane season of 1772, when sixteen landfalls in the Caribbean and Gulf Coast brought destruction on an un-

precedented scale. The cumulative impact of weather-induced crisis during the early 1770s effected a complete revision of Spanish royal economic
policy toward the Americas. The economic structures that had been in
place for decades came under fire, and the changes were nothing short of
a total realignment of trading patterns in the North Atlantic basin.

The modifications of the 1770s grew out of the problems of the 1760s—
such as in Louisiana, where Alejandro O'Reilly arrived in 1769 to quell the
rebellion of the previous year. The rebellion in 1768 and former governor
Ulloa's ignominious flight meant that for order to be restored, Spain's response needed to be unequivocal. As was the case after the fall of Havana
in 1762, Charles III never repeated past mistakes. In 1766, Ulloa was sent
to Louisiana with less than 100 troops and with a conciliatory attitude
toward the resident population. In contrast, O'Reilly returned in 1769
with no fewer than 2,000 men and with an uncompromising stance that
brought swift and severe punishment to the ringleaders of the rebellion.
Yet his intractable position toward commerce with foreigners had to be
tempered by the reality of the situation on the ground. O'Reilly and his
successor, Luis de Unzaga y Amézaga, coped with food shortages caused
by ecological crisis and exacerbated by imperial restrictions. By winter
1770, the colony once again faced starvation, and O'Reilly was obliged to
grant permission to traders in New Orleans to receive flour from farmers
upriver in the Illinois territory.[4]

Always sensitive to challenges to its privileges, the monopoly was outraged when it learned that O'Reilly had allowed Louisiana's residents to
trade outside the imperial system, and once again the directors petitioned
the Ministry of the Indies for permission to relax the regulations under
which they were compelled to operate.[5] Their arguments were twofold.
First, they complained that the quantity of flour that they were allowed to
import was tied to the number of slaves, and they requested an increase
in that ratio from one-to-one to two-to-one. Second, they requested that
they be allowed to trade directly with foreign ports. They argued that their
responsibility was to provide enough flour to feed the slaves, but because
of the shortages in shipments from Mexico, their only recourse was to turn
to Philadelphia or New York. They rationalized that if they traded with
foreigners, they would be doing the same thing as when Caribbean governors authorized voyages to nearby colonies, only they would not bring
in other contraband goods such as china and cloth concealed in barrels of
flour. If they had violated the law in the past, their intent had not been de-

liberate or as egregious as what the locals committed on a daily basis. The Asiento's directors reminded the Council of the Indies that the company was made up of Spanish citizens of the highest caliber who had never engaged in any other contraband activities. Yet they did succeed in convincing the Ministry of the Indies to prohibit the export of North American flour (and other luxury items such as beer) from New Orleans to Havana.[6] Even more significant, they convinced the monarchy that the best course of action was to allow them to purchase flour from British North America. In 1769, the Council of the Indies legitimized the Asiento's contract with Arturo Moylan and Company to provide 24,000 barrels of flour annually from Philadelphia, to be distributed among the cities that the Asiento served.[7] The flour was supposed to be sent to Cádiz and then be shipped to the Caribbean—an arrangement that failed even before the first shipment left port—and payment was secured through bills of exchange redeemable in Europe.[8]

In short order, representatives of each of the signatories traveled to ports throughout the Atlantic basin to guard the interests of their respective employers. In Cádiz, Moylan secured the services of James Duff, a long-time resident with extensive local ties who, along with his partner John Welsh, acted as broker for several mercantile interests.[9] Moylan's Caribbean factor, James Monson, arrived in Puerto Rico via Dominica in February 1769, and his reception was considerably more cordial than that of his predecessor, the unfortunate John Kennedy.[10] In another strategic move, Stephen Moylan, one of James Moylan's sons, traveled to Philadelphia, and he quickly entered the city's economic and social life.[11] Backed by the wealth of his family business and bolstered by their contacts throughout the North Atlantic world, young Moylan became one of the city's most influential merchants.[12] The Spanish monopoly also sent representatives to the Caribbean. In 1769, Gerónimo Enrile, son of one of the principal investors in the Asiento, traveled to Puerto Rico and then on to Cuba to inform the governors of the new contract.[13] Joaquín Pover replaced Vicente de Zavaleta as the Asiento's factor in Puerto Rico, and in July, Manuel Félix Riesch, who had previously served as Antonio de Ulloa's secretary when Ulloa was assigned to Peru, took over the monopoly's operations in Havana.[14]

Beginning in March 1770, Philadelphia merchants sent four shiploads of flour to Puerto Rico.[15] The two largest cargoes went on vessels partially or wholly belonging to Philadelphia's most prominent merchant house,

Willing and Morris. The last shipment, which cleared port on 6 September, was owned by a consortium of five investors: Stephen Moylan, Robert Morris, Thomas Willing, W. Marshall, who had captained boats carrying two of the three previous shipments, and the former San Juan factor, John Kennedy, who had recovered from his bout with yellow fever and now lived in London.[16] For its part, the Asiento reported that the new system was working well. In September 1769, in a moment of candor, the directors reported to the Ministry of the Indies that "no harm will be done by obtaining flour [from the Americans] because, as everyone knows, all of the flour that [goes] from Cádiz to the Spanish colonies is of foreign origin."[17] The Asiento was so pleased that it requested an increase in the amount of flour that could be brought in, from one barrel per slave to two, a request that was quickly approved.[18] The Asiento factors in Puerto Rico and Havana were also more attentive to their responsibility to send regular shipments of flour to the satellite cities, and when a fresh shipment arrived in 1769, Bucareli immediately sent 423 barrels of flour to Santiago.[19]

Indeed, all of the signatories to the new provisioning arrangement seemed satisfied. Only one powerful, dissenting voice rose in protest: that of the British government. In 1770, Britain and Spain were on the verge of going to war over the Falkland Islands (Malvinas),[20] and even though the threat passed, the British crown saw the actions of the Philadelphians as trading with the enemy. In November 1770, official notification went out to Moylan and Company to stop sending flour to Cádiz. Moylan's factor, James Duff, was ordered to cease and desist receiving flour from Philadelphia for sale to the Asiento. By December 1770, the bad news had made its way to the Caribbean, and even Duff's promise that he would tell his suppliers to continue to ship their cargoes until the break occurred did little to reassure the governors of Cuba and Puerto Rico that they could count on a regular supply.[21] Restrictions were reiterated, and provisioning from Jamaica and the Bahamas returned to its illegal status.[22]

The impending crisis in the supply of provisions meant that the locals simply returned to their old bad habits of smuggling with the British islands, and in one of many instances, a sloop carrying replacement troops to Santiago de Cuba captured a small boat of English smugglers on the south coast near Bayamo, in April 1769.[23] Two years later, in 1771, a commissioned coast guard cutter under the command of Clemente Pérez intercepted another sloop belonging to a local captain, Manuel Manresa, with a cargo of flour and salt on its way to a rendezvous somewhere on the north coast of Cuba. In the ensuing battle, Manresa was mortally wounded and his crew was arrested and taken to jail. Yet they did not remain in detention for long.[24] A well-placed bribe of 400 pesos to a co-operative jailer allowed Manresa's second in command and his supercargo to escape and melt into the countryside.[25] At the same time, six Jamaican smugglers were not as fortunate when they were apprehended near Bayamo on the southern coast. They were taken to Santiago de Cuba and interrogated by the authorities about military activity in Kingston before being transported to Havana in December 1771.[26] Still, enforcement successes were sporadic, and on the rare occasions when perpetrators such as the six Jamaicans were caught and convicted, they were usually sentenced to hard labor in the Spanish penal colony, Ceuta, in north Africa.[27]

The open defiance of imperial laws prompted Bucareli to admonish Santiago de Cuba's new governor, Juan de Ayans y Ureta, about the number of fugitives and criminals that roamed free in the area under his authority.[28] The provincial governors of Cuatro Villas and Trinidad also worried how they could prevent the *vegueros* (tobacco farmers) in their areas from continuing their contraband trade with Jamaican smugglers. In March 1772, English sloops unloaded flour at Sabana on the south coast and Viana on the north in return for hides.[29] In May, the smugglers returned with flour and now also with salt.[30] For the most part, these transactions were friendly and conducted with the cooperation of local residents, except for one disturbing incident in late August 1772. The lieutenant governor of Puerto Príncipe, Juan de Lleonart, reported that the hacienda Vertientes, belonging to Lieutenant Francisco de Agramonte, was attacked by an armed party of English raiders, who stole several head of cattle and took two of the hacienda's tenant farmers hostage. Even more disturbing was the reason for the raid. The attackers claimed that they had come in retaliation for being cheated in a previous transaction. The raiders mistook the hacienda of Agramonte for that of their collaborator on the island, but when they realized their mistake, they set the captives free and fled with the cattle.[31]

As in the past, the local population was simply responding to scarcity, but at the same time, the incident emphasized how desperate the conditions on the island were. Local farmers struggled in vain to provide food for their communities, because by 1770 the hallmarks of the El Niño/La Niña sequence were once again evident throughout the region. By late summer 1769, drought in Oriente province brought hunger and misery to the town of Santiago del Cobre when no meat provisioner could be found. Manuel Varela, the commander of the troop detachment, was allowed to offer the incentive that any supplier who came forward could reduce the established weight from four pounds of meat per *real* to three pounds as long as the product was of good quality.[32] Similar to the events in the 1750s, the drought spread westward toward the interior of the island, and in October 1770, the ranchers of Santa Clara requested relief from their obligations to provide meat for the town's slaughterhouse because the severe drought had significantly reduced the herds of livestock in the area.[33]

Food shortages also contributed to the onset of epidemic diseases, and in April 1771, the residents of Santiago del Cobre and of another small town in Oriente, Holguín, were plagued by a wave of fevers. Now royal officials were faced with yet another dilemma; both Holguín and Santiago del Cobre were isolated villages with few royal officials assigned to care for the local populations. Each town had but one doctor, one priest to administer the last rites to the dying, and one notary to record last wills and testaments. When these men succumbed to the fevers, the process of working through death came to a halt. At the onset of the epidemic, the lieutenant governor of Holguín requested help from Bayamo, asking not only for medicine and food, but also for replacements for the officials who had died. When Bayamo's city fathers did not reply, he appealed directly to Bucareli in Havana, and he, in turn, forwarded the urgent request to the governor of Oriente.[34] On one hand, neighboring towns and villages were obligated to respond to the emergency needs of another. On the other, contact with the afflicted populations ran the risk of spreading the disease. Ayans ordered Bayamo and other towns to send help as quickly as possible, but if the epidemic showed signs of spreading, they were to take all necessary precautions, including quarantining the sick populations to halt the spread of the fevers.[35]

Meanwhile, drought extended to the mainland, resulting in crop failure in Mexico, Cuba's ostensible provisioning ground. The situation was so desperate in Yucatán that the authorities from Campeche sent a boat to Louisiana pleading for corn or any provisions. An English boat laden with flour had just arrived in New Orleans, and in an extraordinary move, Unzaga authorized it to proceed to Campeche as an emergency measure to prevent complete starvation.[36] The catastrophe in Mexico was accompanied by a similar failure of the corn and yuca crops throughout the north coast of South America in the viceroyalty of New Granada. Towns

Figure 4.1 The *Postillión de México*, dismasted in the hurricane of January 1771, the same storm that forced the Marqués de la Torre en route to Cuba to return to port. Source: Ingenios y Muestras, number 259, Audiencia de Santo Domingo, Archivo General de Indias, Seville, Spain.

along the northern littoral tried to follow imperial dictates and trade with each other, but disappointed captains returned with the news that no legal port had any crops even for their own residents. In desperation, the town council of Cumaná asked for and received permission from the captain general, Felipe de Fonsdeviela, the Marqués de la Torre, to send a boat with a cargo of mules to Martinique to obtain provisions to satisfy their "most urgent need."[37] De la Torre's experience in coping with the crisis brought by drought in the viceroyalty of New Granada would stand him in good stead when he received orders to transfer to Cuba to replace Antonio María Bucareli, who had been promoted to the viceroyalty of Mexico. De la Torre set out for Havana in January 1771, normally a safe time to sail since the threat of hurricanes was at a minimum. But an uncharacteristic winter hurricane struck the western Caribbean, and the packet-boat belonging to the monopoly that carried the marqués to Cuba was forced to return to La Guaira (in present-day Venezuela) to wait out the storm.[38]

At the same time, in Puerto Rico, shortages and uncertainty had returned while the battle between political and commercial interests escalated. Marcos de Vergara, the first governor to challenge the Asiento's business dealings, had died in early 1769, and a new governor with decades of experience had been appointed to take control, Miguel de Muesas, the former commander of El Morro in Santiago de Cuba during the war with Great Britain in 1762. Few royal officials came to their positions with the degree of experience of Muesas. By summer 1771, the twin problems of ecological stress and political intransigence threatened to bring starvation to his jurisdiction. Never prone to indecisiveness, Muesas invoked his emergency authority and permitted a registered ship to sail to St. Thomas and St. Eustatius to find food. Predictably, the Asiento's factor, Joaquín Pover, lodged several protests against the governor's actions, to no avail.[39] The arrival of the Spanish ship did not escape the notice of watchful eyes in St. Eustatius, and local factors reported to their suppliers in Philadelphia that the Spanish had come and purchased all of the available flour on the island.[40]

The drought lingered into 1772. Beginning in January and continuing for the first six months of the year, royal officials in Santiago de Cuba took drastic measures to provide water to the main fortress that guarded the port.[41] In Havana, de la Torre and the chief naval officer in charge of the port, Juan Bautista Bonet, debated what to do about the diminishing supplies for the troops and naval forces.[42] In desperation, they requested that Louisiana send some of the flour that it had obtained from the Americans, and Unzaga complied by sending one boatload of the valuable commodity to Havana. Unfortunately, the military men failed to consult with Félix Reisch, the Asiento's factor, so when Reisch complained, the intendant in Cuba had no choice but to confiscate the vessel and its cargo.[43] Their actions, of course, gave the monopoly justification to renew its barrage of complaints about the Caribbean governors. In one particularly acerbic letter, the Asiento's directors railed against Muesas's actions in Puerto Rico the previous year. They complained that the governor had no authority to import 6,000 barrels of foreign flour onto the island and complained of the "violence" that had been done to their interests.[44]

Yet the perceived violence done by Puerto Rico's governor would pale in comparison to the real violence in store for the Caribbean populations during the extraordinary and destructive hurricane season of 1772.[45] That summer marked a turning point at which metropolitan and local officials were forced to invoke emergency measures that would prove to be a permanent arrangement. Problems started in June, when a registered packet-boat on its way to Puerto Rico, the *Amable*, was blown off course by a storm at sea and ultimately found safe harbor in Santiago de Cuba.[46] A second hurricane developed in the Lesser Antilles in mid-July and ran westward along the north coast of Puerto Rico before turning northward into the Atlantic.[47] At the beginning of August, a third hurricane developed in the Caribbean and moved westward along the south coast of Hispaniola.[48] On 4–5 August, it struck Santiago de Cuba and Bayamo and continued directly across the island, hitting Havana and the western provinces on 6–7 August.[49] Later that month, a fourth major system began in the outer islands, passed over the south coasts of Puerto Rico and Hispaniola, and wobbled on a west-southwest track.[50] Over water it gained strength, but it passed between the north coast of Jamaica and the south coast of Cuba. Protected by mountains, Santiago de Cuba and Bayamo were spared a direct hit, but more rain fell on the eastern region on 31 August. Continuing westward, the hurricane made landfall on the Yucatán peninsula of Mexico on 1 September.[51] After causing considerable damage, it exited into the Gulf of Mexico, regained strength, turned northward, and came ashore once again between Mobile and New Orleans on 2–3 September.[52] Simultaneously, the outer islands of the Caribbean became the victims of a large and powerful hurricane, the fifth of the season, that caused catastrophic damage in the Virgin Islands, St. Eustatius, Dominica, and Antigua.[53] This storm devastated an already-weakened Puerto Rico and became the final straw that compelled officials to abandon any semblance of compliance and to request emergency rations directly from Philadelphia.[54] A few days later, the hurricane compounded its destruction by dumping copious amounts of rainfall on the north coast of Cuba before turning northward and causing additional damage on the outer banks of North Carolina.[55]

By early September, the magnitude of the catastrophe that engulfed the Caribbean basin was such that no mitigation policy—however enlightened—could provide a remedy.[56] After the first storm, the governor of Santiago de Cuba reported that the hurricane had destroyed the plantains, rice, corn, and yuca. By the end of the month, after the effects of the second storm, he was pleading with the captain general to send help quickly, as all of the crops in his jurisdiction were totally ruined.[57] Similar reports came from other towns and villages in eastern Cuba such as Puerto Príncipe and Baracoa.[58] The town of Santa Clara suspended the sale of provisions to other areas because the hurricane in late August had ravaged the countryside in its area and destroyed all of the crops.[59] Reports from the heavily populated towns around Havana echoed those from other areas of the total destruction of the subsistence crops in that area.[60] Barely six months earlier, the captain general and the monopoly's factor were at loggerheads over bringing Louisiana's flour into Havana. The dispute evaporated in the wake of months of continuous rain, and after the first series of hurricanes, Louisiana governor Unzaga managed to put together three shipments of provisions for Cuba. En route, two of the three ships were caught in the late August storm and sank at the mouth of the Mississippi River, with considerable loss of life and the total loss of the cargo of emergency supplies.[61]

The destruction in Puerto Rico was also catastrophic. After the first storm, the governor of Santiago de Cuba sent the ill-fated *Amable* to Puerto Rico with supplies, but the packetboat was caught in the second hurricane and never made it to its destination.[62] Continuous bad weather in San Juan ruined all of the flour that awaited distribution to other Caribbean cities. In the aftermath of the third hurricane, Muesas and the Asiento's factor were in complete agreement that the situation was critical, and no protest was lodged when several captains were granted permission to travel throughout the Caribbean to try to purchase provisions. In this moment of crisis, the Puerto Rican factor told his captains to "make all haste to any port" to find food to bring back to alleviate the suffering. Pover advised the men that he had commercial contacts in neighboring islands such as Santo Domingo, St. Thomas, St. Croix, and St. Eustatius, and even though he and they were aware of the prohibition on trading with foreign islands, the situation was so grave that the circumstances warranted his actions.[63] It was not long before one such captain returned with the bad news that nowhere in the Caribbean was there any food to be found. As other disappointed captains returned to San Juan harbor, several men reported that the other islands in the Lesser Antilles were all suffering as much, if not more, than Puerto Rico.[64] In the free ports, such as St. Croix, St. Thomas, St. Martins, St. Eustatius, and Dominica, where normally there were abundant supplies of provisions, the residents were in a state of shock as they coped with the realization that entire villages had been swept away.[65] A captain of the Spanish fleet on his way from La Coruña to Puerto Rico reported that as he sailed through the passage between Virgin Gorda and Anguilla he came across fragments of houses, beams, fences, broken masts from ships, bodies, and other vestiges of what once had been a settlement on St. Eustatius.[66] A young Alexander Hamilton lived through the "horror and destruction" on St. Croix, and he described cowering in his house as the storm raged outside.[67]

By September, the magnitude of the catastrophe, along with the folly of adhering to existing policy, was apparent. The first line of defense was to call upon other towns on each island to aid the stricken communities, but the futility of that strategy soon became obvious, especially after the series of storms in late August. The second line of defense was to call upon other

areas of the Spanish empire, but no help was forthcoming from the mainland, as Mexico was struggling with drought in grain-producing areas that
had destroyed the wheat that was destined for the Caribbean.[68] Eventually, the chilling realization set in that even the last line of defense against
starvation, the British islands from which smugglers could normally be
counted upon, could provide no relief, as they, too, struggled to make up
their own losses. The despair was almost palpable in the proclamation
issued by Sir Ralph Payne, governor in chief of the British islands, who
spoke of the "melancholy prospect of an approaching famine" and called
upon fellow citizens to dispatch vessels to the islands with provisions as
rapidly as possible.[69] By all accounts, by mid-September, all Caribbean
colonies were sharing in the misery of their neighbors, and no island could
spare its food supplies in order to help another.[70]

Once the extent of the damage became indisputable, Spanish officials
rapidly took steps to alleviate the suffering in their colonies. Left with no
choice, the authorities in Puerto Rico went outside the Spanish imperial
system and established direct contact with Philadelphia.[71] Later, the two
antagonists, Muesas and Pover, would each write a lengthy report justifying their decisions and explaining to their superiors that the successive
hurricanes had destroyed all of the flour destined for Puerto Rico and for
distribution to other colonies.[72] The news that reached the minister of
the Indies, Julián de Arriaga, from Cuba mirrored the reports from Puerto
Rico.[73] By the end of August, de la Torre informed the court in Spain that
"not only were we hit badly, but in all of the Leeward Islands especially in
the foreign colonies, the destruction has been terrible."[74]

The unprecedented crisis once again highlighted the problems inherent in the Spanish commercial system. As the drought that was emblem-
atic of the El Niño sequence spread to Mexico and destroyed that region's
wheat crop, it became even more obvious that the existing system could
not provision the islands of the Caribbean. Over the winter of 1772–73, the
Ministry of the Indies, influenced by a steady stream of correspondence
bemoaning the crisis in the Spanish Caribbean, acted to remedy the terrible situation. In an effort to find out what had gone wrong, the unpopular Asiento came under sharp scrutiny.[75] Forced to defend themselves
in front of the most powerful group of men in the empire, the primary
directors of the company, Francisco de Aguirre and José de Arístegui, produced a long lament on how the inhabitants of Cuba had gotten rich while
the poor company had suffered.[76] Clearly referring to Willing and Morris

(although never mentioning the company by name), the monopoly complained about the privileges granted to the individuals who now provisioned Puerto Rico, especially the concession that freed them from the obligation to import slaves.[77] As a last resort, Aguirre and Arístegui asked to be released from their contract if the Ministry would not grant them comparable consideration.[78]

Indeed, among the company's most vociferous complaints was the perverse illogic of the regulation that tied the amount of flour that could be imported to the requirement to bring in a corresponding number of slaves. The company was forced to bring in what the Caribbean did not need—more mouths to feed—while time and again it was prevented from providing sufficient quantities of the mainstay of the residents' diet. Even though it had been granted an increase in the flour-to-slave ratio in the wake of the crises in 1768, the company still faced huge losses.[79] More often than not, the company could not sell the slaves it brought to the island, and humanitarian concerns forced the government to take possession and sell the slaves to whoever would buy them at discount prices and on credit. In one instance, mirroring the situation of the 1750s, the royal tobacco factor of Santiago de Cuba bluntly informed the governor that if the monopoly continued to bring in more slaves than the island's market could absorb, "[his office] would no longer buy [the slaves] to distribute to the tobacco farmers."[80]

The Ministry of the Indies was unsympathetic to the Asiento's ostensible plight. That body appointed two men to evaluate the benefits of retaining the monopoly as well as to investigate the reasons for the chronic food shortages in the Caribbean—Tomás Ortíz de Landázuri, the treasurer for the American colonies, and the Ministry's general counsel, Manuel Lanz de Casafonda. The report they submitted in summer 1772 was scathing. To begin, they concluded that the company was absolutely inept, and they termed it "an embarrassment to itself and to the Spanish nation."[81] One important impediment was the animosity among the current directors, Aguirre and Arístegui; the faction who supported them; and José María de Enrile, the most competent member, who also had the most experience in commercial matters. Ortíz and Lanz concluded: "[This organization] cannot continue as it has up to now."[82]

Yet, in spite of its serious drawbacks, the Asiento remained an important link in the supply chain because it had established connections to merchants in foreign colonies who could purchase flour from North

America on a moment's notice. Such linkages were important not just in terms of normal supply but also in times of crisis when food needed to be procured without delay. Rather than abandoning the idea that one monopoly should retain the privilege of trading with the Spanish Caribbean, the existing company was reorganized, with Enrile at its head. His son, Gerónimo Enrile, was named as its factor at Havana.[83] A lengthy new set of regulations accompanied the reorganization that Enrile and his partners agreed to observe.[84]

The counselors still faced two fundamental problems: one, the long-term issue of maintaining a regular and dependable supply and, two, how to deal with the extraordinary demand that occurred in a disaster's aftermath. The present regulations allowed one single shipment, which brought in a sufficient quantity of flour to maintain a reserve of 2,000 barrels of flour on hand at all times. Ortíz and Lanz concluded that this was a bad idea because flour spoiled quickly in the tropics. It was not possible to maintain it for more than four to six months, and in four months the inhabitants could not consume more than a third of the 2,000 barrels. This invariably led to conflicts, because when the monopoly demanded that the bakers buy the rancid flour anyway, the enraged citizens dumped the barrels into the harbor.[85] The officials acknowledged that flour was a "vital and indispensable commodity" and that they would have no choice but to authorize regular but smaller shipments into Caribbean cities.[86] In September 1772, they recommended that the quantity of flour to be warehoused at any given time be reduced from 2,000 barrels to 600.[87]

By early 1773, the Ministry of the Indies had settled on a radical new strategy to provide provisions to the Spanish Caribbean, implementing a series of modifications that effectively opened the Spanish colonies to foreign produce. First, the requirement that flour and slaves be brought in together was abandoned, and the flour-to-slave ratio was increased once again to a ratio of three-to-one.[88] Second, the Asiento was granted one of its long-cherished goals: permission to travel without restrictions to foreign colonies to purchase slaves and flour. Under the new regulations, its ships could sail directly from Santiago de Cuba and Havana to foreign ports without having to stop over in Puerto Rico. Captains were permitted to take hard cash from the royal treasury in Havana; they could pay up to 180 pesos per slave and eight pesos per barrel of flour; and they could bring in both duty-free. Buyers would be allowed to take out loans in the towns where the slaves were sold, and if they defaulted, the company could reclaim the slaves. In these instances, royal governors were instructed to expedite any suits for payment.[89]

Still, because of the propensity to engage in contraband, royal advisers harbored a strong aversion to permitting foreign suppliers into Spanish American ports, which explains a final concession, which permitted foreign ships, particularly those from Philadelphia, to import their provisions only into certain Spanish peninsular cities. From there, flour was shipped on to the Caribbean by the Asiento in Spanish ships.[90] Most flour entered through the port of Cádiz, but northern cities such as Bilbao, Ferrol, and La Coruña received lesser quantities in fewer shipments.[91] The change exacerbated a rivalry within the Spanish imperial system among peninsular port cities as other mercantile groups fought tenaciously against the Asiento's prerogatives.[92]

The extraordinary shift in royal policy can be directly attributed to the environmental crisis in the Hispanic Caribbean. Throughout their deliberations, the counselors justified their actions because of the immediate crisis at hand.[93] Meanwhile, residents and royal officials struggled to cope with the aftermath of the disaster at the local level. With the people who depended upon them reduced to eating putrefying plantains and yuca, de la Torre in Havana, Ayans in Santiago de Cuba, and Muesas in San Juan had no choice but to act to meet the residents' urgent needs. As the crisis deepened in Santiago de Cuba, Ayans granted four sequential licenses to local captains to go in search of food without waiting for approval from Havana.[94] The circular order from the captain general that permitted such commerce with foreign colonies crossed paths with Ayans's letters justifying his actions and begging for relief.[95] De la Torre reiterated how desperate the situation was throughout the island, particularly in Oriente, when he notified his superiors in Madrid that he had permitted any ship that had survived the hurricanes to go outside the Spanish system to find food.[96] At the same time, measures were taken to prevent anyone from taking advantage of the victims. Near Havana, the constable of Melena accused the overseer of a local plantation of price gouging in providing meat for the garrison. The captain general ordered the overseer to return the excess that he had charged, with the warning that if he was ever caught again he would not get off so easily. The incident prompted a decree that anyone guilty of a similar infraction would be fined eight *pesos fuertes*, the equivalent of one month's pay for an ordinary worker.[97]

Gradually, the recovery efforts began showing results, and in late October, a boat laden with 200 barrels of rice and 200 barrels of *chicharos* (chitterlings) left Havana for Santiago de Cuba, along with a promise that another boat was on its way with meat and pork.[98] By November 1772, western Cuba's immediate need for food had been met, and three more ships were loaded with surplus provisions and sent to Santiago, which was still struggling to recover from the sequential storms.[99] Yet as late as February 1773, local officials there and in Bayamo still had to cope with shortages of flour and rice.[100] Provisions continued to flow from Philadelphia to Puerto Rico throughout the winter of 1772–73, and some of the precious North American flour was shipped on to Havana.[101] Clearly, the Caribbean-wide emergency had transformed activities that were illegal in July 1772 into legal commerce by February 1773.[102]

Once the immediate threat of starvation had passed, the residents faced the challenges of repairing the damage to the infrastructure that stretched from one end of the island to the other. The town of Batabanó on the southern coast, which had been so seriously affected by the hurricane of 1768, was once again devastated. In 1764, during the Conde de Ricla's tenure, a cavalry troop was permanently assigned to the town to guard against smuggling and to escort tobacco shipments from the wharf in Batabanó to the tobacco factory on the outskirts of Havana. The detachment was doubly needed in such a remote location, because after the previous storm, it became the first responders to the crisis. In 1772, though, the first responders became the first victims when the barracks constructed for their lodgings gave way under the continuous rainfall.[103] At the far eastern end of the island, the lieutenant governor of Baracoa told a similar tale when he informed de la Torre that the barracks housing his men had collapsed.[104] The militarization program near Havana suffered additional setbacks when the barracks for the Catalonian Riflemen and the military hospital of San Ambrosio were badly damaged.[105] Private and public buildings near the capital that sustained serious damage included houses and shops, the slaughterhouse, and even the aqueduct that provided water to the city.[106] South of Havana, the villages of Jesús del Monte and Bejucal reported that many structures were destroyed and that the rebuilding was hampered because of a shortage of palm thatch for roofs.[107] Making matters worse, the *capitán del partido* of Bejucal, Esteban Rodríguez del Pino, began to suffer from the effects of the seasonal fevers that routinely set in after the passage of a storm.[108]

One of the most urgent tasks facing mitigation teams was clearing obstructions in the harbor. Havana was the most important port for ships returning to Spain, and even under normal conditions, keeping the channel and the harbor open was an ongoing task. This became even more crucial after passage of a hurricane to facilitate emergency provisions and supplies arriving and being dispatched to other areas. Salvage divers Nicolás Marín and José and Simón Montenegro and their crews were already hard at work trying to remove debris from the *Neptune*, the *Europa*, and the *Asia*, which had been beached in the hurricane of 1768. On 5 August, when weather signs gave warning of the approaching storm, naval commander Juan Bautista Bonet pulled the divers off the salvage operations in order to have enough men to crew a rescue boat. After the storm had passed, the men tried to resume salvage operations six times on 9 August, but the runoff into the bay made the bottom too obscured to work.[109] On the twelfth day after the storm, salvage operations were postponed indefinitely, and the divers and their crews were transferred to clearing the wreckage of a sloop that had gone down near the principal wharf.[110] Bad weather continued to impede their work for over a month. A frustrated captain general complained to his superiors in Madrid that from 23 August through 12 September numerous thunderstorms had stirred up the mud in Havana harbor, obscuring the bottom. For more than a month, the diving crews were prevented from accomplishing anything until the inclement weather subsided.[111]

Faced with the task of keeping the port open, Commander Bonet acted on his own authority, which provoked a conflict with the head of the engineers, Silvestre de Abarca. Both men claimed the use of the only barge in the harbor to carry out repairs on the buildings that housed their men. Although his authority was supreme on the island and he could have resolved the argument in favor of one man or the other, de la Torre opted to pass the matter on to Madrid. Charles III's reply, swift and unequivocal, was a strong statement about the metropolitan attitude regarding how colonial officials should behave in the aftermath of disaster. The jealousies, rivalries, and competition that were endemic among royal officials would no longer be tolerated. Henceforth, cooperation was to be paramount. The king reiterated that keeping Havana harbor open was crucial, and he authorized royal officials to use whatever means necessary to support the recovery process.[112]

Armed with the unequivocal orders of the monarch, de la Torre formed a junta composed of himself, Bonet, the two intendants—Montalvo of the navy and Altarriba of the army—and chief engineer Abarca, and during the winter these men met regularly to discuss their options. The junta debated one of the most pressing issues—that of sand being carried into the harbor by the incredible amount of rain.[113] The runoff caused *bancos* (sandbars), which were a danger to ships.[114] Abarca and Bonet agreed that a new barge was critical to sound the bay to find out where the sandbars had formed. The junta called upon naval contractors Josef Chenard and Vicente Morand, who had constructed a similar vessel to dredge the harbor in Veracruz. In December 1772, Chenard and Morand received a contract for a barge like the one they had built in Mexico, which would cost nearly 50,000 pesos.[115] That same day, Montalvo submitted an estimate detailed down to the last real (48,503 pesos, one-half *real*) to his superiors in Madrid.[116]

Another suggestion was to pave the city streets to minimize the amount of soil washed into the bay.[117] As was customary, residents were allowed to submit bids for the contract, and shortly thereafter Agustín de Piña and Manuel de Brito were awarded the contract to pave the city streets. The contract came with a grant of the labor of 100 criminals and deserters, men who were sentenced to forced labor as a consequence of their transgressions.[118] Over the winter of 1772, de la Torre's reports to the Council of the Indies tracked the progress of the recovery and the growing cooperation among royal officials in Havana.[119] By spring 1773, deliberations among royal officials became commonplace, and Charles III relayed his satisfaction via his minister of the Indies, Julián de Arriaga.[120]

Clearing the rural roads of mud and debris was also hampered by continuous rainfall. The most serious obstacle was the collapse of the bridge over the Cojímar River east of Guanabacoa. The original bridge had been built of wood, but over a period of years it had been repaired several times, sometimes by simply patching strong, new wood on top of the older, deteriorating structure. After weeks of uninterrupted rain, the dilapidated old bridge gave way, dealing a serious blow to commerce and communications between Havana and Matanzas to the east. Replacing the bridge would require a large outlay of funds, especially since de la Torre was adamant that the reconstruction be done in a substantial manner. On 6 October, the captain general appointed Domingo de Lisundia, the Marqués de Real Agrado, to oversee the project. Real Agrado was the *regidor* (magistrate or alderman) of Havana and also among the most influential sugar producers in Río Blanco, an agricultural district east of the capital. Chosen for his "caution and zeal," he was given complete authority over the new bridge's location and was charged with completing the task "in the manner least grievous to the public."[121] Subordinate officials included José de San Martín, adjutant from the second battalion of Havana stationed in Guanabacoa, and three lieutenants from local militias, including Domingo Santaya, Josef Portillo, and Andrés Visiedo. These men were responsible for hiring the laborers and overseeing the day-to-day progress on the project, which provided employment for a number of local day laborers and skilled masons.

Now de la Torre faced the question of how to pay for the bridge in "the most equitable manner," and he implemented a solution introduced by Bucareli after the hurricane of 1768. The residents were assessed on a sliding scale based upon the extent of their properties, their ability to pay, and the benefit they would gain from the reconstruction of the bridge. De la Torre determined that those who would derive the most benefit were the sugar plantation owners, and on 11 November, he compiled a list of eighteen landowners of seventeen *ingenios* (large sugar plantations) and one *trapiche* (small mill) and sent it to Real Agrado for collection.[122] Most plantation owners were assessed twenty or twenty-five pesos for their large tracts of land, but Real Agrado, perhaps because he chose to set a good example since he was in charge of collecting the tax, contributed fifty pesos, double that of any other plantation owner. Town dwellers in Guanabacoa paid between a half-peso (4 *reales*) and six pesos per property.[123] Lieutenants Santaya, Portillo, and Visiedo shared the responsibility for collecting the assessments from the owners of modest agricultural properties such as *estancias*, *vegas*, and *potreros* (cattle farms) in eight villages and districts, who were charged two pesos per *caballería* of land.[124] The first round of taxation began in November 1772, raising nearly 1,191 pesos to begin the project.[125] A year later, good progress had been made on the impressive bridge of stone and mortar, but the overseers were faced with a shortfall of nearly 1,900 pesos. De la Torre ordered a second round of taxation, raising the assessment for owners of small farms by a modest one *real* per *caballería*. Similar proportional increases were mandated for the sugar plantations.[126] By March 1774, just eighteen months after the disaster, the bridge was complete.

The reconstruction of the Cojímar bridge provides a case study in public attitudes toward civic responsibilities. Virtually all of the residents accepted the burden of rebuilding the roads and bridges in their district, as long as they believed that the taxation was implemented in an equitable manner and that they would benefit from its reconstruction. Only three landowners refused to pay their assessments when requested by the collectors. Citing the order promulgated by Bucareli after the hurricane of 1768, Antonio Martínez argued that his ecclesiastical status exempted him from being taxed, and widow Angela Barba claimed that the farm (hacienda) that she occupied, one of the largest in the area, actually belonged to her son, who served on active military duty, therefore being exempt.[127] The sugar plantation owners accepted the valuations without complaint, and only Domingo Garro, who owned a half interest in a small tract, vociferously resisted paying the assessment levied upon him.[128]

West and south of Havana there was less damage from the hurricane itself, but like the rest of western Cuba, that district had suffered from continuous rainfall since the beginning of the summer. Already known by the designation "Vuelta Abajo," which in the following century became synonymous with Cuba's exquisite tobacco, the region was the center of that crop's production. The roads of Vuelta Abajo were under continuous traffic from the mule teams that brought the crop from the area's *vegas* and from the wharf in Batabanó to the tobacco factory located just outside the city walls to the west of town. Even before the series of hurricanes struck, the area's residents struggled to keep ahead of the damage caused by the rain. As the rainfall accumulated during the August hurricanes, one by one, the local constables voiced their frustrations that as a part of a road was opened, another was destroyed. A sympathetic de la Torre advised them to wait until the water receded before trying to open the roads and instead to concentrate their efforts on repairing buildings, especially militia barracks, as soon as possible.[129] Yet as late as November the roads to the south of Havana were still impassable, according to Nicolás Duarte, the constable of Batabanó.[130]

Frustration flared into open disobedience as time wore on. As hurricane fatigue took its toll, the task of maintaining the peace and public order fell on the shoulders of these local authorities. In July, after days of continuous rain and even before the first major storm struck the island, Nicolás Rodríguez del Pino, the constable of Güines, warned de la Torre of discontent among his citizens.[131] As his patience with their obstinacy ran out, he put one of the leaders in jail, hoping that his show of strength would serve as a warning to the rest. After spending a few days in jail, the ringleader was released.[132] The villagers of San Miguel del Padrón voiced similar complaints, and when some refused to work, an exasperated de la Torre instructed local officials to impose a "penalty appropriate to their disobedience" (*pena condigna a su inobediencia*).[133] Unlike the instance in Güines, where the leaders were jailed, in Managua, most punishments involved the loss of property. The constable confiscated three slaves from different individuals and used their labor to rebuild the barracks in Río Hondo. In another instance, a pair of oxen were taken from their owner, and de la Torre ruled that the oxen were to be used for eight days in transporting materials. If, at the end of the eight days, the owner still refused to contribute, the oxen were to be sold at public auction and the money contributed to paying the accumulated fines of the stubborn man.[134]

One particular center of opposition was the tobacco-producing village of Jesús del Monte, south of Havana, already notorious for its recalcitrance during a five-year rebellion earlier in the century.[135] The main road from Batabanó to Havana ran right through Jesús del Monte, and it already was in poor repair from the heavy traffic carrying tobacco to the warehouse.[136] After the August storms, Jesús del Monte's residents complained to Constable Francisco José Roxas Sotolongo that they were being forced to maintain the road but that it was the mule drivers and their mule teams from the southern villages who were causing the damage. In December 1772, the ringleader of the resistance, Domingo del Castro, offered a variety of excuses as to why he was not liable for road duty.[137] In cases of public disobedience, the authority of the district constables was usually sufficient to maintain order, but Roxas Sotolongo could also rely upon a cavalry company of regular soldiers headed by Lieutenant Commander Martín Navarro, who had the power to mobilize the local militia companies under his command.[138] Perhaps this explains why, less than two months later, del Castro and several of his co-conspirators were under arrest and on their way to Havana facing sentences of hard labor—on the same road-building project to which they had refused to contribute in their home village.[139]

Constables, military units, and local militias were aided by a cadre of men of proven loyalty and courage who received special commissions from the captain general to perform specific police functions, for example, to capture deserters, fugitives, or runaway slaves. The ability to award these commissions originally rested with the local ayuntamientos, but in 1763 Ricla appropriated the privilege and appointed Juan Antonio Cabrera as captain of the *vagos* (vagrants) because of his previous "distinguished service," and Bucareli issued a similar commission to Antonio Rocabuena in 1768.[140] De la Torre, who had come to the island with a no-nonsense approach, greatly expanded the practice. Between April and May 1772, he granted three additional commissions, one to Andrés Oliveres for criminals and deserters in Havana; another to Patricio Enríquez for similar troublemakers in Matanzas; and a third to Josef Gil, whose authority extended throughout the island.[141] Enríquez was selected because he, like the others, was "an intelligent subject [of His Majesty] of known honor, responsibility and good behavior, who possesses all of the qualities that [this position] requires."[142] These men were allowed to behave in ways that were prohibited for ordinary citizens. A commission granted to Sebastián de Espinosa, for example, allowed him and two trusted companions to search for deserters in any part of the island. They were also allowed to carry weapons of any kind—a truly extraordinary concession—and they were obliged to report on any person, houses, or activities that appeared suspicious.[143] The men were paid according to the number of fugitives they captured and from how far from Havana the criminals had to be transported.[144]

After the passage of a storm, such extraordinary commissions were vitally important. To begin, the normal police functions of the *capitanes del partido*, the regular troops, and militia members were shifted to rescue and recovery. As long as regular police units were diverted to disaster mitigation, this ensuing confusion provided an opportunity for slaves, criminals, and conscripts to escape to the woods. Predictably, a group of slaves escaped from the plantation of Gabriel Beltrán de Santa Cruz to the densely forested hills south of Jaruco and established a runaway (*cimarrón*) community.[145] Another group of runaway slaves stole oxen, cattle, other animals, and firewood from the sugar plantation and cattle ranch, Divina Pastora, near Guatao, which belonged to María Basabe.[146] Areas to the west of town were less affected by the storm, but even so escaped slaves formed another *cimarrón* community on the far reaches of the plantation of the Marquesa de Cárdenas de Montehermosa near Bahía Honda.[147]

Plantation owners, heavily invested in the labor of slaves, had a keen interest in keeping order in their areas. Five influential sugar planters from the Managua district (Pedro Calvo de la Puerta, the Conde de Buena Vista; Félix de la Torre; Ignacio Loynaz; Félix de Acosta y Riana; and José de la Guardia) brought the concerns of their community to the captain general asking that order be restored to their district. This committee recommended that Josef López, a retired militia captain, be designated as the official bounty hunter for their area, while María Basabe requested that her overseer, Guillermo Rodríguez, be granted the same authority.[148] The "fear and disorder" that coursed through the communities on the west side of the bay led to Francisco Bello's appointment on the recommendation of his employer, the Marquesa de Cárdenas de Montehermosa.[149] Over the winter of 1772–73, a substantial rural police force emerged in the towns and villages circling the capital, as special commissions were granted to an increasing number of experienced men authorized to capture runaway slaves, deserters, criminals, and vagrants engaged in a variety of types of criminal behavior.[150]

Given the scope of the Caribbean-wide catastrophe, the long-term recovery proceeded smoothly, and ironically, the reorganized Asiento contributed to its success. The original decree permitting commerce with foreign colonies was published on 1 May 1773. In a few short months, the pattern of intra-Caribbean trade had changed dramatically. Beginning as soon as the news arrived in summer 1773 and continuing through August 1776, an average of one Asiento ship per month sailed directly from Cuban ports to Jamaica and Barbados, with occasional voyages to St. Eustatius. Within weeks, the ships returned with their cargoes of slaves and flour. During the hurricane season, Asiento ships remained in port, but once the danger had passed, the monopoly increased its sailings (in January and February, for example) to compensate for any shortage that remained from the fall. Another important change involved removing the imperial dictate to trade solely with the authorized port for each island. Under the old regulations, provisions for the secondary cities had to be cleared in San Juan or Havana, respectively; only then could foodstuffs be shipped on to Santiago de Cuba, Bayamo, or Baracoa. If the ordeals of the 1760s had taught royal officials anything, it was that the highly perishable flour must be shipped and distributed quickly to avoid spoilage. To that end, the monopoly assigned the sloop *Industry* specifically to a triangular route from Havana to Kingston and then on to Santiago de Cuba, carrying a smaller number of slaves and flour to the eastern city.[151] Another change involved anticipating scarcity. Even before the hurricane season of 1774, Jamaican merchant Peter Barral was in contact with Havana's factor, Gerónimo Enrile, suggesting that Enrile let him know in advance how many barrels of flour the monopoly would need for the coming months so he could order them from Philadelphia or New York.[152]

Some general conclusions about the quantity of flour that was imported may be gleaned from the extant data recording the number of slaves brought into Cuba. Most Asiento ships brought in between 100 and 200 slaves, so logically they carried at least twice that number of barrels of flour. Thus, between 200 and 400 barrels of "superfine" North American flour per month were available to Havana's and Santiago de Cuba's consumers. Regular arrivals further guaranteed that the product was always fresh and that the Asiento would no longer suffer the losses associated with rancid flour. Ironically, such abundance brought new and unanticipated problems. Cuba's consumers were accustomed to scarcity, but now there was no shortage of superfine flour to bake into fresh white bread. This abundance created rivalries among the bakers who had access to the flour. In order to protect their markets, the most influential bakers of Havana organized into a guild (*gremio de panaderos*) and asked the crown to approve their actions. The crown was willing to grant such a privilege but not without imposing some reciprocal obligations on the part of the bakers, especially since the public was vehemently opposed to another monopoly on vital provisions. The compromise was that the *gremio* in Havana was allowed to form, as long as it accepted the obligation to contribute to the fund for providing uniforms for the militia of Havana.[153] In the interest of the public welfare, the captain general was given the authority to set the price that *gremio* members could charge for bread, based upon the price of flour that was imported into the city.

The evolution of a coherent post-disaster recovery program was a long process, but one whose worth was proven time and again in the subsequent years. The effects of the extraordinary hurricane season of 1772 were still felt as late as the following spring 1773. In Oriente, the island's largest river, the Río Cauto, raged outside its normal course long into January.[154] At the same time, the inclement weather led to the loss of the schooner *San Vicente*, which was carrying tobacco from Santiago de Cuba to the monopoly's warehouses in Havana.[155] By November, the circumstances had suddenly reversed—the classic El Niño/La Niña signature—and a provisions contractor in Matanzas was unable to fulfill the terms of his contract to provide meat to the woodcutters for the royal navy because the drought was killing the cattle in that province.[156] Bad news also came from the Asiento's representatives in New Granada, who reported that the harvest there had failed as a result of the unusually dry winter.[157]

In late October 1774, the cycle reversed again, and the western end of

Cuba received the brunt of a storm of medium intensity. Towns between Havana and Matanzas were the most seriously affected, and predictably, in the villages surrounding Havana, all of the primary food crops were destroyed.[158] Maritime interests also suffered. One example was the frigate *La Perfecta*, which was seriously damaged in the storm and limped back to Havana to remain in drydock well into December.[159] Matanzas was particularly hard-hit. Rivers spilled out of their banks, destroying crops and drowning livestock in record numbers. In January 1775, de la Torre sent Lieutenant Luis de Toledo on the first of what would become a series of reconnaissance missions to gather information about the destruction that the province had suffered.[160] Vicente de Fuentes, who rented land for a cattle ranch from the Marqués de Justíz de Santa Ana, reported the dreadful conditions to his landlord, who passed the news on to de Toledo. The majority of his cattle had drowned, and those that survived were afflicted with an unknown disease (*peste*) that took an additional toll on his herd. Consequently, de Fuentes begged to be relieved of the obligation to supply cattle to Matanzas. Another dismaying letter told of how the starving people were cutting down centuries-old trees to get the honey in beehives located high up in the branches.[161] Drought returned the following spring (1775), forcing the supervisor of Havana's sawmill, which drew its water supply from the aqueduct, to suggest that the cisterns be filled at night and covered during the day to minimize losses from evaporation.[162]

As always, when torrential rainfall followed severe drought, the impact was exponentially greater because the ecosystem was already in disequilibrium. When a devastating storm struck eastern Cuba in late August 1775, crisis conditions for months thereafter were certain, in spite of advances in royal policy since 1772. The first missives to Havana of Oriente's governor, Juan de Ayans y Ureta, warned that after any serious storm a shortage of provisions was inevitable and that the needs of the victims would exceed what the area could produce. The governor wrote of the "considerable ruin" of most of the food supply and that in a very few days the residents "would feel the full effects of the calamity."[163] The full magnitude of the disaster, however, was not apparent until reports from the countryside had been compiled and sent to the governor. The towns and villages reported almost total destruction of their food supply. In the Cauto district to the east of Santiago de Cuba, over 2,300 baskets (*serones*) of corn had been destroyed, and flooding had drowned cattle, swine, and other farm animals such as chickens and goats. Every hog, every head of cattle, and every other farm animal perished in Caney. As the hurricane winds scoured the countryside, entire stands of banana trees were destroyed, down to the roots, damage that would take years to recover from.[164] The leveling effect of disaster made misery widespread and indiscriminate in its effects, and even families of high status were forced to sow tobacco seeds alongside their slaves.[165]

For the second time in as many years, catastrophe threatened eastern Cuba, and the misery was compounded by the international political atmosphere. Like the decade previous, Spain was on the verge of going to war, this time with Portugal over a boundary dispute in South America between Portuguese Brazil and the Spanish viceroyalty Río de la Plata (present-day Argentina). Royal governors, lieutenant governors, and other military men offered their assessments that the munitions and troops under their command were dangerously unprepared for a potential conflict. After the hurricane in September 1775, Ayans worried that the rain had damaged the powder for the cannon in Santiago de Cuba and the fortresses surrounding the city.[166] He also feared that Baracoa was totally indefensible. "Baracoa could be taken with two navios by closing the passage between Cuba and St. Domingue with cruisers out of Jamaica. Ultimately, the enemy could close the entire Spanish naval routes from Campeche, Vera Cruz, and finally the entrance to Havana."[167] He opposed allowing British ships to enter the harbor at Santiago de Cuba or Guantánamo Bay, citing the precedent in the 1740s when British ships were given safe harbor in Cartagena (present-day Colombia), only to return months later to lay siege to the city.[168]

By the winter of 1775–76, conditions in Oriente were eerily reminiscent of the disastrous winter of 1761–62. No one could forget the circumstances just over a decade earlier that ultimately had resulted in the fall and occupation of Havana. At the same time, policy makers and strategists in Spain could not postpone preparations for war in the Americas. In February 1776, a royal troop transport from Spain arrived in Santiago de Cuba with the regiment of España en route to its new post in Havana. Widespread sickness aboard ship forced the captain to stop in the eastern city to transfer 131 ailing soldiers to the new military hospital.[169] The soldiers faced a long recovery period, and the majority of them could not continue their journey until May.[170] Thirteen of the soldiers of the regiment were so ill that they remained behind in Santiago de Cuba, and one such soldier, Pedro de la Peña, was not healthy enough to travel to Havana until

August.[171] Five members of the regiment died.[172] Even after a lengthy recovery in Oriente, upon arrival in Havana some men had to be readmitted to the hospital in Guanabacoa to recuperate.[173]

The unidentified fever that arrived on the troop transport quickly spread among the residents of eastern Cuba. Shortly after the sick soldiers were housed in the city, a senior official in the tobacco monopoly, Mateo de Echavarría, fell ill and requested permission to retire to the country to convalesce from the effect of fevers.[174] In March, Ayans also wrote that his health had deteriorated to the point that he was not able to carry out his regular responsibilities.[175] Juan Manuel de Rebollar, a lieutenant colonel of the second battalion, received similar permission for medical leave in April, and he moved to the rural village of Jiguaní in the hope of restoring his health.[176] At the same time, a soldier in Rebollar's unit, Joseph Antonio Rolando, spent nearly three months in the hospital as a consequence of the fevers.[177]

As had happened in 1762, the ruinous weather continued from early spring throughout the summer, with the situation worsening as time went on. As storms continued to wrack the area, two British ships sought safe harbor in June; both had been caught in storms at sea between Cuba, Hispaniola, and Jamaica in May.[178] Another arrival in July, the British ship *Firebird*, had been separated from its fleet near the island of Grenada. It made one emergency landfall in Santo Domingo, only to be blown off course once again on its way to Jamaica. The next time the captain and crew set foot on dry land was when they limped into Santiago de Cuba harbor seeking shelter, days overdue and miles away from their original destination.[179] Spanish forces were similarly burdened. Juan de Lleonart, the former lieutenant governor of Puerto Príncipe, survived one of the most harrowing ordeals. By 1776, he had been promoted to the rank of sergeant major of the second battalion in Santiago de Cuba. At the beginning of August, he and twenty-five men under his command left the city in the company of two other ships en route to Havana. Two days out, the convoy encountered a terrible hurricane at sea, and Lleonart and his men were shipwrecked on the north coast of Cuba near Holguín. Rescue appeared impossible, so the sergeant major and his men set out on foot to return to Santiago de Cuba. Along the way, many fell sick with the usual variety of fevers, but the company pressed on. Finally, on 10 October, the survivors made it back to Santiago de Cuba, to the surprise and delight of their families. The two ships that accompanied them were never heard from again.[180]

Yet, as if the ecological crisis and seasonal fevers were not enough, the disease environment went from bad to worse in June when smallpox was confirmed in Santiago de Cuba. This highly infectious disease is caused by a virus that is spread through direct human contact, through the air, through cloth such as sheets, blankets, or other bedding, or through the clothing of an infected patient, even long after the patient has died or recovered. Victims feel the first symptoms within a ten- to fourteen-day incubation period, although the first symptoms are often confused with those of influenza or other fevers. Shortly thereafter, a rash develops that produces pustules and scabs that appear over the victim's body. Mortality rates range from a low of 7 percent to a high of nearly 100 percent in instances of "virgin soil" epidemics, which occur when populations have had no previous exposure to the disease.[181] The most vulnerable are the very young, the very old, and pregnant women, and poor diet and malnutrition contribute to higher mortality rates.[182] Those who survive are immune for life, although they often bear visible scars from the disease. Older persons are often immune, and smallpox claims a disproportionate number of children, who have no immunity.[183]

Characterized by symptoms of hemorrhaging and skin eruptions, the disease that struck the townspeople of Santiago de Cuba was most likely *variola major*, the deadliest form of smallpox, with an average mortality rate of 25 to 30 percent. At the first signs of the illness, a municipal officer, a physician, a surgeon, and a notary went house to house, looking for evidence that the occupants were infected.[184] Well-to-do residents were confined to their homes, and the less fortunate were expelled from the city to a quarantine area. When a patient died, all items in the house were burned. The room was disinfected by removing the plaster on the walls and the floor tiles and then washing the walls with lime. Even jewelry was melted down to remove any trace of infection.[185] In addition to a scientific approach, town leaders relied upon the time-honored comfort of faith. As soon as the epidemic was confirmed, Santiago de Cuba's town council called for a public procession and prayers to be held in the streets of the city.[186] Ironically, the show of faith probably spread the disease further as persons already infected but showing no visible signs of the disease came into direct contact with healthy residents.

In spite of all precautions, followed to the letter by municipal officials, by the end of June, the news was grim. The inspections revealed that the disease had spread to all of Santiago de Cuba's barrios and that the number of deaths had reached levels "never before seen" in the city.[187] Mortality struck all ranks of the citizenry and created havoc at all levels of government. On 24 June, the aged and already debilitated governor Juan de Ayans y Ureta died in his residence, and the lieutenant governor, Estéban de Oloríz, took control of the province.[188] Among the casualties was Lieutenant Colonel Rebollar, who also had been debilitated by the wave of fevers during the previous winter.[189] Smallpox claimed the ranking treasury official, Juan de la Passada, along with many less notable residents of the city. Deputy Treasurer Antonio Rocabruna also fell ill, but he survived.[190] As late as September, the normal functions of government were still impeded, because so many members of the town council were too sick to leave their houses and a quorum could not be convened for meetings.[191]

Blame for the epidemic was placed on the Asiento—Santiago de Cuba's angry residents accused the company of not holding the slaves in quarantine for the required forty days.[192] Years of frustration and resentment against the monopoly boiled over in the townspeople's belief that the company had ignored their welfare.[193] Their accusations were confirmed by the news that came from the neighboring town of Trinidad. In late May, when an Asiento vessel unloaded the cargo of slaves it had transported from Santiago de Cuba, local officials found that two slaves showed signs of smallpox. Both were immediately put in quarantine.[194] The most likely candidate for bringing the disease to Cuba was the frigate *Minerva*, which had docked in Santiago de Cuba on 4 May with 167 slaves, 13 of whom were unhealthy.[195] Or perhaps the disease was introduced even earlier, on the sloop *Industry*, which had arrived in early April with 130 slaves from St. Eustatius.[196] Then came the news that the sister city in the Asiento's supply chain, San Juan in Puerto Rico, was also afflicted by the disease, and the suspicion that the slave trade was the culprit grew even stronger.[197] Certainly the slave trade seems the most likely explanation as to how the disease arrived in Cuba, but other explanations are possible. The disease could have been present onboard one or both British ships that arrived seeking shelter from severe storms on 2 June. Another potential carrier was a launch on patrol that had encountered another British ship in distress and had towed it into Santiago de Cuba harbor.[198] The town council first reported the disease on 14 June, and since smallpox has a fourteen-day incubation period, a potential carrier on either of these ships would have shown no signs of the disease upon arrival.[199]

The accusations that the slave trade was responsible for introducing smallpox to Spain's Caribbean cities was perhaps the final straw that compelled Charles III's inner circle of advisers to confront the problems that still existed in the Caribbean. The reform measures implemented in 1773 allowed the Asiento to continue its existence, but the members of the king's council were less than pleased with the outcome. Much of their unease originated as a consequence of complaints from cities assigned to the Asiento along the northern littoral of South America, such as Cartagena, Cumaná, and La Guaira. The city fathers of Cumaná were especially critical since they had suffered a similar outbreak of smallpox in 1769 that was also traced to the slave trade. Now the king's counselors faced the realization that slaves brought by the Asiento from Jamaica were the likely carriers of the disease, and the most likely way that smallpox had gotten to Jamaica was via the United States.[200] A contemporaneous outbreak had raged in North America since fall 1775, causing great mortality in the thirteen colonies and having serious consequences for both sides in the American Revolution.[201]

The steady stream of complaints from the Caribbean basin cities prompted the Ministry of the Indies to take action. In late 1775, it authorized an inspection visit to examine the account books and operating practices of the Asiento in San Juan, Santo Domingo, and Santiago de Cuba. The chief treasury official in Havana, Juan José Eligio de la Puente, was chosen for the assignment, a sensible choice since the Eligio de la Puente family was synonymous with royal service. Their bureaucratic pedigree was already generations old when in 1733 his uncle Francisco had authorized and paid for a printed compilation of his and his family's services to the Spanish crown in 1733.[202] In 1764, the younger de la Puente was the ranking official in the royal treasury, and in that capacity, he supervised the transfer of the civilians, the garrisons, and the materiel from Florida in 1763.[203] He also supervised extensive networks with ship captains from the exiled Florida community, making him the natural choice when a trusted official was needed to coordinate espionage activities against the British. Equally important, however, were his family ties to one of the most influential international merchants in Cuba, Juan de Miralles, who was married to Eligio de la Puente's sister, Josepha. (See the next chapter for a discussion of Miralles's significant contributions.)

Armed with direct authority given to him by the minister of the In-dies, José de Gálvez, in late 1775 Eligio de la Puente sailed to the major port cities of the Spanish Caribbean. In each city, he audited the treasury accounts and made a detailed examination of how the economy in each area functioned. Among his responsibilities was to offer suggestions about how the Asiento and other provisioning systems could function more effi-ciently with fewer losses to the treasury from the contraband trade.[204] He arrived in Santiago de Cuba in February 1776 at the height of the epidem-ics and thus became a firsthand observer of the misery and suffering of the residents. By the time of his arrival, the number of deaths and absences from illness had compromised the effectiveness of government. Eligio de la Puente remained in the pestilential city for several months because there were so few trusted and competent officials to take over the reins of government.[205] The deaths of most competent treasury officials had made chaos of the accounts, and it took him months to put the books in order.[206] During his time in Santiago de Cuba, he had the opportunity to observe how the Asiento functioned, and the report he sent to de la Torre reinforced the pessimistic observations of the Council of the Indies in 1773. He suggested that any treasury official assigned to the satellite cit-ies should be knowledgeable of the taxes and customs duties that had to be collected. Among the deputy's responsibilities would be to learn about the internal conflicts among the local merchants, but this person also had to recognize legitimate grievances when merchants incurred losses.[207]

For the second time in two decades—both instances on the verge of war—a deadly combination of disastrous weather and epidemic disease combined to create catastrophic conditions in Cuba. Yet there was a sig-nificant difference between events in 1762 and in 1776. By 1776, metropoli-tan bureaucrats in Charles III's group of advisers had confronted imperial problems with a maturity and modernity rarely attributed to Spain. Sub-sequent seasons were plagued by the continuance of the El Niño/La Niña cycle, which from 1773 through 1776 reinforced in bureaucrats' minds that the current commercial structure of the empire was unsustainable. The gathering storm in the thirteen colonies allowed the proponents of independence to take advantage of the Spanish empire's demand for pro-visions, which would lead to their decision that Great Britain was a greater hindrance than help. The consequences for Spain, Cuba, and the United States will be addressed in the next chapter.

CHAPTER FIVE

In a Common Catastrophe All Men Should Be Brothers

B Y THE SUMMER of 1776, the disenchantment so pronounced in the correspondence between Captain General de la Torre and treasury official Eligio de la Puente was symptomatic of the problems that would compel a new approach toward colonial affairs. The question of how to deal with the economic malaise generated by bad weather and the ensuing environmental crisis, the consequences of which had spread throughout Spain's empire, occupied the full attention of royal officials.[1] The political setting exacerbated rather than alleviated the problems at hand. In 1775, Spain suffered another and more significant military defeat in Algeria in North Africa. Poor planning and poor leadership caused the deaths of over 1,500 young Spanish soldiers, and the resulting public outrage forced a reorganization in Charles III's advisory councils.[2] The reorganization brought some of the most talented, experienced, and visionary men to positions of influence, among them José de Gálvez, who became the minister of the Indies in February 1776.[3] That same year, Spain and Portugal were on the verge of going to war, which threatened to involve Great Britain in defense of its Portuguese ally.[4] This potential conflict was hastily settled when the queen of Portugal, sister of Charles III, made an unusual journey to Madrid to consult with her brother, after which Spain retreated from its belligerent position.[5] Before it was certain that war would be averted, Spain began a troop buildup in the Americas, and as always, Cuba was affected by the arrival of another wave of peninsular soldiers, increasing the demand on the already-overtaxed provisioning system.[6]

As had occurred in the 1760s, the weather did not cooperate with European plans. The suffering and misery during the first half of the 1770s dragged on into the latter half of the decade as the El Niño/La Niña cycle did not subside, as it had done in decades past. The turbulent weather during fall 1775 through summer 1776 that ravaged eastern Cuba (discussed in the previous chapter) was no less severe in and around Havana. In early June, storms wracked the southern coast. The hamlet of Batabanó, which had sustained considerable damage in the hurricanes of 1768 and 1772, was again devastated. The barracks and stables built to lodge the men and their horses as part of the heightened defense posture became the first casualties. Wind and rising water completely destroyed the structures so recently rebuilt after the hurricanes of 1772, and the floodwater surged inland, ruining subsistence crops in the area.[7] The intendant of the army, Juan Ignacio de Urriza, ordered the captain of the closest area not damaged by the storm to transfer provisions and rebuilding supplies to Batabanó, "so that the troops and horses do not suffer any losses or discomfort."[8] The same storm that damaged the south coast of Cuba caused catastrophic results in Jamaica at a time when conditions on the British island could not have been more desperate. Still not spent, the system regained strength, curved northward through the Gulf of Mexico, and struck Havana's satellite city, New Orleans, causing the Mississippi River to overflow its banks—flooding the lowlands, destroying the crops, and placing additional pressure on the provisioning system.[9]

As winter 1776–77 set in, drought returned to the center of the island, especially in the provinces of Puerto Príncipe and Cuatro Villas. The dry winter gave way to spring, and as the rainy season returned, reports came from Havana's neighboring province, Matanzas, that the bridge over the San Juan River was in danger of collapsing.[10] Continuous rainfall throughout the summer made the roads impassable and prevented José de Alvarado from returning to Havana from Trinidad. Alvarado was stranded in the port on the southern coast when he could not engage any boat to take him to Batabanó because the captains feared sailing during the autumnal equinox.[11] At the end of October, eastern Cuba once again was wracked by a hurricane, which appears to have passed between Cuba and Saint Domingue. Even though it was on the western fringes, Bayamo, nonetheless, suffered considerable damage.[12] Santiago de Cuba, closer to the eye of the storm, felt the full fury of the wind and rain, and for the third time in as many years, reports from that unfortunate town told of the misery of its residents and the scarcity of provisions.[13] As always, the first line of defense was to sail to the closest ports, but Saint Domingue also fell victim to the storm.[14] Upon learning of the disaster, the new captain general, Diego José Navarro, authorized Urriza to transfer provisions to the east and to otherwise help in all ways possible.[15] By 1778, the continuous bad weather was having an impact on areas that did not suffer a direct hit. The barracks in Jubajay, originally damaged in the hurricane of 1774, finally collapsed, after suffering continuous assaults from wind and rain.[16] One by one, storage areas in the numerous fortifications, tobacco warehouses, and barracks and stables deteriorated slowly and steadily under the forces of nature.[17] The population also suffered from a shortage of provisions, when bad weather ruined subsistence crops and bad harvests reduced the total amount of provisions. By the end of the year, the hospital administrator in Havana sent an urgent request to the constable of the small town, San Miguel del Padrón, to send eggs and fowl for his patients. The constable replied that wild game birds had been hunted out because there were so many poor and desperate residents in his area.[18]

The epidemic and environmental conditions, so reminiscent of 1762, must have sent chills of apprehension through the war planners in Madrid, who had learned from hard experience that they could not ignore local conditions as they had in the past. Among the hallmarks of Charles III's reign was that he and his advisers viewed his empire as an interconnected whole. When they debated if and when to enter the war, his counselors analyzed imperial issues in global terms. From their perspective in Spain, the outlook would not have been reassuring. Reports from other governors and royal officials told of similar dire conditions throughout Spanish America. From 1771 through 1779, drought was pervasive in every major city on the mainland from Mexico to the southern tip of Chile.[19] The regions obligated to provision Cuba, the agricultural provinces of Mexico, suffered repeated environmental crises from 1765 through 1778.[20] Likewise, the northern littoral of New Granada fell victim to decades of turbulent weather that ruined its agricultural potential.[21] These "synchronous events" throughout Spanish America placed all provisioning networks developed over the previous decades in jeopardy.[22]

Such were the empire-wide environmental conditions that influenced metropolitan planners in Spain. To their credit, however, this time royal advisers recognized that the need to supply provisions to troops could not be left to chance or to the inefficient methods of the past. Charles III and his war counselors were prepared to abandon old ideas and restrictions to assure that the military units sent to the Americas were well supplied. Natural disasters directly affecting the Spanish empire, as well as the collateral effects of such phenomena, were fundamental in changing imperial policy concerning the commerce in comestibles, which, in turn, forever altered economic patterns in the Caribbean and the North Atlantic basin.

The Domino Effect: 1774–1777

During the mid-1770s, as misery and scarcity plagued the Spanish Caribbean, Great Britain faced its own problems in dealing with a rapidly escalating confrontation between itself and its thirteen North American colonies. Far away from the epicenter of the crisis, the opportunity to market their flour, rice, and lumber to communities in the "greatest state of misery" would provide an irresistible temptation for North American merchants. Meanwhile, from October 1774 to November 1775, the political relationship between North America and London deteriorated significantly. Amid growing confrontations between the metropolis and the colonies, a representative body of delegates, the Continental Congress, met in Philadelphia in October 1774 and enacted an embargo on sending provisions to the British West Indies, to take effect in September 1775. In the intervening nine months, the delegates debated their options. Faced with a precarious economic and political future within the British empire, the need to secure foreign markets was obvious. The delegates knew that their survival depended upon their ability to market their goods. Without trade, the nation could not survive. If actual combat began, the insurgent army would need weapons, ammunition, and gunpowder. The colonies also needed money, and Congress's good credit had to be maintained. Without trade, the colonies could not get intelligence about British troop movements. The delegate from Georgia summarized the situation: "If we must trade we must trade with somebody and with somebody that will trade with us, either foreigners or Great Britain. If [we must trade] with foreigners, we must either go to them of they must come to us."[23] In October 1775, as the embargo against Great Britain took effect, the Continental Congress enacted a resolution that authorized exports to foreign ports in exchange for arms, ammunition, gunpowder, and cash.[24] It was simply a matter of survival. Congress cut out the middlemen in Jamaica, Barbados, and the Bahamas, and within weeks American merchantmen were authorized to sail to any port that would welcome them.[25]

While events unfolded in the thirteen British colonies, Charles III, his ministers, and the Asiento itself took decisive measures to ensure an adequate supply of provisions for the Caribbean. All participants were aware that if combat actually began, the provisioning capability of North American merchants would be seriously jeopardized. The monarchy needed to look no further than the recent experiences of the early 1770s, especially the Caribbean-wide crisis during 1772, which brought home in painful fashion the realities of interconnected commerce. No one doubted the obvious: Spain depended upon foreign—preferably North American—flour. One solution was to quietly increase the volume of trade from North America to Spanish peninsular ports, primarily Cádiz, but also to the northern cities of Bilbao and La Coruña. From July 1774 to July 1775, during the interim waiting period when it was uncertain whether the embargo would be implemented, at least twenty-six ships laden with flour cleared Philadelphia harbor for Spain.[26] The trade involved virtually all of the city's most prosperous merchant houses, many of which had benefited from the brief opening of trade to Havana in 1762.[27] As was customary, the trade in Spain was conducted by factors, who were well placed to receive the cargoes and to sell them at the most advantageous price. The most prominent of the Philadelphia merchant houses, Willing and Morris, dealt with several different men who were long established in their respective European cities.[28] In Cádiz they used the services of Duff and Welsh and on occasion those of the firm Noble and Harris.[29] A long-time contact and staunch supporter of the American cause was Etienne Cathalan of Marseilles, who transshipped Willing and Morris's provisions to Barcelona.[30] In northern Spain, the Philadelphians dealt with José de Gardoqui of the firm Gardoqui and Son of Bilbao.[31] These transactions were not always without complications, and there was no small amount of resentment against the Americans by the entrenched Spanish firms. The Philadelphians were entering a fiercely competitive system characterized by personal and regional rivalries between and among family networks in northern Spain, southern Spain, the Mediterranean coast, and Madrid.[32] As a consequence, the reception the Philadelphia merchants received was less cordial than that which they received in the Caribbean. In 1776, for example, Willing and Morris had difficulty marketing one of their cargoes when the merchant guild in Cádiz refused to accept the shipment. It appeared the ship would be turned away, but at the last minute, Willing and Morris's agents, Duff and Welsh, succeeded in convincing a Spanish factor to take responsibility for the flour on board.[33] Philadelphia's flour rarely remained long in Spain since the Asiento quickly transshipped it to the Hispanic Caribbean.[34]

These new realities of international commerce occupied a central place in the deliberations at the Spanish court, especially in debates about whether the Asiento provided any benefit to the empire. In 1775, the Asiento was no more popular and no more profitable that it had been two years previous. Shortly after the first Continental Congress met, Charles III once again formed a council of experienced men to evaluate the monopoly's performance. In February 1775, royal advisers Marcos Ximéno, José de Gálvez, Tomás Ortíz de Landázuri, and Manuel Lanz de Casafonda gathered to begin deliberations on how to secure provisions for their Caribbean cities. All veterans of inner-court politics, they were ill disposed to allow any more concessions to the highly unpopular monopoly. Their conclusions were reached, in part, because of complaints from the mainland cities about the high prices charged by the Asiento, accompanied by the persuasive argument that the client cities were perfectly capable of sending ships to foreign ports and purchasing their own provisions.[35] At the same time, even the most loyal and dedicated officials in the colonies voiced complaints about the Asiento's performance. Two of the highest-ranking officials in Cuba, the head of the mail system, Armona, and the intendant of the army, Urriza, commented on the stranglehold that the Asiento held on the commerce of the mainland cities. Together they wrote to the council criticizing the monopoly's inability to prevent contraband and opining that in order to remedy the abuses inflicted by the monopoly and to supply the residents, "nothing can [be] better nor quicker than *comercio libre*."[36]

Perhaps in anticipation of a negative assessment, the monopoly itself sought to modify its contract with the crown that established its privileges, this time expanding its sphere of influence by accepting investors outside its immediate circle.[37] For the most part, the operational mechanisms of the Asiento remained unchanged. José María Enrile retained his position as manager in Cádiz, and day-to-day operations in Havana were still supervised by his son, Gerónimo, and by the factor, Manuel Félix Reisch. Behind the scenes, though, the inclusion of a new group of interests marked a momentous shift. The new plan brought an unidentified group of French mercantile interests represented by Pierre Agustín Caron de Beaumarchais into the enterprise.[38]

The reorganization of the Asiento's board of directors in 1775 with Beaumarchais as a member was a critical step. Beaumarchais, a man of extraordinary ability best known for his literary talent, is also renowned

as one of the most outspoken supporters of the American cause. His linkages to the Spanish monarchy and to the Asiento have been identified in
different contexts, yet the full import of his relationships with members
of the Asiento's board of directors along with his vigorous champion-
ship of the American Patriots' cause have never been studied. During the
previous decade, Beaumarchais was always near the epicenter of international commerce. In 1765, he, along with other merchants, including Juan
de Miralles (whose signal importance will be discussed in detail later in
the chapter), submitted competing proposals to form a monopoly company to supply the Caribbean, the contract that ultimately was awarded
to Aguirre and Arístegui. By 1773, the inept performance of the original
company, coupled with significant financial losses and ten years of unending complaints from its client cities, forced the Asiento to reconstitute its
board of directors. During the investigation in 1773, which led to the first
reorganization, Charles III's ministers were unrelenting in their scrutiny
and unrestrained in their harsh conclusions. More significant, though, by
1775, the Spanish crown had come to a gradual realization that monopoly
companies were outdated and inefficient entities that did more harm than
good. That Charles III allowed the Asiento to continue its existence with
Beaumarchais a part of a new board, in spite of the evidence of its ineptitude, is clear evidence that his presence and his politics had tacit—if not
outright—approval at the highest level.[39] Equally important, given Beaumarchais's politics and the subsequent decisions taken by the Spanish
crown, it is clear that as early as 1775 Charles III intended that the Asiento
would be the conduit through which hard currency could make its way to
the rebellious Congress in Philadelphia.

During the first months of 1776, events moved very quickly. British
cruisers made the Atlantic crossing perilous for American vessels, while
suppliers and consumers in the Hispanic Caribbean scrambled to adjust
to the changing realities of transatlantic trading networks.[40] In January
1776, Robert Morris's friend Etienne Cathalan attempted to ship a load
of muskets, bayonets, and powder to the insurgents on an English ship
via Jamaica, but the captain refused to carry the cargo. Cathalan devised
an alternate plan to get the vital military supplies to the Patriots. He purchased a schooner and outfitted it as a French ship bound for the French
West Indies. The Marseilles merchant rationalized that not only would
they be sailing under French colors with the protection of the French
navy, but they would also save the costs of higher duties if the products

were kept within the French imperial economy.[41] In response to Cathalan's plan, Willing and Morris altered its marketing strategies, abandoning
the practice of shipping flour to Europe and installing its own commercial
agents in the French Caribbean ports of Mole San Nicolás and Cap Français (Guarico), both on Hispaniola, and in St. Pierre on Martinique.[42] The
agents' task was to receive flour and other comestibles and to exchange
them for gunpowder, weapons, supplies, medicine, and, most important,
hard cash to meet their expenses.[43]

Meanwhile, Beaumarchais set up dummy companies in France to supply the Patriots with arms and ammunition supplied by the French king,
all the while serving as adviser on the board of directors of the Asiento.
Simultaneously, his fellow board member, the elder Enrile, warned Gálvez
of the threatened disruption in supply from Spain and the options that
were available to the monopoly. In early 1776, no one was certain that the
North American embargo would be effective, and the Asiento was still
operating under the assumption that boats would be able to obtain provisions in Jamaica and/or Barbados. In February, Gálvez issued an order to
the captains general in the Caribbean basin to intensify their efforts to
obtain information about the events unfolding in North America.[44]

Upon receiving Gálvez's orders, de la Torre drew upon an espionage
network that was already in place. One of the most effective networks
was centered in the exiled *floridano* community. In spite of the cession of
Florida to Great Britain, *floridano* captains continued to sail northward to
familiar waters on both sides of the peninsula. Outside the fortified cities
of St. Augustine and Pensacola, British control was tenuous. A few Spanish citizens remained in Florida as agents provocateurs, others traded with
Indian groups along the Suwanee River, and others operated a cattle ranch
in Apalachee, unmolested throughout the period of British rule.[45] Captains of fishing vessels maintained contact with these Spaniards in British
Florida through regular visits to a settlement of Minorcan Catholics south
of St. Augustine at New Smyrna. These men carried messages back and
forth, thus providing information about the status of the British garrison
in St. Augustine and about the condition of family properties.[46] Equally
important, the ship captains served both as agents and as ambassadors
by promoting contact and commerce with Florida's west coast Indian
nations, who were implacably hostile to British rule.[47] The information
provided by these networks was collected by the ever-versatile Eligio de la
Puente, who passed the information on to war planners in Madrid.

At the same time, de la Torre redoubled his efforts to gather information, sending spies to other key areas with which they were familiar from previous visits. Among the most experienced of these men was Pedro Trujillo. During the 1760s, Trujillo had served as a captain of the *guardacostas*, charged with intercepting smugglers along Cuba's southern coast.[48] In that capacity, he had created a network of confidants in Jamaica. By 1776, honesty and loyalty had earned him the rank of lieutenant on a naval frigate, and so he was a natural choice for de la Torre to send to Jamaica under the pretext of conducting illicit commerce.[49] Another agent, Joseph Carrandi, traveled westward to the settlement of Filipinas, an especially vulnerable area because of its strategic position observing the maritime traffic in the sea route returning to Europe.[50] One particularly effective agent was naval officer José Melchor de Acosta, who regularly sent copies of the British press from his position in Guarico in Saint Domingue. At the eastern end of the island, the governor of Baracoa, Rafael de Limonta, collected information from fishing captains and other persons who may have had contact with neighboring areas. Because of Limonta's intelligence-gathering network, the Cuban captain general was among the first officials to learn that the thirteen colonies had declared independence in 1776.[51]

The most studied of these missions was carried out by Miguel Eduardo, the bilingual public interpreter of Havana.[52] In February 1776, Eduardo was given a special assignment to gather information for the political administrators in Cuba and in Spain and a separate charge to purchase flour for the Asiento.[53] The original instructions specified that Eduardo was supposed to go to Rousseau in Dominica, because, according to Enrile, "Rousseau is closer than Philadelphia, and we do not have any contacts there."[54] At the same time, though, he was warned: "You are to maintain the most religious secrecy so nobody learns of your true destination."[55] The Havana factor acknowledged that he did not know anything about Willing and Morris, except through intermediary contacts, but that the merchants had provided flour to Puerto Rico in the past.[56] In May 1776, Eduardo left Havana with letters of introduction to Willing and Morris in particular, and to several other commercial houses in free ports such as St. Eustatius and Barbuda, with 12,000 pesos, which was earmarked to purchase flour, and an additional 1,000 pesos to cover his expenses.[57]

The turbulent weather that plagued the Caribbean in spring and early summer 1776 undoubtedly provided credibility for his cover story that he was headed for Rousseau. The ship in which he traveled, the *Santa Bar-*

bara, one of the dozens of vessels caught at sea, was blown far off course and was forced to seek shelter at the first landfall, at the mouth of the Delaware Bay. The vessel was captured by the British navy, Eduardo was made a prisoner, and the 12,000 pesos was confiscated. The Spanish ship was declared a prize, and the captain and crew were ordered to proceed to Chesapeake Bay and to present themselves to the court of Admiralty. While a captive, Eduardo compiled a diary of events, which became a model of intelligence gathering, and so the mission was deemed a success.[58]

In the end, Eduardo's mission was a spectacular intelligence-gathering accomplishment and a spectacular commercial failure. The misadventure resulted in a net loss of 12,000 pesos, an enormous sum. Worse still, it did not solve the problem of providing flour to Cuba. Conversely, the mission was less important for its military success or commercial failure than for what it signaled to the Continental Congress. By sending a messenger to Philadelphia in search of provisions in early 1776, it was clear that Spain intended to continue purchasing flour from the North Americans, regardless of where the product was marketed. The decision had been made in Madrid as early as February (and probably earlier, given Beaumarchais's presence on the Asiento's board of directors), and that intention was communicated to Havana no later than April. By early May, Eduardo was on his way to effect the purchase, and whether or not he was intended to purchase Philadelphia flour in Rousseau or in Philadelphia, the Patriots were aware of the Spanish intentions no later than May 1776. Thus, a month before he signed the Declaration of Independence, in July 1776, Massachusetts delegate John Adams stated with confidence: "There will be little difficulty in trading with France and Spain."[59] That information became public knowledge by October after the *Pennsylvania Gazette* published the news of the *Santa Barbara*'s arrival with "10,000 Spanish milled dollars to procure provision for them. . . . [The captain] says we may expect a number of their vessels to the continent in the course of the winter."[60]

Throughout spring and early summer 1776, as fact and rumor swirled throughout the Caribbean, one Asiento ship a month traveled between Cuban and British Caribbean ports. Each returning captain verified the news of the deteriorating situation between Great Britain and the thirteen colonies. By May 1776, the North American embargo on shipping provisions to the British islands was having an impact. Concerned about his dwindling ability to fulfill the contract with the crown, that month the Asiento's factor in Puerto Príncipe asked for permission to send a boat to Jamaica to ascertain whether vital provisions would still be available from the British island.[61] Consternation gripped his counterparts, the merchants in Kingston, who witnessed their trade with Cuba evaporate within months. Replying to the Asiento's inquiries, in July 1776, a group of merchants that dealt with the Asiento sought to assure the lieutenant governor of Puerto Príncipe, the Conde de Ripalda: "We will do everything in our power to continue to do business with you."[62]

The declining trade between Cuba and Jamaica is evident in the commercial transactions of the Asiento's frigate *Minerva* (the vessel that most likely brought smallpox to Santiago de Cuba), which left Havana for Jamaica on 1 May to purchase flour and slaves. For the *Minerva's* captain, Ramón de la Hera, the voyage to Kingston should have been routine. Hera was one of the most experienced captains in the intra-island trade, having been employed by the Asiento before and after the decree allowing direct contact with the British islands in 1773.[63] In the months since the implementation of the embargo, he had made three trips between the two islands—in December 1775, in March 1776, and the present voyage in May. He always carried a considerable quantity of Spanish hard currency, averaging 60,000 pesos per trip, but even though money seemed to be no object, his return cargoes revealed the declining conditions in the Jamaican market. Even more disturbing, the slaves that he had been able to purchase increasingly showed signs of illness upon arrival, culminating in the outbreak of smallpox in Oriente in June.[64] Other ship captains sailing the intra-island route to Jamaica reiterated the litany of misery that had descended on the neighboring island. By June, the Patriots' embargo had effectively cut supplies.[65] To make matters worse, the same early-season hurricane that had caused so much damage near Havana made a direct landfall on the north coast of the British island. Already vulnerable because of the miserable conditions caused by the embargo, the suffering of the residents increased exponentially.[66] As had occurred in the 1760s, when political authority broke down in the wake of catastrophe, a familiar pattern set in. Roughly two weeks after the storm struck, slaves on the northern provinces rose up in rebellion. Returning captains reported that over 25,000 slaves from 70 plantations were killing their white masters and overseers and setting fire to buildings and cane fields.[67] Other missives told of the great scarcity of provisions and the confusion that had set in as a result of the catastrophe and the uprising.[68]

While the British West Indies markets stagnated and the residents starved, a festival-like atmosphere prevailed in the entrepôts created in the French islands, especially in neighboring Saint Domingue. In the harbors of Guarico and Port au Prince, international merchant ships and North American corsairs lay side-by-side at anchor, while traders on shore made deals and executed contracts for future purchases. Commercial houses, among them Willing and Morris, maintained public warehouses loaded to the rafters with provisions. According to the returning Asiento captains, "any type of merchandise, even munitions and arms," could be purchased in any port.[69] Agents representing the thirteen colonies assured their employers in Philadelphia that the plan to conduct business via the French colonies was a success. In August 1776, William and Morris's agent in Mole San Nicolás, John Dupuy, assessed the situation as being "warlike," but that the conditions worked to the Patriots' benefit. Competition had increased the supply of vitally important gunpowder, and as a consequence, the price had fallen from 6 French *livres* per hundredweight to 4.25. The French government sent three frigates to cruise along the north coast to protect Patriot merchant ships that attempted to make the run to the island. Likewise Spain sent frigates to patrol along the coast of Santo Domingo, extending their cruising area to the east to Puerto Rico.[70]

In early September 1776, the *Minerva* again cleared Havana harbor for Kingston with a new captain, Miguel Gonzáles, and with 80,000 pesos to purchase slaves and flour. Upon arrival in Jamaica, Gonzáles found chaos, so he sailed on to Guarico, where he managed to purchase 700 barrels of flour for consumers in Cuba. While in port, Gonzáles was approached by one Stephen Ceronio, who introduced himself as the "official representative of the thirteen colonies."[71] Displaying a keen knowledge of Spanish commercial regulations, of Eduardo's unsuccessful mission to Philadelphia, and of the fact that the Asiento was seeking to purchase flour, Ceronio offered a solution in a letter addressed to Enrile in Havana. Since Spanish vessels were not able to enter ports in the thirteen colonies without risk of being intercepted by the British navy, the agent offered to send the boats directly to Spanish Caribbean ports with flour, rice, and other commodities. He acknowledged that direct commerce was prohibited at present but that such contact would benefit both nations. Gonzáles passed Ceronio's letter on to de la Torre and Enrile, but they could do little more than forward the missive on to José de Gálvez and the monarch.[72]

In spite of relentless diplomatic pressure by U.S. ambassadors in Spain, by October 1776, Charles III had settled on an official policy of middle-of-the-road neutrality.[73] Citing international law that guaranteed freedom of the seas and the obligation to provide asylum for ships in distress, Charles III declared that North American ships could be admitted into Spanish ports in times of urgency (over the protest of the British ambassador) but that they could not engage in any commerce. They would be received with hospitality, and they could pay for their emergency relief in specie, in bills of exchange, or in slaves.[74] On the other hand, it would be foolish to reject the opportunity to secure provisions that were so close at hand, so at the same time the Asiento was given permission to purchase flour in Cap Français and Port au Prince.[75]

Colonel Antonio Raffelin, the commander of the mounted cavalry in Havana, the most prestigious military unit on the island, was chosen to convey the Spanish policy decision to Ceronio. Like Eligio de la Puente, Raffelin was a sensible choice. A native of Paris, he had joined the Spanish military in 1742 and had distinguished himself in the European campaigns in the 1760s before transferring to the Caribbean theater during the Seven Years' War.[76] He had participated in the defense of Havana in 1762, and in 1763, while being held prisoner of war in Jamaica, he had conducted a similar intelligence-gathering mission on the orders of the Conde de Ricla.[77] Upon his return to Cuba in 1765, he elected to remain on the island, and shortly thereafter, he was commissioned into the mounted cavalry, where he rose to become its commander.[78]

In February 1777, Raffelin left Havana for Saint Domingue aboard the *Santa Barbara*, the same vessel that had taken Miguel Eduardo on his journey to the Chesapeake. He carried specific instructions to gather information and to meet with Ceronio to inform the agent that U.S. ships could make port in an emergency but that no trade would be allowed.[79] His trip was supposed to last but a few days, but the *Santa Barbara* proved unseaworthy and had to be careened. Stranded in Saint Domingue, Raffelin had ample time to assess the commercial atmosphere and the potential for regular trade with North America via the French islands.[80] At a time when the ability to secure flour and other provisions in Jamaica was seriously compromised, Raffelin reported on the number of "*Bostoneses corsarios y mercantiles*" (American corsairs and merchantmen) in the harbor. More important, he concluded that the Asiento captains would be able to purchase flour as per their contract, with no problems.[81]

At the same time, Raffelin began collecting intelligence on the conduct and progress of the war. He wrote to José de Solano, the president (equivalent to the captain general) of Santo Domingo, informing Solano of his mission and of the change in Spanish commercial regulations. In April, he communicated that the British navy was unable to stop the North American corsairs and that hunger was widespread in Jamaica. The embargo forced the Jamaicans to depend on French smugglers, who brought flour and other life-saving provisions. At the same time, the Jamaicans were adamantly opposed to the Americans and resented the French for the help that they openly gave to the rebellious colonies.[82] By late April, other captains verified that the market in Jamaica had collapsed, that there was no longer the potential to buy flour or slaves, and that even the secondary goal of acquiring information was no longer a possibility. Finally, after six weeks' delay, Raffelin secured passage back to Havana on a French ship.[83] By his own estimation, his stay was only a modest success, but his six weeks in Cap Français were far more productive that Eduardo's costly (and celebrated) mission the previous year. Raffelin had paved the way for Asiento captains to secure flour in Saint Domingue, and he had also established a communications network in which Ceronio would send monthly reports on conditions in the insurgent colonies to Solano via Montecristi.[84]

By March 1777, propelled by scarcity and need in Cuba and by the economic decisions of the merchants in Philadelphia, the French ports in Saint Domingue had replaced Kingston and Barbados as the regular ports-of-call for Asiento ships. The shift in commercial relations was again evident in the records of the Asiento sailings. In the three years immediately after the monopoly gained the privilege to trade with foreign islands (1773–76), Asiento boats went almost exclusively to the British islands, overwhelmingly to Jamaica. After the rupture between Great Britain and the thirteen colonies, a transition period began, from July 1776 through mid-1777, during which time Spain adopted a wait-and-see policy. Captains continued to sail to Kingston, but invariably they returned with discouraging news, few provisions, and slaves who showed increasing signs of illness. After March 1777, the Asiento abandoned voyages to the British colonies entirely, in favor of going to Saint Domingue. Another significant change was the de facto uncoupling of the requirement to bring in a fixed number of slaves in relation to the quantity of flour. From 1773–76, the licenses granted clearly stated that both slaves and flour were the goal of the voyage, while after May 1776, captains were granted permission simply "to buy flour for the benefit of the community" (*para la compra de arinas a beneficio de éste público*).[85]

At the same time, though, neither the crown's advisers nor the Asiento were willing to abandon completely traditional avenues to secure flour for Cuba. The monopoly tried to maintain commercial relationships with Jamaica, and in March 1777, Enrile again sent a boat to Kingston to make contact with the merchants. Like the previous voyages, the captain returned home without success.[86] Another avenue was opened by the new governor of Louisiana, Bernardo de Gálvez, nephew of the minister of the Indies. In response to flooding in New Orleans after the hurricane strike in summer 1776, Gálvez initiated direct contact with Philadelphia via Robert Morris's agents in the Louisiana city, François DePlessis and Oliver Pollock.[87] A few months later, in March 1777, commercial relations between New Orleans and Philadelphia were solidified when a packetboat under the command of Bartolomé Toutant Beauregard sailed northward to purchase flour and to gather news of the rebellion.[88]

With every report from governors, commercial agents, and spies, royal ministers in Madrid weighed the positive and negative aspects of the interaction. The letters exchanged between de la Torre and Juan José Eligio de la Puente after the latter's residence in Santiago de Cuba during the horrific spring and summer of 1776 were among the evidence that guided Charles III's counselors in formulating policy for the empire. Neither the captain general nor the highest-ranking treasury official in the Caribbean could offer a positive assessment of the situation. The men agreed that the existing restrictions worked against local merchants because regulations were formulated in Spain by bureaucrats who had no idea of the conditions in Caribbean cities. Even though the Asiento had a deputy in Cuba ostensibly to run the operation efficiently and fairly—Gerónimo Enrile—he could do nothing to satisfy the complaints of consumers because he was little more than a representative of the merchants in Cádiz and was obliged to promote the Asiento's interests over the welfare of its client cities.[89]

Before reaching a final decision, His Majesty, prudent as ever, sought the advice of another man with years of experience in Cuba, Lorenzo de Montalvo, the Conde de Macuriges. Now elderly, Montalvo was well aware of the problems that had plagued the island during his forty-nine years serving his monarch. While the empire was reeling from the consequences of the British victory in 1762, he had remained in the city acting as an intermediary between the British forces and the local population. After the British left, Montalvo had been rewarded with his noble title and placed in charge of the royal accounts as the naval intendant of Havana. Among his many duties was dealing with the shortage of provisions, especially in the two critical years after the return of Spanish rule. During the horrible winter of 1772–73, Montalvo was a key member of the *junta de guerra*, the group of advisers who struggled to cope with the aftermath of hurricanes and food shortages, which were unprecedented in their experience. By the time Charles III called upon him for one last service, Montalvo had been among His Majesty's most trusted advisers in Cuba for nearly a half century.

Montalvo's report, submitted in October 1777, revealed how much ideas about the existing economic system had shifted in a decade. His first conclusion reiterated the obvious: the current system was unsustainable because Mexico could not supply Cuba. The bakers' guild in Havana was required to obtain its flour from Veracruz, but more often than not, the flour that did arrive was already rancid or full of weevils. On the other hand, the Asiento was able to purchase North American flour from Jamaica and from Spain. To no one's surprise, Cuban consumers preferred the Asiento's flour because it was fresher and the monopoly sold it at a better price. Montalvo did not advocate abandoning the supply route from Mexico, but he did suggest abolishing the bakers' guild, even though it contributed 23,000 pesos for the militia uniforms. Instead, he suggested taxing the entry of wine, aguardiente, and cacao. After presenting and evaluating the options available, Montalvo offered a not-so-stunning conclusion: the only solution that would guarantee an adequate supply of flour to Cuba would be free trade.[90]

By the end of 1777, the movement toward free trade had become a tidal wave. Many studies attribute the change to José de Gálvez, but the most vocal proponent of *comercio libre* was his even-more-powerful colleague, Tomás de Landázuri, who for over ten years had firsthand experience attempting to remedy the shortcomings of the Asiento.[91] When Gálvez assumed control of the Ministry of the Indies, he supported Landázuri's assessment that the imperial economic system must be reformed. The opinions of advisers such as Armona, Urriza, de la Torre, Eligio de la Puente, Montalvo, and Gálvez's nephew Bernardo, men who were firsthand observers of the commercial environment, undoubtedly influenced the decision. Only one obstacle stood in the way—Spanish imperial law—so beginning in February 1778, the minister took steps that would lead to the gradual reform of Spanish commercial regulations. The first area to be affected by concessions was the viceroyalty of the Río de la Plata (present-day Argentina), and by August, Gálvez's protégé, Francisco de Saavedra, began working on mechanisms to implement empire-wide changes.[92] In October 1778, these measures came together in one of the most important documents in Spanish economic history, the *Reglamento para el comercio libre* (1778).[93]

Few legislative documents have generated so much scholarly interest.[94] Certainly, the *Reglamento* was one of the milestones in promoting the economic development of the Río de la Plata, Cuba, Puerto Rico, and Cuba's satellite cities. The powerful merchant guild in Mexico was implacably hostile to the reforms, predicting—correctly—the deterioration of commerce between its area and the rest of the empire. Some merchants, anxious to perpetuate their commercial influence, actually proposed returning to the system in place in 1720, the *flota*.[95] The Asiento was also unhappy with the new legislation. The opening of free trade sounded the death knell for the monopoly, and according to the leading expert on the monopoly's business operations, by 1780, the Asiento ceased to function.[96]

While few would doubt the importance of the *Reglamento* to Spain and its colonies, no study has identified its importance in solidifying trade between the Spanish Caribbean and the thirteen insurgent colonies. In April 1777, after Raffelin's mission to Cap Français, royal officials in Havana and other cities acted under the premise that flour and other provisions would be shipped regularly from Philadelphia to the French islands. Thus, the obstacles to provisioning Cuba, Puerto Rico, and Santo Domingo appeared to have been cleared. Then two events disrupted the supply. In June 1777, the British army occupied Philadelphia, and the Continental Congress was forced to flee west to the safety of York. Many rural farmers who supported independence refused to bring their produce into the city, and by fall 1777, scarcity and famine plagued the residents of the capital.[97] To make matters even worse, over the winter of 1777–78, the normally abundant farms in the hinterland near Philadelphia fell victim to bad harvests from drought. Conditions were so desperate that by November one observer wrote: "Almost everything is gone of the vegetable kind; butchers [are] obliged to kill fine milch cows for meat, mutton and veal not even heard of."[98] Facing scarcity at home, the Continental Congress limited the amount of provisions that could be exported to the French ports, while North American agents in the Caribbean such as Ceronio began to fall seriously behind in their obligations.[99]

Events in the thirteen colonies jeopardized Spain's plans for feeding Cuba, and in late 1777 Charles III decided to send emissaries to the insurgent colonies to assess the political and commercial conditions. The monarch commanded the captain general to choose two knowledgeable men who knew the area and were well informed about commercial policies and the current restrictions on trade. No one was more qualified in fiscal matters than Eligio de la Puente, and in November 1777, Navarro formally nominated the treasury official to undertake a mission to British Florida. The second man chosen was Eligio de la Puente's brother-in-law, merchant Juan de Miralles, a "wealthy resident from a family who is recognized in Havana."[100] Miralles's nomination coincided with his own proposal to undertake a secret mission to North America to get flour and slaves. Citing the benefits that it could bring to Havana and to the royal treasury, he proposed to sail to North America in the spring. He further proposed to employ the services of a good captain familiar with the voyage up the east coast, *floridano* Antonio Pueyo, a member of Eligio de la Puente's well-oiled espionage network. For secrecy, Miralles proposed to use the alias Pedro Payan.[101]

Although his political activities are well documented, mostly regarding his time in Philadelphia, Miralles's selection to represent Spain was the culmination of a lifetime of commercial success and related service to the Spanish crown.[102] As early as the 1750s, the Miralles family had established itself among the trading networks in the transatlantic commercial world.[103] During the Seven Years' War, Miralles was on board a ship returning to Havana from Jamaica when the British laid siege in 1762.[104] After the war, he was one of several competitors, including Beaumarchais, who submitted proposals to import slaves and flour into Caribbean cities, the privilege that ultimately was awarded to Aguirre and Arístegui in 1765. With the creation of the new Asiento in 1773, Miralles lost his investment of 70,000 pesos, but he was probably relieved when the crown allowed the original directors to bow out of their contract with only significant financial losses as punishment.[105]

The Eligio de la Puente and Miralles clan was a formidable force, not only because of their influence but also for their knowledge and expertise, which bridged the worlds of politics, law, economics, and international commerce. The men and their families moved in the highest circles of Cuban and Spanish society, and they enjoyed the favor of Charles III's most trusted officials. The warm tone of the correspondence between de la Torre and Eligio de la Puente demonstrated the regard that the rank-ing official in the Caribbean had for the treasury official.[106] De la Torres's successor, Captain General Navarro, reinforced the favorable impression when he praised Miralles's expertise in fulfilling other assignments and stating that he would be a "worthy representative" to the thirteen colo-nies.[107] By the 1780s, the family's influence was such that the hero of the Spanish army, Bernardo de Gálvez, served as a sponsor at the wedding of his daughter, Josefa de Miralles.[108]

Still, the appointment of Eligio de la Puente and Miralles generated resentment and jealousy in some circles of Charles III's court, especially within a group of partisans of a network revolving around Alejandro O'Reilly, the disgraced general responsible for one of the most significant military failures in Algeria in 1775. Prior to his fall from grace, O'Reilly had gone to Cuba to restore Spanish rule in 1763. In 1765, he wrote a negative assessment of Miralles, which very likely resulted in the merchant's failure to secure the Asiento privilege in 1765.[109] Miralles's appointment in 1779, received directly from José de Gálvez, was a vindication of the attempted character assassination of him and his family by the O'Reilly faction. It was also a tangible rejection of the rival clan, which would come back to haunt Cuba in the 1790s when O'Reilly and his brother-in-law, Luis de las Casas, briefly returned to power in the court of Charles IV.[110]

In December 1777, while Eligio de la Puente traveled to familiar terri-tory in St. Augustine, Miralles left for the United States. His first stop was Charleston, South Carolina, where he met and formed fast friendships with many leaders of the new nation.[111] In March, Miralles purchased and shipped 292 barrels of Carolina rice to Havana onboard Antonio Pueyo's schooner, the *San Antonio*.[112] By June 1778, the merchant was in Edenton, North Carolina, and when the British army evacuated Philadelphia, he left for the capital.[113] Philadelphia welcomed Miralles, and he plunged into the commercial and social life of the city.[114] He rented a magnificent man-sion that had been the property of Joseph Galloway, a Loyalist who had left the city when the British army evacuated in 1778.[115] He entertained lavishly while in Philadelphia, as befitting his rank as a wealthy gentleman. He rode through the city in a fine carriage, and he was known for sump-tuous dinners and soirées, during which the house was alight with can-dles.[116] His inner circle included the French ambassador, Sieur Gerard, and the Chevalier de la Lucerne. Although his official status was ambigu-ous, Miralles was welcomed in the highest circles of the new government.

In December 1778, when a new Congress was chosen, Miralles and his friend Gerard were welcome guests at the festivities.[117] Soon the Spaniard was involved politically with most prominent personalities of the new nation, including George Washington and the commander in chief's aide-de-camp, Alexander Hamilton. Among Washington's closest associates was Stephen Moylan, the young man who had been sent to Philadelphia in 1770 when the first commercial contract between Moylan and Company and the Asiento was executed. In the intervening years, Moylan had become established in Philadelphia, and with the declaration of war in 1776, he had been appointed quartermaster general of the Continental Army.[118] Six years later, in Philadelphia, the close association of the two men—one of the most influential investors in the old Asiento and the scion of the firm that was contracted to provide flour to the monopoly—brought the Atlantic world connection full circle.

Meanwhile, throughout 1778, Spain maintained its official neutrality in the face of growing pressure, especially after France entered the war on the side of the Patriots.[119] At the same time, because Charles III was determined to maintain an overwhelming military presence in the Caribbean, Spain was committed to trade with North America due to the inability of the Spanish imperial system to provide food for the troops. In 1779, Gálvez stated the monarch's position unequivocally. He was adamant that the supply of provisions for his forces in the Americas would be sufficient to withstand any exigency. Writing to Manuel Antonio de Flores, the viceroy of Santa Fe, about possible military actions, Gálvez observed: "The intemperate climate, the ease with which provisions go bad, the few resources that the area can provide for subsistence and other local conditions work against the success of any invasion."[120] The conclusion was obvious: a sufficient supply of food was necessary and the military forces could not rely upon local provisioners or on Mexico for their needs. Trade with North America was unavoidable.

The same intemperate climate that plagued the Caribbean also threatened to derail Miralles's mission. In summer 1778, strong storms ravaged the East Coast from the Gulf of Mexico to New England. Charleston, the primary southern port and a secure source of rice, was hit by a hurricane on 10 August 1778, during which many of the ships in port were sunk.[121] A severe storm also came ashore near Newport, Rhode Island, wrecking the British ships that were pursuing a French fleet but possibly averting an invasion.[122] Coming on the heels of the drought during the winter of

1777–78, the cumulative effects of weather seriously reduced the harvests throughout the new nation. In September 1778, Congress had no choice but to suspend exporting foodstuff in order to compensate for the shortages that lingered from the previous year.[123]

Securing the same provisions for the Spanish army, of course, was Miralles's task, and in October 1778, shortly after his arrival, he entered into a partnership with the financier of the Revolution, Robert Morris. Within days of the official announcement of *comercio libre*, the Spanish envoy and the Philadelphia businessman began the first of their many commercial ventures. At the end of October, they outfitted a schooner and sent it to Havana laden with food and other necessities, a voyage that was now perfectly legal under the new *Reglamento*, since Miralles was a Spanish citizen. Although the Continental Congress had officially suspended the export of provisions, in November a confident Miralles wrote: "I believe that if Havana or any other part of the dominions of our august Sovereign were in need of such provisions, the Congress would permit the export [of them]."[124] As if to verify his observation, just weeks later, in November, Congress permitted a shipment of flour to go out to the French islands.

"The Distractions of War"[125]

And so, in May 1779, confident that his ministers had foreseen as many pitfalls as possible, Charles III entered the war against Great Britain.[126] In preparation for an attack on British territory, more than 7,700 men made their way to Havana, and on the eve of the first expedition against Pensacola in 1779, 5,300 army personnel alone were lodged in Havana and in the barracks.[127] By April 1780, 11,752 infantry soldiers were stationed in Havana, and even more men made up the ranks of the naval forces.[128] With almost 12,000 more mouths to feed, Miralles's mission in Philadelphia became even more vital, and now American ships sailed legally and frequently to Havana.[129] The success of the Miralles and Morris partnership and the new legislation was evident in the increasing amount of Philadelphia flour that made its way to the Spanish army, and by 1779, the number of barrels of flour imported from the United States equaled that from Mexico.[130]

With Cuba as the keystone, the primary theater of war became the Gulf Coast, but inclement weather conditions plagued the war effort from the outset. On 31 August 1779, a strong storm passed over present-day Pinar del Río, destroying all of the subsistence crops before continuing toward the Gulf Coast, where it did significant damage to Spanish preparations to move against the British.[131] Anticipating shortages because of the rigors of war, the merchants in the town of Santa Clara raised their prices, earning them a quick reprimand.[132] Desperation was still the rule in the British islands, resulting in raiding parties coming ashore on the north coast of Cuba, where the men stole cattle and small boats and often took local residents hostage.[133] On the south coast of the island, the citizens of Trinidad maintained their time-honored tradition of relying on contraband to provide for their needs.[134]

Guiding the war from Spain, the counselors chose Bernardo de Gálvez to lead the military assault against the British, a decision that did not meet with the approval of his fellow officers in Havana. Many senior officers, including Captain General Navarro and the head of the naval squadron, Juan Bautista Bonet, were jealous of the Gálvez clan's rise to power. Beginning in 1779, Navarro and Bonet began offering excuses as to why they could not send reinforcements from Cuba to support the strategists' plans along the Gulf Coast.[135] In spite of their obstructionism, made worse by the hurricane that came ashore near New Orleans and seriously compromised the effectiveness of his forces, Gálvez went ahead with a surprise attack on British settlements upriver, with stunning success.[136] The victories expelled the British forces from the eastern shore of the Mississippi River, and afterward, Bernardo de Gálvez hoped to strike quickly against Mobile and Pensacola, while a secondary force in alliance with Patriot forces would move against East Florida.[137] Early in 1780, Gálvez sent his trusted comrade and friend Estéban Miró to Havana to arrange for military units to be transferred from Cuba to aid in the conquest of Mobile. Instead of receiving a warm welcome and complete support, Miró was forced to confront the problems caused by the dissension in the Spanish forces.[138]

For all combatants, 1780 was a pivotal year, not for military victories but for the sequential disasters caused by extraordinarily bad weather. Historical climatologists rank the hurricane season of 1780 as the deadliest in terms of the loss of human life, estimating that over 20,000 people perished in the hurricane of 10–16 October in just one landfall in the Leeward Islands.[139] Yet the nightmare that spanned the Caribbean basin by November was but a continuation of the disastrous weather that persisted from the previous year. On the night of 22 February, a rare winter

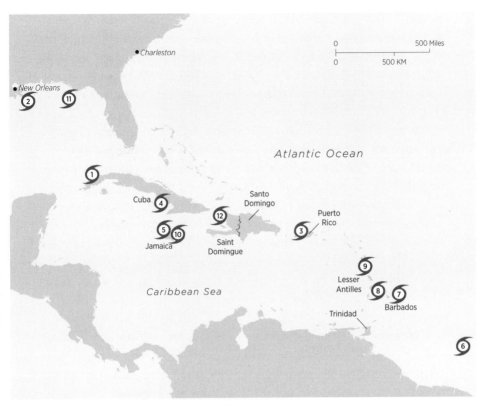

Map 5.1 Hurricane strikes in the Caribbean basin, February 1780 and from July to October 1780. Sources: see appendix 2.

storm began pounding the north coast of Cuba near Havana. The furious storm continued its destruction for most of the next day, only subsiding at sundown. At the height of the storm, naval commander Bonet and the men under him struggled to save the ships from the damaging effects of a storm surge exceeding eighteen feet.[140] Bernardo de Gálvez's envoy, Miró, recorded in his diary that on the nights of 22 and 23 February the strong wind made departure impossible; Bonet observed that in his fifty-three years of service he had never experienced such a strong gale in the winter. The naval commander went on to opine that if they had been at sea, all aboard would have perished.[141]

Daybreak on 24 February brought a realization of the extent of the destruction; virtually every structure in town had suffered at least minor damage. The signature fortress in Havana, El Morro, and the barracks for the mounted cavalry inside the fort were among the casualties of the

storm.[142] The most impressive new fortification was the recently completed Castillo de la Cabaña. The hurricane-force winds blew in the doors
of the artillerymen's barracks, and flying debris shattered the beautiful
stained glass windows over the altar in the chapel.[143] Wooden *barracones*
(barracks), built on the military parade ground just outside the city wall
(Campo de Marte) to accommodate the transient units, sustained considerable damage.[144] Likewise, the responsibility for repairing damaged
barracks and stables in the villages rested with the commanders of the
local forces.[145] At the height of the storm, the eighteen-foot storm surge
pounded into the bay, damaging three large ships that were being prepared
to join the planned convoy to Mobile.[146] And of course, the precious provisions intended to feed the troops were in danger of spoiling from the
torrential rain that leaked through holes in the roofs.[147] By some good
fortune, on 24 February, a lieutenant of the volunteers appeared at the
doors of El Morro with eight *tercios* of salted beef, vegetables, and hardtack, which would carry the troops over until regular provisions could be
secured.[148]

Given the extent of the damages, the recovery efforts were nothing
short of miraculous. Within a week, feverish efforts had fixed all three
boats and replaced the provisions that the troops had consumed on the
days that they waited on land for their ships to be readied.[149] Still, the
senior commanders in Havana delayed sending reinforcements, citing
the "inconstant" weather that threatened any ship that put out to sea.[150]
A small force under Miró's command finally left Havana to reinforce
Bernardo de Gálvez's forces, and on 14 March, Mobile fell to the Spaniards.[151] When the official notification of the young man's victories arrived in Spain, the exultation was tangible. The correspondence between
uncle and nephew reflects their jubilation: "His Majesty has received the
news with the greatest satisfaction," wrote José de Gálvez.[152] On the other
hand, the senior leaders in Havana received a stern reprimand for their
dilatory tactics.[153] The disruptive atmosphere in Havana jeopardized the
success of future military campaigns along the Gulf Coast, and in June
1780, Charles III decided to send José de Gálvez's protégé, Francisco de
Saavedra, to Havana as his personal representative to Cuba, with the direct
authority to implement the monarch's wishes.[154]

In spite of its positive effect on the morale of the Spanish forces, the
capture of Mobile created additional problems for the war planners. Now
the already-burdened army had to care for the British prisoners of war

taken in the action.[155] In April, Navarro issued a circular letter to all of his commanders ordering that "all prisoners are to be treated with all the humanity and gentility that is possible within the context of the security of the island."[156] The following month, Bernardo de Gálvez requested that the authorities in Havana mobilize all their resources to send flour and other provisions to his forces. He also suggested that once the ships arrived in Mobile, the prisoners of war be transferred to Veracruz, presumably to shift the burden for feeding them onto the viceroy of New Spain.[157] Now Miralles's assignment became even more vital, and together with his business partner, Robert Morris, he redoubled efforts to send provisions to Cuba. In early March 1780, the two men purchased another ship, the *Golden Rule*, to increase the volume of trade to the island.[158] Shortly thereafter, Miralles left for the countryside in Morristown to join a group of his friends who were visiting Commander-in-Chief Washington and enjoying the hospitality of his camp. Misfortune struck when Miralles contracted a fever. He died suddenly on 28 April. His death was an immense setback to the careful measures put in place to provide food for the troops in Gulf Coast expeditions, but in the ensuing months, Miralles's partner, Robert Morris, continued to send provisions as per his agreement with his late friend.[159] Four of Philadelphia's ships, the *Fox*, the *Buckskin*, the *Lincoln*, and the *Havana*, maintained the commercial links between Havana and Philadelphia, bringing vital provisions for the troops and civilians.[160]

The "Common Catastrophe"

Summer 1780 began poorly for the garrison, officers, commanders, and civilians, who sought to make the best of the bad situation. As late as April, New Orleans still struggled to recover from the extraordinary storm in February. Most important, the warehouses that stored arms, ammunition, and provisions remained inadequate, in spite of the efforts of the new governor, Martín de Navarro, to rebuild the structures.[161] Meanwhile, just as the war council in Cuba was burdened with additional mouths to feed— the prisoners of war from the Mobile campaign—the central region of Cuba that supplied fresh beef for the garrison was stricken with a severe drought. In late April, local officials in Puerto Príncipe and Santa Clara were notified that the increased number of troops necessitated that each town make "the strongest efforts" to send as many head of cattle as possible. Puerto Príncipe's leaders responded that it would be difficult because of the severe and extensive drought that had descended on the region.[162] Urriza sent an emissary and a stern warning to the lieutenant governor to facilitate the purchase of 8,000 head of cattle, and ten days later, 58 steers left the country on their way to Havana.[163] Then, without warning, in July conditions reversed, hurricane season began anew, and continuous rainfall throughout the summer and fall plagued the local populations. In July, Navarro sent three carts to the rural region near Managua to purchase palm thatch to reinforce the roof on the leaky barracks on the Campo de Marte.[164] By August, the continuous rainfall once again threatened to wash away roads and bridges, so the residents of Arroyo Arenas petitioned the captain general for permission to build a bridge over the river so that they could bring their produce to town.[165]

On 4 October, at six o'clock in the morning, the telltale steady winds of an approaching storm began to blow in Puerto Príncipe. The wind came first from the southeast. Around noon the direction shifted to coming in from the northeast, and it continued to blow until midnight. Drenching rain soaked the province until the morning of 6 October. As soon as it was humanly possible, the lieutenant governor, Ventura Díaz, left his residence to survey his jurisdiction. He began by posting a declaration that detailed the "usual measures [to be implemented] for the common relief."[166] Predictably, all of the crops in the ground were ruined because the heavy rain fell on a parched landscape suffering the effects of a severe drought, a worst-case scenario, and property losses were estimated at over 100,000 pesos.[167] Soon after the hurricane, a series of fevers descended upon a population that had endured so much over the previous year.[168] As late as January of the following year, the residents were only surviving because of the provisions and other help sent to the province from Bayamo and Holguín.[169]

Less than a week later, the most destructive hurricane to impact the Caribbean in recorded history made its first landfall in the Lesser Antilles. In Barbados, more than 4,000 slaves died immediately, and perhaps 1,000 more perished of disease in the aftermath.[170] To the surprise of eyewitnesses, even well-built houses of stone could not withstand the winds.[171] The British navy suffered significant casualties because its fleet had just arrived in the Caribbean and had stopped over in St. Lucia before continuing on to Jamaica. Numerous other ships, both military and merchant, were caught at sea in the storm. A Spanish vessel, the *Diana*, carrying Francisco de Saavedra to Cuba, had sailed with a convoy of French warships, but the

Diana's captain chose to part company with the convoy near Tenerife in the Canary Islands. The French fleet sailed on to Martinique, where the entire fleet was lost in the hurricane. The *Diana*, sailing a more southerly course, was also caught up in the storm. The captain and crew fought for forty-eight hours to save the vessel, and ultimately they made their way safely to Margarita.[172] In total, the hurricane of 10–16 October caused over 20,000 deaths and resulted in incalculable property losses.[173]

Meanwhile, in Havana, preparations were under way for an assault on Pensacola. On 10 October, the expedition under the command of José de Solano, the former president of Santo Domingo, sailed out of Havana and into the full force of yet-another hurricane, which likely developed in the northern Gulf of Mexico.[174] By 1 November, the absence of information warned military planners in Havana that something drastic had occurred to the fleet en route.[175] A week later, the *junta de guerra* still had no definitive news of the fate of the expedition, so contingency plans began to be formulated.[176] By the end of the month, reports of the disaster had made their way to Havana. Solano's fleet had almost made it to Pensacola when the hurricane scattered the ships throughout the Gulf of Mexico. Several vessels were thrown ashore on the coast of the Yucatán peninsula. Some survivors were picked up at sea and taken to New Orleans; others were never seen or heard from again.[177] One survivor spoke of his ordeal, clinging to debris among "fragments of ships and cadavers floating in the water."[178] In Havana, military planners scrambled to outfit the few remaining ships in drydock to send on a rescue mission to search for survivors.[179] For the second time in a year, the shipyard of Havana worked feverishly to patch together careened ships to provide emergency relief after a devastating storm.

In spite of the disaster, the business of war could not be suspended, and the forces now occupying Mobile still needed to be supplied.[180] The loss of all the provisions for the Pensacola expedition was a serious setback, and in November, Urriza passed on the bad news that the supplies of hardtack, flour, and meat were completely inadequate for the demand. Flour posed a particular problem, in spite of a shipment that had just been secured from the "*Ingleses Americanos del Norte*."[181] Military officers were charged with scraping together anything that they could spare, while at the same time notices arrived that the provisioning grounds in Puerto Príncipe had been destroyed a week before.[182] Unaware of the disaster that had just devastated the neighboring islands, the *junta de guerra* worried that the British would take advantage of their weakened and vulnerable position.[183] Extra watches were ordered at all of the towers on the island after the lieutenant governor of Pinar del Río reported that he had spotted a ship with ten or twelve bronze cannon that appeared to be British.[184] The weather, of course, did not cooperate with military strategists. Continuous rainfall ruined all of the scarce remaining gunpowder in the armory inside El Morro.[185] Their French allies in Saint Domingue were also hurt by the bad weather. In one coastal city, Mole San Nicolás, it had been raining continuously for fifteen days.[186] Misfortune had also befallen Charles III's personal emissary, Saavedra, who was captured at sea and made a prisoner of war in Jamaica for two months. In spite of the hostilities between the two nations, he was well treated by his captors and was allowed to roam freely throughout Kingston upon his promise of good behavior. English soldiers from the doomed British fleet who had been shipwrecked on Martinique also were treated humanely. The governor of Martinique wrote to his superiors that he had freed the English sailors who had been shipwrecked on that island because, "in a common catastrophe, all men should be brothers."[187]

To royal officials taking stock in Madrid at the end of 1780, the outlook for 1781 was bleak.[188] With the exception of the victory at Mobile, Spain's forces could only count failures, not successes. The severe winter storm in February, the near-mutiny of senior officers in Havana, Miralles's death in April, and the horrible weather throughout the summer—culminating in the hurricanes in October that hit Puerto Príncipe and the expedition to Pensacola—all had drained precious resources with no benefit to show for all of the efforts. Nevertheless, as early as November, preparations were being made to organize another expedition against Pensacola.[189] In January, the troops that were scattered throughout the Gulf Coast returned to Havana for regrouping.[190] Throughout the western end of the island, from Matanzas to Cabañas, local constables were ordered to conscript caulkers, ship's carpenters, and axmen for transfer to Havana.[191] At the same time, the island's militia units were put on alert, while the regular forces prepared to depart on the second expedition against Pensacola.[192] By the end of February, even the remotest hamlet on the island knew that hostilities were imminent, when a circular order arrived announcing that as of 6 March, "no boat, large or small would be allowed to put out to sea until further notice."[193] Bernardo de Gálvez's second attempt to capture Pensacola had begun.

The result was a resounding success that eliminated the British threat in the Gulf of Mexico and turned the tide of war in Spain's favor.[194] Buoyed by victory, the Spanish government chose to press the war effort on several fronts. In Central America, the father of Bernardo, Matías, led an expeditionary army against British interlopers in present-day Nicaragua. Another front was opened against the Bahamas, also British territory, and a planned expedition against Jamaica was organized in Saint Domingue in 1782.[195] Rewards for Bernardo's bravery and audacity followed on the heels of each victory. After the Mississippi River valley victories, his post as governor and captain general of Louisiana was separated from the authority of the viceroyalty of Mexico and granted complete autonomy.[196] After Mobile and Pensacola, he was awarded the title of Conde de Gálvez and made captain general of Cuba. His brief tenure in Cuba preceded more honors, when, upon the death of his father, Matías, in Mexico City, Bernardo was chosen to succeed the older man as viceroy of New Spain, in 1785.[197]

"To Supply That City with Everything That It Needs"

The Spanish capture of Pensacola gave a different atmosphere to the war, and the decision to expand the conflict made securing provisions for the troops even more important. The definitive declaration arrived in Cuba in early 1781, charging the military strategists with securing whatever resources they needed to ensure the success of the expedition against Pensacola.[198] The expedition in Central America led by Matías de Gálvez received the same concession.[199]

By April, the question about where the provisions would be purchased was settled. Because of the "effects of the shortages," Havana harbor was thrown open to North American boats with the charge from José de Gálvez "to supply that city with everything that it needs." In addition, merchants received tax concessions on imports and exports. Each shipment arriving would be assessed at half the regular import duties, and outgoing cargoes would also be taxed at half their value. José de Gálvez also relayed Charles III's sincere gratitude to "the honorable Robert Morris," who, in the uncertainty after Miralles's death, had sent his ships in ballast to the Danish islands to purchase flour for the Spanish troops.[200] Throughout spring and summer 1781—to the "complete satisfaction of His Majesty"—an increasing number of North American vessels sailed southward to Havana bringing vital flour, rice, and lumber to supply the war effort.[201]

By late summer 1781, the commercial atmosphere in Havana was transformed into the festival-like conditions that had prevailed in Cap Français just a few years previous. Like weeds in a field, North American trading houses sprang up overnight. The head of one such business was Joseph Grafton, of Salem, Massachusetts, who, in summer 1781, decided to investigate the potential for "profitable ventures" in Havana. Prior to his departure, he secured letters of introduction to people in Cuba and throughout the Atlantic world. Grafton arrived in Havana onboard the brigantine *Romulus* on Sunday, 1 September 1781. Obviously well aware of the protocol required to conduct business, he came with the appropriate "gifts" for the town constable—a yard of calico fabric and three pairs of silk stockings.[202]

The profitable ventures that Grafton had hoped for materialized almost immediately, and by the end of the summer, he and his family were fully engaged in the Cuba trade. In 1781, Havana became the staging ground for the French fleet in its preparations for the expedition against Yorktown, and Grafton's and other American merchant ships unloaded barrel after barrel of flour, rice, potatoes, and other comestibles. On 12 November, they received the news of Lord Cornwallis's surrender to Washington. Two days later, the Americans staged a victory party, attended by over fifty people, including French officers and Irishmen in Spanish service. With each toast that was drunk, the brigantine *Schuylkill* discharged its cannon in celebration.[203] Their unqualified welcome was assured because the weather remained treacherous. In February 1782, a Philadelphia brigantine was wrecked off the coast of Matanzas in a winter storm that pounded the north coast of the island, and in July, a relatively weak hurricane passed over the western end of the island, causing considerable losses to livestock and crops but, miraculously, no loss of human life.[204]

Grafton's commercial success in Havana would last for three years, but on Friday, 23 April 1784, North American commercial activities came to an abrupt halt.[205] That morning, he was conducting routine business onboard the schooner *Betsy* when the town constable came with five soldiers to tell him to leave the city. Grafton first appealed to Oliver Pollock, Morris's former agent and now the commercial representative for the United States in Cuba, who secured permission for him to stay, but the soldiers on the ship refused to allow him to leave. The next morning, the *Betsy* set sail with Grafton on board. He had not even been granted permission to retrieve his laundry from his washerwoman's house.[206]

After the expulsion of foreigners in 1784, savvy merchants simply shifted their centers of operation and created international trading networks reminiscent of the 1760s and 1770s. By the mid-1780s, Philadelphia, Baltimore, and New York had replaced London as the hub where deals were made and cargoes were loaded for shipment to Havana. Getting around the prohibition against foreigners in Havana proved less of a problem than it appeared to be at first, especially since the miserable weather conditions of the previous four decades showed no signs of abating into the latter half of the 1780s and the 1790s. Still desperate for provisions, in the interim, the returned province of East Florida provided an opportunity to comply with the spirit and the letter of the law. After the expulsion of 1784, the merchants of St. Augustine capitalized upon their heritage and serendipitous geographic location to become the conduit through which North American provisions continued to flow to the consumers of Cuba. Their success and the subsequent solidifying of commercial relations with North America will be explored in the following chapter.

CHAPTER SIX

The Tomb That Is the Almendares River

IN LATE JUNE 1791, St. Augustine captain Don Antonio de Alcántara sailed into Havana harbor at the helm of his schooner, the *Santa Catalina*.[1] A decade earlier, his arrival would have been unthinkable because his port of origin was in British hands, and Great Britain was at war with Spain. In the subsequent years, however, after the Floridas returned to Spanish rule, captains such as Alcántara found themselves in an advantageous position. Such men and their families capitalized on their sailing expertise and their status as Spanish citizens to establish commercial linkages among Havana, East Florida, and the United States. Their circumstances became particularly profitable after North American merchants were expelled from Cuba in 1784 and commerce with the island was restricted.[2] By 1791, Alcántara had become the head of a respected clan of merchants and captains who sailed as far away as New England to the north and Montevideo to the south. Just days after unloading his cargo, Alcántara cleared Havana harbor unaware that one of the most destructive hurricanes in the island's history was poised to strike the northern coast of Cuba and the Straits of Florida.[3] At home in St. Augustine, Alcántara's wife—the *Santa Catalina's* namesake—Catalina Costa, waited in vain for her husband to return. What remained of the schooner probably washed ashore on the Florida peninsula; the fate of her captain and crew was never officially determined.[4]

The hurricane of June 1791 came at a critical juncture in Cuban history. Lingering over western Cuba for several days, it was the cause of a flood that claimed over 3,000 lives, drowned over 11,000 head of cattle, and caused incalculable property damage.[5] It also marked the region's entry into its fifth—and possibly its worst—decade of environmental crisis. The crisis in Cuba was exacerbated because a new and inexperienced monarch, Charles IV, sat on the Spanish throne. The new king was ill-equipped to deal with matters of government, and his policies vacillated from one extreme to another.[6] Within the maelstrom of Spanish court politics, he promoted a rival faction that for fifteen years had been alienated from the center of power.[7] International politics, including the increasing influence of the United States, the French Revolution, and its colonial counterpart, the rebellion on Cuba's neighboring island of Saint Domingue, posed challenges to the monarchy, and the situation was complicated further because Spain faced an impending financial crisis. By January 1789, the Spanish empire was in a state of confusion, and Cuba's economic position in the empire was problematic at best.[8]

Regardless of the political turmoil and economic uncertainty, thirty years of pragmatic and beneficent rule meant that by 1788, the Cuban people were accustomed to a positive response when disaster occurred; they were understandably unprepared for a royal reversal after Charles III's death in December of that year. Led by the captain general, Luis de las Casas, partisans of the rival political faction in Spain came to power in Cuba. This new group of bureaucrats represented metropolitan attitudes that had changed from a response that sought to implement disaster mitigation measures to almost complete indifference to the sufferings of the people.[9] Royal authority on the island became a near-textbook example of inadequate response to disaster. Having to cope with a series of weather catastrophes in the wake of hurricanes that struck the island four out of six years of Las Casas's tenure in Cuba (1791, 1792, 1794, and 1796) directly contributed to political chaos on the island.

NEARLY A DECADE EARLIER, in the celebratory atmosphere after the victory over Great Britain in early June 1784, frenetic preparations were under way throughout the Spanish Gulf Coast and the Caribbean in anticipation of the return of Spanish control to territory lost at the end of the previous war. In the barracks and on the wharves of Havana, men, materiel, and the newly appointed governor, Brigadier Vicente de Zéspedes, prepared to embark for East Florida.[10] In Philadelphia, Father Thomas Hassett, appointed as East Florida's principal Catholic priest and ecclesiastical judge, packed his belongings for the trip southward to his new congregation.[11] In Havana harbor, Captain Pedro Vásquez readied his brigantine, the *San Matias*, and the other ships under his command for the important responsibility of carrying the governor and his entourage to their new assignment.[12] Although not part of the expeditionary force, the members of the regiment of Asturias celebrated their part in the victory while they waited for their ship, the *San Cristóval*, to be readied to carry them home to Cádiz.[13]

Far out to sea, the telltale counterclockwise circulation and dropping barometer warned that a deadly storm was brewing in the tropics. As it organized and gathered strength along the twenty-fourth parallel, it skirted the north coast of Cuba and bore down on the Straits of Florida and the peninsula. Like many early-season storms, though, the cooler landmass of North America deflected the brunt of the tempest away from the mainland. Recurving northward, the worst of the storm stayed out to sea, although violent winds battered and copious amounts of rain drenched Havana province.[14] Anxious to arrive at his destination, Governor Zéspedes waited impatiently for the weather to clear.[15]

At last, on 19 June, Zéspedes and the 500 men who accompanied him departed for St. Augustine on the *San Matias*. Sailing on a fresh wind that trails the passage of a strong storm, the convoy made good time and arrived off the city in seven days. Yet the hurricane that had frustrated Zéspedes's departure also frustrated his arrival, for upon entering the harbor, pilot Joaquín Escalona conveyed the news that the main channel leading into St. Augustine had been silted over from the strong winds and tides of the storm. Zéspedes was again forced to wait until the following day, when Escalona returned in his shallow-draft launch and ferried the governor into the city.[16] Unable to cross the sandbar, which drew only seven feet of water, Vásquez and the fleet of ships under his command proceeded north to the port of St. Marys to complete the disembarkation of men and materiel.[17]

For Zéspedes, the storm was just an inconvenience, but for other members of the expedition the dangers were far greater. Father Hassett made his way southward toward East Florida aboard the *Santa Ana*, captained by Miguel Ysnardi. The hurricane caught the *Santa Ana* on 28 June, and at the height of the storm's fury, she foundered on the reef of Arogüito Key in the Bahamas. Badly injured, Hassett and the other survivors made their way ashore. There they repaired one of the *Santa Ana*'s boats, and he and twelve other men sailed to Havana, where the authorities were notified to send a search party for the remainder of the *Santa Ana*'s crew.[18] The same fate befell the regiment of Asturias. The *San Cristóval*, sailing north in the Gulf Stream between the Florida peninsula and the northern Bahamas, also foundered on a reef. The ship and eight soldiers were lost, but the majority of the regiment, along with the ship's crew and captain, made it to a nearby island, where they too were rescued and brought to the safe harbor at St. Marys.[19] But nature was not yet finished with the expeditionary force. In early July, the high winds and rough seas of another early-season storm caused many boats anchored in St. Marys harbor to lose their anchor cables and crash into one another. The *San Matias* collided with the *San Antonio de Padua* and suffered considerable damage above deck, although she escaped any structural damage below. At last, after recuperating in Havana, Father Hassett arrived in St. Augustine, and in early August the shipwrecked regiment of Asturias was able to depart for Spain.[20]

The situation that faced Vicente de Zéspedes in Florida in 1784 was virtually identical to the one that had faced Antonio de Ulloa in Louisiana in 1768. Zéspedes also confronted a surly population of foreigners with a treasury that was inadequate for the costs of administration.[21] Zéspedes was sent to St. Augustine with the ridiculously small sum of 40,000 pesos, which was hardly enough to pay ordinary operating expenses, much less to cover the extraordinary costs of the aftermath of the storms—additional boats were needed to take the regiment of Asturias to Cádiz,[22] and crews from the damaged vessels had to be transferred to the ships that would be returning to Spain, an additional expense.[23] While waiting for the equipment and munitions to be unloaded, the captain of the regiment of Asturias and thirty of his soldiers were lodged aboard the *San Matias*.[24] St. Augustine's treasury bore the entire cost of their maintenance, since they were not permitted to set foot on shore, and ultimately it was saddled with all of the transportation costs.[25] On a smaller scale, Father Hassett, who had lost everything in the shipwreck, petitioned the crown for restitution, and upon Zéspedes's recommendation, he was awarded 400 pesos in 1786.[26]

Zéspedes, a veteran of decades of service in the Hispanic Caribbean, was well aware of the monarch's wishes concerning disaster mitigation. For the past two decades, royal policy had sought to do everything possible to lessen the effects of catastrophe on the stricken population. It was apparent that funds in his treasury were inadequate, so on numerous occasions he pleaded with Bernardo de Gálvez, by then the captain general of Cuba, and with Juan Ignacio de Urriza, the intendant in Havana, to send him more money.[27] The problem was exacerbated because the Mexican treasury, which supplied the Florida *situado*, was experiencing its own difficulties because of poor harvests caused by drought.[28] Faced with popular riots due to food shortages caused by a drought of historic proportions in the mid-1780s, Mexican officials limited the amount of money they sent to Havana for St. Augustine. Worse still, when the *situado* did arrive, Urriza took a percentage of the monies before shipping the remainder to East Florida.[29]

Determined to not repeat the experience of his predecessor, Zéspedes quickly implemented emergency measures at the local level.[30] He authorized Spanish ships to travel to foreign ports to purchase provisions, foreign ships were allowed to enter St. Augustine harbor if they carried food, and foodstuffs could be imported duty-free.[31] Ironically, his actions directly violated the spirit of the expulsion of North Americans from Cuba; nonetheless, his measures met with the approval of his superior officer, Captain General Bernardo de Gálvez, who confirmed Zéspedes's decisions in 1786.[32] Zéspedes's response subsequently won approbation at the highest levels of government, from Minister of the Indies José de Gálvez.[33] Captains such as Alcántara, now sailing under Spanish colors, would become eager accomplices, while the United States, hard-pressed for currency and prohibited from trading directly with Cuba, would come to use St. Augustine as the gateway to the island.[34]

Already-well-connected families had little difficulty in capitalizing upon hardship and amplifying their established maritime networks, and with the opening of trade with both northern and southern ports, St. Augustine—like Kingston and Cap François before—took on the function of an entrepôt. St. Augustine's captains sailed north to Charleston, Philadelphia, New York, and New London, where they purchased food and other goods in unlimited quantities. Then they returned to St. Augustine, where the products were unloaded, redesignated as "*frutos del país*," and reloaded onto ships bound for Havana. One such beneficiary was Miguel Ysnardi, the master and captain of the *Santa Ana*, which had been wrecked in the 1784 storm. As a member of a kinship and commercial network (reminiscent of the Moylan clan of the 1770s) that linked East Florida to Philadelphia, Havana, and Cádiz, Ysnardi set up operations in St. Augustine where he enjoyed a meteoric rise to power and prominence. The Havana branch of the operation was managed by his wife, Juana de Torres, who established a permanent household there in 1791.[35] The Ysnardi clan expanded its commercial contacts to Baltimore through contracts with merchants John and Margaret Frean.[36] The family seat remained in Cádiz, where the patriarch of the family, José Ysnardi, and another son, Tomás, were the conduits through which the clan traded throughout the Atlantic world.[37] The most prominent member of the family was yet another brother, José María Ysnardi, who served as proconsul to the United States in Philadelphia, even as he supervised the northern terminus of the commercial enterprise.[38] In his post as proconsul, José María Ysnardi would become one of the most influential men in the Atlantic world trading networks, and within a year of its first ventures in St. Augustine, the family enjoyed so much success that it was forced to contract merchandise out to other captains.[39]

Some former British residents, such as Thomas Tunno, with established connections in New York remained in East Florida after the change in sovereignty, acting as collection agents for debts owed to departing British citizens.[40] In the wake of the crisis in 1784, Tunno's ship, the *Swift*, was one of the first to transport life-saving provisions into East Florida.[41] By 1785, he had acquired an agent, Juan de Aranda, and under Aranda's stewardship Tunno's cargoes did not comply with the spirit of imperial regulations, much less the letter—merchandise was not even perfunctorily unloaded before it was shipped on to Havana.[42] In 1787, Tunno left East Florida, possibly to reconfigure his interests in the United States, and by 1789, he had returned to St. Augustine and had established even more lucrative commercial contacts with Cuba.[43]

Meanwhile, after the expulsion of 1784, North American merchant houses cautiously resumed business in Havana. By 1785, it appeared that the draconian measures adopted by the captain general the previous year would be tempered by a reversal in policy generated by the goodwill established between Philadelphia and Bernardo de Gálvez during the war.[44] By the mid-1780s, Philadelphia's monopoly on the Cuban trade was challenged by commercial firms in New York, Baltimore, and Charleston, and other cities began shipping their agricultural products, especially lumber for building and rebuilding. One such merchant house was run by John Leamy of Philadelphia, and another was the firm of Lynch and Stoughton in New York.[45] In spite of their success in Havana in the early 1780s, the Grafton family of Salem never recovered its former prominence, and by 1787 the family had abandoned trade with Cuba, leaving several lawsuits pending against factors on the island.[46] By the end of Charles III's reign, Spanish proconsuls in Philadelphia such as José María Ysnardi and Diego de Gardoqui, scion of the trading house Gardoqui and Sons, which had been instrumental in funneling money to the Patriots, were given the authority to issue licenses that made U.S. ships' voyages to Havana perfectly legal.

As a result, as the 1780s wore on, the island grew more dependent upon flour from the United States, and after Charles III's death, the new monarch retained his father's commercial policy of neutral trade, which had originated out of the necessity of war.[47] The reauthorization of neutral trade sent shock waves through a sector of the mercantile community in Havana, which had hoped that its lobbying at the new court would exclude the North Americans from the island. As the battle over who would supply provisions to the island spilled over into the 1790s, it became no less acrimonious as time passed. When Charles III had promulgated the *Reglamento para el comercio libre* in 1778, he sought to alleviate potential shortages and avert any injustice by including a provision that any city that did not have a merchant guild (*consulado*) would be permitted to form one.[48] At the time, neither Havana nor Santiago de Cuba had taken advantage of the concession, but in 1783, Santiago de Cuba petitioned for the privilege, which was granted in 1789.[49] The merchants of Havana soon followed with their own request, and the Consulado de la Habana was created in 1792, followed quickly by complementary associations with overlapping membership, such as the Sociedad Económica de Amigos de País, the Sociedad Patriótica, and a Junta de Fomento (Council to Promote Business) formed from the membership of the Consulado.[50] Most members of the new groups were partisans of the faction that had been excluded from power during the tenure of the Gálvez family and the reign of Charles III.[51] Their leader was the captain general, Las Casas, to whom many traditional historians have attributed the commercial modernization of the island.[52]

With the reauthorization of neutral trade in 1790, probably the most disappointed member of the Consulado was Cuba's most famous creole intellectual, Francisco de Arango y Parreño. Residing in Madrid, for months Arango had been lobbying the new regime, as outlined in his *Discurso sobre la agricultura*, first presented to the Council of the Indies in 1789.[53] The much-heralded tract was little more than a litany of grievances of the Havana ayuntamiento, the Consulado, and the Junta de Fomento, which had invested heavily in sugar plantations. During the 1770s, they suffered through poor harvests and the indignity of having officials refuse to transport their sugar to Spain on royal ships during the war with Britain.[54] The final insult came when the local currency (*moneda macuquina*) was devaluated and recalled at a fraction of its value.[55] Other agricultural enterprises on the island such as tobacco farming and coffee cultivation had received concessions from Charles III, who had sought to diversify Cuban agriculture in his own way by not favoring one industry over another. Planters who had invested heavily in sugar mills watched in growing resentment as the Mexican military subsidy (*situado*) arrived in Havana, only to be spent on their rivals in the military faction or be transshipped north to finance the thirteen colonies' struggle for independence.[56] The motivation for Arango's original proposal, then, was to move the economic base of the island away from military spending and subsidiary industries and to stimulate sugar production in its place. His argument was couched in the nationalist rhetoric that Spain needed to recapture economic sovereignty over the island. His suggestions to "promote agriculture" would directly benefit the empire by making Cuba profitable so that such foreign dependence would be unnecessary.[57]

In one sense, Arango's nationalist rhetoric received a warm welcome in Charles IV's court, and after 1789, the new regime introduced measures to address some of the complaints inherent in the *Discurso*. One such measure was the permanent uncoupling of the obligation to provide flour along with slaves. The requirements that governed the old Asiento's commercial dealings had been abandoned in practice as early as 1780. During the 1780s, separate licenses to import slaves were granted to select individuals, who brought in their human cargoes on a case-by-case basis from neighboring colonies.[58] For the most part, during the 1780s, U.S. vessels laden with flour and other provisions such as rice rarely traded in slaves.[59] North American merchants preferred to smuggle in valuable consumer goods in the bottoms of barrels topped off with flour and tried to avoid paying Spanish customs duties by smuggling out large quantities of specie in hidden compartments of ships.[60] Meanwhile, in 1788, a group of commercial interests in Havana petitioned for permission to monopolize the slave trade to the island without the restriction that the new arrivals be accompanied by provisions to support them.[61] Instead of relying upon yet another monopoly company to provide slave labor for the island, a Royal Cédula, promulgated on 28 February 1789, opened the trade to anyone who could engage in it. Historians are unanimous that the new concession of free trade represented a watershed in Cuban history because the liberalization of Spanish laws regulating slave imports removed the obstacles to providing labor.[62] More important, though, the permanent removal of the slave/flour requirements opened the door even wider to merchants from the United States, who eventually would come to dominate the market in Cuba.

This was exactly what the self-appointed new guardians of the Cuban economy hoped to prevent, and they scored a degree of success when, in spring 1791, higher import tariffs were imposed upon North American flour that was brought into Spain. Factors for the American merchant houses in Cádiz wrote in growing alarm about the protective tariffs that had been placed upon flour and rice shipped to the peninsula for re-export to the West Indies. The Spanish customs collectors were forthcoming about the reasons for the tariff: "A new duty of two hard dollars per barrel . . . intended to favour the fabricks of flour in this country" had been laid on imports, according to a contemporary observer.[63] One merchant house had nearly 8,000 barrels in storage on the wharves of Cádiz; another estimated that between 10,000 and 13,000 barrels of flour would go unsold. Entreaties to William Carmichael, the commercial representative of the United States at Madrid, to press for the exemption of the flour already in Spain were to no avail.[64] Then, in June 1791, representatives from Havana scored another victory when they convinced the king to order all foreigners out of the city. Those who failed to comply with the proclamation would be subject to a fine of 500 pesos.[65]

But the second expulsion of foreigners in 1791 would be no more effective nor lasting than the previous decrees. For the next five years, efforts to provide sufficient provisions for the island became a dizzying dance of reversals of one royal dictate after another. Merchant captains who sailed with licenses issued by Spanish consuls in U.S. cities in 1791 arrived in Havana only to find the situation reversed and their cargoes in danger of being confiscated. Barely eighteen months later, in June 1793, the port was thrown open again in response to the exigencies of war with France, only to have the situation reversed in January 1796.[66] Ultimately, though, the ever-changing commercial policies were not in local or metropolitan hands. The evil twins, El Niño and La Niña, became the arbiters of whether and to what extent foreign produce would be welcome in Cuba.

Like the previous cycles, the onset of the El Niño/La Niña sequence began with a severe drought that had its greatest impact on the center of the island. In Santa Clara in February 1791, rancher Andrés Moreno presented his problems to the city council members. He acknowledged that he was responsible for providing one head of cattle a year, but the "calamities" brought on by the weather and the subsequent diseases had ravaged his herd since 1787.[67] The cattle ranchers on the Isla de Piños echoed the reports from Santa Clara that the drought had limited their ability to provide dried, jerked beef (*tasajo*) for the troops at Batabanó.[68] Writing from Guanabacoa, city council members Miguel Núñez and José Pérez de Medina linked the drought to the failure of the major food crops, which in turn had brought diseases to the area.[69]

Meanwhile, in Havana, the suffocatingly hot and dry conditions weighed heavily on Don Jacinto Barreto, the Conde de Casa Barreto, who was the patriarch of a family whose roots had been established in Cuba for at least a century. By summer 1791, the old man was tired and ill, and in late June, he elected to retire to his country mansion outside Havana's city walls on the banks of the Chorrera River (sometimes known as the Almendares River, which is today's common usage).[70] There he hoped the fresh breezes from the river would restore his health.[71] Instead of recovering, however, Barreto took a turn for the worse. He drew his last breaths as the drought ended abruptly and rain began to fall on western Cuba. His household attendants prepared his body for burial, unaware of how rapidly disaster approached. Heavy rains continued as the mourners arrived to pay their last respects before the coffin that lay in the entryway of his house. At last, the saturated ground could absorb no more water, and the Chorrera River and the rivulets that fed it spilled out over their banks. As the water rushed toward the ocean, it scoured the fragile ground and picked up felled trees that were stacked along the riverbanks awaiting transportation to the Real Arsenal (shipyard) of Havana.[72] The rising waters caught the piles of lumber and created a churning wall of water, mud, trees, and debris. Barreto's house was directly in its path. The wall of mud and debris crashed into his house, and in its wake it carried his coffin downriver and out to sea.[73]

This relatively weak storm—probably no stronger than a Category 1—became one of the most costly in terms of loss of human life and property damage. Factors that contributed to the storm's lethal consequences were its duration and the fact that it lingered over western Cuba for days, combined with the topographic characteristics of the countryside surrounding Havana and ecological changes that had been in process for three decades. To the south, southeast, and southwest of Havana, the terrain rises from sea level and ascends to a range of hills running east to west paralleling the coast. In some places the elevation rises abruptly, and there the hilly terrain is punctuated by gullies and ravines in which rivulets and seasonal streams carry runoff northward to the ocean. Villages and hamlets in the interior from Havana were particularly vulnerable because they were located along the margins of these streams and rivulets.[74] Equally important, the rains of June fell on terrain that had been eroded by the twin processes of deforestation to provide trees for the shipyard and urbanization caused by explosive population growth, which had doubled the number of residents in a little over one generation.[75]

In weak storms, the worst damage was done to the plantains, which were blown from the trees, lending the name "banana wind" (*viento platanero*) to the spectacle of thousands of plantains littering the countryside. Normally, residents could salvage some of the crop, but after days of flooding, the plantains on the ground were ruined by contamination from the decaying bodies of dead animals. Such was the case in the village of Calvario, located directly to the south of Havana. Calvario stood at the head of a steep ravine cut by the Luyanó River, which emptied into Havana bay. Among the losses to livestock were two hundred domesticated fowl and a mule that died on the *potrero* of Antonia Gonzáles; the *estancia* of Agustín García lost eighty-six chickens and a cow. Two weeks after the passage of the storm, seven families in Calvario still had nowhere to go. Josef de Mena struggled to survive with his wife and eight children.[76] Downstream in the village of Jesús del Monte, the first casualty reports were submitted by Constable Félix Gonzáles, who wrote that forty logs of an extremely hard wood called "breakaxe" (*quiebrahacha*) had undermined the pilings of the bridge in the village of nearly 2,500 people.[77]

The greatest number of deaths occurred in the small settlements to the west of Havana located within the drainage basin of the Almendares River. The headwaters of the river began in the same hills south of town, and like the Luyanó River, the Almendares coursed out of the hills, flowing north and emptying into the Straits of Florida. Steep sides of a gorge had been carved out of the terrain, which funneled the rainwater into a raging torrent that bore down on the Almendares basin settlements, which were home to nearly 4,000 people. The first village in the floodwaters' path was Santiago de las Vegas, and in its aftermath, the constable, Vicente de Soría, anguished over how he could help the hundreds of victims in his town and the nearby hamlets.[78] Farther downstream, 600 logs were stacked along the banks of the river in Jubajay (Wajay) waiting for transportation to the sawmill.[79] In their path lay hundreds of modest wattle-and-daub houses

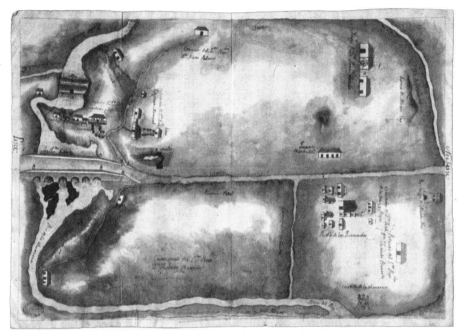

Figure 6.1 View of the Chorrera River basin prior to the storm of 1791. Note Jacinto Barreto's riverside property in the lower left-hand corner. Source: Mapas y Planos, number 527, Audiencia de Santo Domingo, Archivo General de Indias, Seville, Spain.

clustered near a bridge across the river near the village of Prensa on the east bank of the river. Without warning, the churning wall of water and debris crashed into the villages in its path. Where Barreto's house once stood, a crater sixty feet deep was all that remained.[80]

For the survivors of the catastrophe, the horror had just begun. Families who owned small boats rushed into service, rescuing people who clung to the roofs of their houses.[81] Vicente de Castilla, the constable of Prensa, was overwhelmed with the task of retrieving the hundreds of cadavers floating in the receding waters, suspended in trees, or entangled in fences. On the third day after the catastrophe, Castilla asked for a large boat to intercept the bodies floating down the river, which were intermingled with the carcasses of dead animals and debris from destroyed houses—furniture, windows, and beams from roofs.[82] Castilla's counterpart on the west side of the river, the constable of Quemados, Cristóval Pacheco, faced equally dire conditions.[83] Entire families were carried away by the torrent, including María Guadalupe, a free woman of color, whose body

Figure 6.2 Bridge over the Cojímar River built by the Marqués de la Torre in 1774 and destroyed in the storm of 1791. Source: Mapas y Planos, number 561, Audiencia de Santo Domingo, Archivo General de Indias, Seville, Spain.

was found on 27 June so decomposed that the rescuers could only bury her where she laid. The bodies of her five children were not found until six days later.[84] Elsewhere, constables in inland areas that escaped the worst of the devastation coped with the expected flight of slaves taking advantage of the distractions of the disaster. It took three weeks to recapture one fugitive who had hidden himself in the woods surrounding the village of Guayabal. For three weeks, he had committed various assaults and robberies, including killing a horse for food, until captured by civilian patrols and returned to his owner.[85]

Among the hardest hit were the transportation networks. The three most important bridges had pilings undermined by the unceasing rainfall and rising waters, which seriously compromised getting products to market. Especially important was the bridge at Puentes Grandes, the "esophagus" of the city; other bridges on the main arteries into the city, the Pastrana bridge in Luyanó and the Ricabal bridge over the Cojímar River east of town, had been washed into the sea.[86] The June storm also

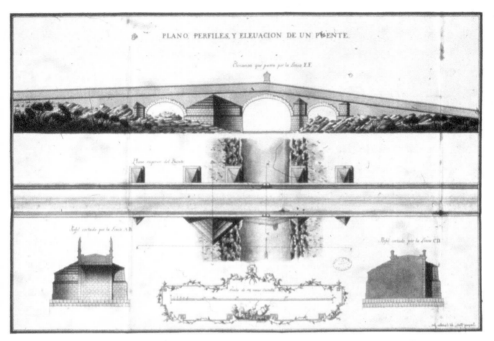

Figure 6.3 Plans for the new bridge over the Cojímar River after the storm of 1791. Source: Mapas y Planos, number 562, Audiencia de Santo Domingo, Archivo General de Indias, Seville, Spain.

caught the shipping industry unaware. Because the early summer was still considered to be a safe time to sail, no warnings were issued as Alcántara steered the *Santa Catalina* northward. The schooner, her captain, and her crew became casualties of the rare early-season storm, but the crew and passengers of the brigantine *El Gallo*, bound for New Orleans with mail, lumber, and tar, which left on 20 June, were more fortunate. The ship encountered the hurricane 120 miles south of the mouth of the Mississippi River. Four passengers and four crew members were lost in the shipwreck, but the survivors were rescued by a passing French vessel and brought into Havana on 6 July.[87] For the remainder of hurricane season, the administrator of the mail system, José Fuertes, took precautions to minimize any losses from what portended to be a difficult year. As a consequence of the wreck of *El Gallo*, Fuertes grudgingly allowed one mail ship, the *Cortes*, to leave for Spain in mid-August. Afterward, no ship was permitted to sail until well into October, after the dangerous period around the equinox had passed.[88]

In the first test of the new regime, Las Casas responded in a manner completely opposite from the humanitarian concerns to which Cuban sensibilities had become accustomed. The horrific reports from his subordinates of famine, sickness, and universal misery failed to move him. Even though local rescuers worked against time and under the worst of circumstances, the captain general was not pleased with their efforts. Upon receiving reports on the ground from the Almendares River basin, Las Casas criticized the constable of Quemados for burying the victims' bodies in place, admonishing him that he should have made every effort to recover the bodies and bury them in the cemetery.[89] Instead of implementing a program of disaster mitigation designed to lessen the burden on the people, the captain general held the villagers responsible not just for their own recovery and cleanup but also for the reconstruction of the island's infrastructure. Meanwhile, Las Casas ordered the constables of each urban and extra-urban barrio near Havana to go house-to-house "soliciting donations" from ordinary people to help aid the stricken residents of the flood.[90]

The cavalier attitude of the new governor, coupled with his demands to rebuild the roads and bridges on both sides of the city, became the flashpoint for confrontation between the ordinary people (*el pueblo*) and royal officials, especially since the policies were so alien to previous policy. The damages in 1791 were no more serious than after the hurricane of 1768, yet the captain general during the earlier hurricane, Antonio María Bucareli, set the example for the behavior of subsequent leaders on the island. After the storm, Bucareli paced in the portico of his residence waiting for the tempest to pass. Afterward, he personally took to his horse and made the rounds of victims. Under Bucareli's mandate, a taxation schedule for rebuilding was implemented under which owners of small parcels paid far less than owners of large plantations. Urban dwellers were assessed based on their ability to pay, with everyone paying at least a token payment. During the crisis of the early 1770s, the Marqués de la Torre used the model formulated by Bucareli in raising money to rebuild the bridge over the Cojímar River from 1772 to 1774; only three individuals resisted paying the taxes, which were applied in an equitable manner.[91] In the wake of the catastrophes in the 1780s, local citizens stepped forward to assume the duties normally assigned to military members, volunteering without coercion to rebuild the damaged infrastructure without complaint.[92]

The hurricane of 1791 struck a population that expected similar consideration and that received none. The constables in the small villages of Quemados and Luyanó were ordered to assess the damage and, more important, to mobilize their male residents to begin cleaning up the roads.[93] Las Casas also demanded that Guanabacoa's leaders mobilize their citizens and begin rebuilding the bridge.[94] The town leaders responded with an absolute rejection of his demand, and they suggested that the captain general call upon the towns that would derive the most benefit, Río Blanco del Norte, Río Blanco del Sur, and Jaruco to the east toward Matanzas.[95] San José de las Lajas echoed Guanabacoa's suggestion to call on sugar producers and sent a list of *ingenios* and the numbers of workers each could contribute.[96] Similarly, the captain of Arroyo Arenas reported that the rain prevented the residents of his jurisdiction from repairing the road through that town.[97]

Faced with popular resistance, in addition to having to cope with the aftermath of disaster, Las Casas temporarily backed away from his hardline position and sent a conciliatory missive to the leaders of Guanabacoa in which he "invited" them to hop on the bandwagon and "contribute" as many workers as they could spare.[98] He also took seriously the demand that the burden be shared by all, and he ordered the captain of Gibacoa, at the eastern terminus of the road from Guanabacoa, to mobilize his citizens to that end. On the western side of the bay, the constables of Govea, San Antonio, Wajay, and San Luis de la Seiba del Agua were commanded to convene the leading residents of their area to decide the "most equitable" way to divide the responsibility for road maintenance. Las Casas suggested that the best place to announce his order was as the parishioners exited from Mass on Sunday morning.[99]

Shortly after the meeting was to have taken place, angry letters arrived at his residence from the captains of San Antonio and Govea complaining that they were the only leaders who had obeyed his decree—the truant leaders blamed the foul weather for their absences.[100] Their disobedience was replicated in the behavior of other constables who neglected to tally the number of able-bodied citizens who were liable for road repair. Felipe Núñez Villavicencio, the captain of San Pedro, was one such official, and his defiance earned him a rebuke from an angry Las Casas.[101] The captain general responded with a furious tirade in the *Papel Periódico de la Havana*, which asked the rhetorical question: "What has the city done for itself lately?" The editorial criticized the people because they had no incentive to work.[102] A month later, in June 1792, Las Casas issued a decree, the *Bando de Buen Gobierno*, targeting the lower classes by sentencing the *"vagos y mal entretenidos"* to the public works projects.[103]

For months after the catastrophe, residents struggled to recover under the most difficult of circumstances. The extensive flooding, filled with the carcasses of thousands of dead animals, had created a sanitation nightmare, and predictably, two weeks after the hurricane, a wave of fevers swept over the region, further debilitating the survivors.[104] In many villages, conditions remained hazardous for weeks afterward. Before the storm, the village of Jesús del Monte could boast that it contained 170 houses, including the summer homes of the Conde de Lagunillas and María de Jesús Aróstegui. After the "horrible event," which washed away the bridge at Puentes Grandes, many families had fled the jurisdiction.[105] Thomas Borrego was even more graphic about the consequences that befell his village, Wajay. Before the storm, Wajay was a prosperous agricultural community of more than thirty homesteads; afterward Borrego wrote of houses full of mud and homeless families fleeing for their lives. In response to Las Casas's call to mobilize the able-bodied men for road duty, Borrego responded that "he could not find a living soul in the region. . . . Now they will only be found in the tomb that is the Almendares River."[106]

Las Casas's indifference to the suffering of the common folk contrasts sharply with the preferential treatment he afforded the ayuntamiento members, who were also affected by the uncharacteristically early storm. Economic disaster threatened if the planters could not get their sugar to market, and from all accounts, the roads leading into Havana were washed away by the flood. The ayuntamiento's leaders offered a solution in a proposal to change the toll schedule on the bridge that would replace the one that had been destroyed at Puentes Grandes. Instead of the current toll schedule, which put the burden on the wealthy who owned carriages, the ayuntamiento suggested that all who used the bridge should contribute to its rebuilding. Las Casas accepted the ayuntamiento's suggestion enthusiastically, and he thanked the citizens for their initiative in offering a solution that "extended the spirit of sharing the burden [*repartimiento*]."[107] As a further reward for its civic responsibility, he offered relief from its obligation to maintain the public theater with a loan of 4,000 pesos "borrowed" from the militia uniform fund.[108] At the same time that they struggled to recover from their losses, the new toll schedule fell heavily upon the small producers to the west, who used the bridge to bring their cattle to market.[109]

Let Them Eat Casabe

In the towns and villages surrounding the capital, agricultural production was severely disrupted, both in the primary foodstuffs and in the economic mainstays of tobacco and sugar.[110] Perhaps, then, it was simply a lack of good judgment for the ayuntamiento and the governor to celebrate the decree expelling foreigners from the island, which could only result in a further reduction in food supplies. To make matters worse, the Consulado introduced a novelty in provisioning procedures. It proposed to substitute locally grown and less costly yuca made into casabe for the traditional bread made from white flour. The measure had a dual purpose: to limit the amount of North American flour that was brought into Cuba and to curb the influence of the greatest consumers of the flour, the garrison and the military bureaucracy, who were also the most outspoken adversaries of the governor and the planters in the Junta de Fomento.[111] During the winter of 1791–92, the ayuntamiento informed the intendant that "it would serve the common good" if he would substitute casabe for white bread in the daily rations of the troops, sailors, low-ranking administrators, hospital patients, and king's slaves.[112] The ayuntamiento hoped to justify its suggestion by enlisting the aid of the Protomédico (chief physician) of Havana, who concluded that casabe was "just as nutritious and just as delicious as white bread," especially when its bitter taste was disguised by preparing it with broth or piling it high with scrambled eggs and lard.[113] The ayuntamiento's scheme to substitute casabe for bread made from wheat flour was a great insult to the European and Europeanized Cuban people, who adamantly refused to change their diet to that of the lower ranks.[114]

Hurricane season of 1792 began within the context of the escalating conflict between the Cuban people and the captain general in alliance with the Consulado. The people resented the forced and wholly unnecessary shortages resulting from the elites' insistence on limiting the supply of U.S. flour brought to Cuba. Meanwhile, the bad weather continued without respite, manifested first in a severe spring drought and followed by torrential rain. Santa Clara was particularly hard-hit. Beginning in

June, one by one, desperate property owners appeared before the town council begging for relief from their obligation to provide cattle: tenant farmer Hilario de León for the hacienda Caguaguas; Manuel de Ayala for the hacienda Maleras; Joaquín Moya for his hacienda Minas Bajas; Juan Antonio for his *potrero*; and José Montenegro for his *estancia*.[115] Seven months later, the owner of the hacienda Calabazas, Tomás Honorio Pérez de Morales, still suffered the effects, which had reduced the number of his pigs by half (from 600 to 309) and left him with only nine head of cattle from his once-large herd.[116]

In the middle of the summer, the drought subsided and the hurricane season began. From far and wide, reports from the local captains detailed how the copious rains made their jobs difficult if not impossible.[117] The season culminated with a storm of moderate intensity that struck the environs of Havana on 29 and 30 October. It brought less rain than its predecessor, but it came with higher winds, and it interrupted or undid the work that had already been accomplished in rebuilding the roads and bridges destroyed in 1791.[118] In addition, the storm created a second set of infrastructure problems. The previous year's hurricane had caused only minor damage to public buildings because they were out of the flood zone, but in 1792, structures that escaped harm the year before sustained considerable wind damage. In Havana, 500 houses that were spared in the previous hurricane were destroyed in this October storm.[119] Heavily damaged were the two primary hospitals in Havana, Pilar, located outside the city walls near the shipyard, and the military hospital, San Ambrosio, in the southern portion of the city.[120] Especially hard-hit were buildings that were used to lodge remote military detachments. Matanzas lost its militia barracks, and Calvario, Güines, Cano, and Managua lost the buildings that served each town as barracks and stables for the mounted cavalry companies assigned to their areas.[121] Finally, even though this was a moderate hurricane, it came on the heels of a month of continuous bad weather, thus compounding the destruction. For example, early in the month of October, the church at Hanábana was hit by a lightning bolt generated by an ordinary afternoon thunderstorm, and it burned to the ground.[122] Two weeks later, the hurricane came ashore, and the rising waters swept away many head of cattle, thereby destroying the economic mainstay of the area.[123] The eastern end of the island did not escape the environmental crisis. In Santiago de Cuba, in November, the town council met in an emergency session to determine how to respond to "the torrent of rainfall that makes walking the city streets impossible" and the related problems in the countryside caused by the impassable roads.[124]

"I Am the Pastor of My Sheep"

By November 1792, all of Cuba had suffered from over two years of summer deluge and winter drought that brought misery, scarcity, disease, and death and heightened tensions within the community. In an effort to calm domestic unrest, the newly installed bishop of Havana, Félix José de Trespalacios, entered the dispute by taking the side of the Cuban people. In a series of published edicts and public pronouncements, he rebuked the captain general and the ayuntamiento for their unreasonable demands. The captain general responded with his own rebuke of the bishop, in effect telling the prelate to confine his activities to ecclesiastical matters and leave the business of government to the secular officials. Las Casas suggested that the bishop could help by saying the customary Mass asking for divine clemency that marked the end of the period of penance.[125] Angered at the governor's interference in traditional ecclesiastical prerogatives, Trespalacios refused to say the Mass. The ayuntamiento, enraged by the delay and eager to get on with the reconstruction, found the friars at the Church of San Francisco to be willing accomplices, but San Francisco did not have any pews to accommodate the worshipers. The ayuntamiento members solved the problem by raiding the Cathedral of Havana (the former Jesuit church), and the people of Havana witnessed the ignominious sight of their leading citizens stealing benches and carrying them through the streets of the city to the Church of San Francisco, where the Mass took place. The outraged bishop fired an angry protest off to the Council of the Indies, but the reply would be delayed at least four months in the transatlantic passage.[126]

During the winter of 1792–93, while the El Niño/La Niña cycle deepened its hold on Cuba, the public war between church and state raged in Havana. In the absence of a ruling from Spain, neither man would bow to the other's authority. Meanwhile, the Cuban people faced food shortages and rising prices for the few commodities that were available. Spring brought another dimension to the dispute with the approach of Lent, during which the faithful were prohibited from eating meat on certain days. Shortages of the primary provisions, especially salted fish, however, meant that substitutes for meat were unavailable, so in early February the captain

general and ayuntamiento jointly issued a decree that "out of necessity and
to avoid the gravest inconvenience," poor people would be allowed to eat
meat four days of the week, "just as they had been permitted during war
time."[127] Without asking permission from Trespalacios, the decree was
printed and published on the door of the Cathedral, in parish churches,
and in the usual places where public announcements were read. Only after
the declaration was made public knowledge did the political officials request ecclesiastical approval.[128]

The response from the bishop was immediate and scathing. In a letter to the captain general, Trespalacios acknowledged that there was a
shortage and that prices were exorbitant, but he admonished the governor: "I am the pastor of my sheep and you have usurped my ecclesiastical
privilege." The affront was even worse because the offensive decree had
been posted at the Cathedral and in the parishes.[129] The bishop was also
incensed that the captain general and ayuntamiento had coerced the official printer for the diocese, Francisco Seguí, into printing the leaflets,
with the justification, "with permission of the government [permiso del
superior gobierno]," usurping ecclesiastical prerogative to grant exceptions
to dogma.[130] In a move obviously designed to recapture ecclesiastical prerogative, on 18 February 1793, Trespalacios issued an edict to the "faithful
citizens and inhabitants." He spoke of how "divine justice has punished
us for our sins the last days of the previous October followed by the great
drought that still continues, that has caused us to lose our crop of grain
and vegetables."[131] Because of the shortage of provisions, the faithful were
granted permission to eat meat on Sunday, Monday, Tuesday, and Thurs-
day, but never on Wednesdays. The bishop also believed that the church
was more effective in carrying out the post-disaster recovery efforts, and
in a measure to replenish the church's funds, he imposed an ecclesiastical
tax on all residents that was reminiscent of the equitable and moderate
efforts of the 1760s and the 1770s. The governor and commandant of the
navy were assessed ten pesos; everyone else, including their wives, was
assessed two pesos each, until such time as there would be 12,000 pesos
in the relief fund. Country dwellers were assessed one peso each until
there would be 60,000 pesos in reserve. Everyone else who was not of
the higher ranks had to contribute two reales (a quarter-peso), regardless
of status.[132]

Not to be outdone, ten days later, Las Casas responded with his own
tax of ten reales per caballería on all landowners, regardless of rank, for the

rebuilding of damaged buildings.[133] The familiar demand to convene a meeting of village leaders to determine how to implement and enforce the tax was reiterated. For areas that were tempted to drag their feet and delay work on public buildings, Las Casas initiated proceedings to confiscate private residences to substitute for the destroyed barracks and stables.[134] Instead of inspiring the people to a spirit of civic responsibility, however, the net effect was to set the residents of rural towns against each other. A dispute began among Jaruco, Río Blanco del Norte, and Río Blanco del Sur over the obligation to construct and maintain a road over the hill that separated the towns.[135] Güines and Matanzas, both independent villas, argued over the territory claimed by each in collecting the tax of ten *reales* per *caballería*, and as a consequence of that dispute, the village of Naranjal requested its release from Matanzas's jurisdiction.[136] The constable or *capitán del partido* of Managua, José López, pointed out the injustice in taxing the residents and then distributing the proceeds according to the quantity of land in each jurisdiction. After counting the number of *caballerías* in his area, he complained that Managua contained over 1,400 *caballerías* of land, and Calvario, where the barracks was located, contained but 300. If the tax were collected, the citizens of Managua would be forced to pay for the barracks that benefited the neighboring village. López closed his letter to Las Casas with a request for advice on how to resolve the problem to avoid the "excited exclamations" from the residents of his town.[137]

Meanwhile, popular resistance passed to a more rebellious stage. The constable of Gibacoa reported "serious disorders" among the small holders, who defied his demands and the entreaties of their parish priest that they cooperate with the decrees.[138] The leaders of Guanabacoa fought tooth and nail against assuming the costs of building the bridge over the Cojímar River, and the captain general threatened to remove them from office and replace them with men who were more amenable to his wishes.[139] In Jesús del Monte, where the storms of 1791 and 1792 produced long lists of human casualties and property damage, not only were the residents forced to work on the roads and bridges, but after the storm of 1792 they became financially liable for the reconstruction of the cavalry barracks in town.[140] It took more than a year of heated debate for a team of engineers and craftsmen to come up with an estimate of 158 pesos for the barracks' reconstruction, and even then, the town leaders could not convince the residents to contribute to the building's reconstruction.[141]

Catastrophe, Commerce, and International Conflict

After three years of bad weather and worse government, Cuba could not have been less prepared to cope with the external pressures being brought to bear upon the Spanish empire from several fronts. The revolution that began in France in 1789 spread to its Caribbean colony, Saint Domingue, and escalated into a conflict tinged with elements of racial inequality between supporters of the French king, Louis XVI, and those who supported the republican National Assembly. The consequences of the French Revolution caused grave concern in Spain and in her Caribbean colonies, particularly in Santo Domingo, which shared the island of Hispaniola.[142] The conflict in the neighboring colony became more complicated when a slave rebellion erupted in Saint Domingue on 23 August 1791.[143] The Spanish governor, Joaquín García, put his forces on the highest state of alert when he learned of the rebellion that had broken out near Guarico, the closest major city to the frontier. But instead of turning eastward, the violence spread west and south toward Saint Domingue's major ports.[144] In May 1792, the regiment of Cantabria was transferred from Puerto Rico to guard the border, while refugees fleeing the wrath of rebellious slaves poured into Santo Domingo.[145] In January 1793, after the execution of Louis XVI, Spain declared war on republican France, and the frontier campaigns shifted from a position of containment to an aggressive stance intended to strike a blow against the republicans.[146] The regiments of Caracas, Maracaibo, Mexico, Puebla, Havana, and Santiago de Cuba were mobilized to be sent to the war front.[147] Spain recruited black soldiers led by Juan Francisco in the north and Jorge Biassou and Toussaint L'Ouverture in the south, gave them military status as "auxiliaries," and agreed to provide them with clothing, munitions, and provisions.[148] At the same time, an unlikely alliance was forged with Great Britain, and that nation's forces led the assault on the cities on the west coast of the island.[149]

War brought the question of provisioning the troops to the forefront. The Spanish crown had ordered reinforcements from neighboring Caribbean cities to Santo Domingo, but the financial obligation to provide fresh provisions for these regiments fell upon the treasuries of their respective garrisons.[150] Now the drain on Havana's treasury became more acute and the battle over flour became more acrimonious. While the Cuban regiments prepared to "follow their flag" into battle, the captain general and the Junta de Fomento sought to undermine their potential for victory. The painful lessons of the past forty years about the importance of a well-fed army were cast aside, and to the veteran Cuban troops, such actions were nothing short of treasonous.[151]

In May 1793, the third and fourth companies of the second battalion of Cuba from Santiago de Cuba, under the command of veteran colonel Juan de Lleonart, sailed for Santo Domingo.[152] Back home, not waiting for royal policy to catch up with local needs, the governor and residents in Santiago de Cuba took matters into their own hands to make sure that they would not go hungry. In a scene all too familiar, in June, an American boat appeared in the harbor requesting "shelter and hospitality." Onboard was a quantity of flour, dried fish, and other comestibles. Juan Bautista Vaillant, the governor of Santiago de Cuba, a veteran with a lifetime of service in Cuba and a political enemy of the captain general, permitted the captain to land his cargo. Vaillant justified his decision that the exigencies of war had suspended the shipment of provisions to his jurisdiction and that the flour that was in Santiago de Cuba was already spoiled. Citing the "extreme necessity" in the town, his decision to permit the sale of flour to avoid the "clamoring" of the residents was obvious.[153] Shortly thereafter, the notification that Charles IV had reopened neutral commerce to the Hispanic Caribbean arrived in Cuba.[154] Meanwhile, the news had already spread like wildfire up and down the Atlantic coast, especially after the Spanish consul in Philadelphia, Diego de Gardoqui, requested that the American government repay its debt to Spain, left over from the War of Independence, in provisions.[155] In short order, American ships appeared in the major Hispanic Caribbean ports, expecting to be able to sell their cargoes of flour, meat, and vegetables.[156]

In late June 1793, the Spanish army of operations in Santo Domingo, cobbled together of companies from several Caribbean cities, scored its first victory at Juana Méndez (also known as Oüanaminthe), the closest French fort to the major Spanish outpost on the northern coastal plain, Dajabón.[157] There, a royalist French garrison under the command of Pedro LaFevilliez surrendered to the forces led by Gaspar de Casasola, colonel of the fixed regiment of Santo Domingo; afterward a jubilant García wrote to Spain describing the capture of the fortification in glowing terms. The "victory" at Juana Méndez was followed by an orgy of reward in which every officer—many of whom had never before seen combat in their entire careers—received a commendation for merit.[158] Not until years later, in 1799, would the true story of the battle of Juana Méndez be revealed, when the former commander broke his silence. LaFevilliez confessed that with the execution of Louis XVI, the French forces in Saint Domingue were faced with a dilemma: the majority were royalists who were outraged by the execution of the legitimate monarch by the usurpers of the Convention (the revolutionary assembly in Paris). They could not support the government that claimed to represent France, and so, when presented with either surrendering to a royalist army or continuing to fight in favor of the illegitimate government in Paris, he and his subordinates unanimously agreed to surrender.[159] Barely a shot had been fired.

The capture of Juana Méndez yielded the priceless intelligence that the enemy was plotting to attack the Spanish outposts at San Miguel, San Rafael, and Hincha, which lay southward across the mountain range that separated the north coast from the central valley. The garrisons in those isolated forts were seriously undermanned, and so the high command at Dajabón posted the first Havana company under colonel Matías de Armona to San Rafael, while the third and fourth companies of the second battalion of Santiago de Cuba were ordered to reinforce the garrisons at San Miguel and Neyba. The timely arrival of healthy soldiers from Cuba bolstered the effective forces on the island, and in August the Spanish armies under Joaquín Cabrera, colonel of the regiment of Cantabria, were able to rout an enemy force that had occupied San Miguel, recapturing several pieces of artillery as well as provisions and pack animals.[160]

In spite of the successes at Juana Méndez and San Miguel, the victory came at a high price. As a consequence of the campaign, 507 men were afflicted by fevers; among them was the commander of the western armies, Casasola, and in addition, the Spanish army inherited fifty sick French prisoners of war. Across the mountains, fever also decimated the troops attempting to defend the scattered outposts in the central valley. Half of San Rafael's defenders were incapacitated, and even with the arrival of additional men of the first Havana company, García warned that it "was not enough to overcome the epidemic."[161] The lack of provisions, of course, simply compounded the problem. García complained of the "extreme necessity" for fresh rations, and at the same time the defender of Oüanaminthe, LaFevilliez, admitted that one of the considerations that led to his surrender was the lack of provisions for his men.[162]

As disease took its toll on the regiments of Santo Domingo and Cantabria, García called upon his neighbors in the Caribbean basin to send more reinforcements to replace his rapidly dwindling forces.[163] Meanwhile, the commander of the Spanish squadron in the Caribbean, Gabriel de Aristizábal, sailed for Caracas. Arriving in July, he requested that the governor, Pedro Carbonell, transfer the province's infantry and artillery units to his command, mobilize the province's civilian militia, and at the same time, initiate a forced draft—even going so far as to clean out the jails.[164] By August, the veteran infantry and artillery companies were on board Aristizábal's ship en route to Santo Domingo, and late in 1793, his squadron blockaded the port city, Bayajá (Fuerte Delfín), on the north coast.[165] The city offered no resistance, and on 29 January 1794, Aristizábal accepted the articles of capitulation from the three senior commanders of the French army on board his flagship.[166] As had occurred in Juana Méndez, the Spanish army captured the French fortification without firing a shot.

In late January 1794, García, who had overseen the successful campaigns from the safety of his residence in the capital, traveled to the frontier to take personal command of the war effort.[167] Once in Dajabón, he went to the hospitals to visit the troops and inspect the conditions under which they were living. Later, he went to Juana Méndez, where he praised LaFevilliez for opting to surrender. While in Dajabón, García received the news of Aristizábal's capture of Bayajá, so the captain general traveled to the port city to meet with the naval commander and to plan subsequent military operations. A few days later, García convened a war council (*junta de guerra*) made up of the commanders of the units from several Caribbean cities to strategize how best to carry out the campaign against the republicans entrenched in Guarico.[168] Dajabón, safely behind the lines in Spanish territory, was designated as the primary center of operations, while Bayajá became the vanguard city where an invasion would be organized. A defensive cordon was established along the western and southern frontier that ran through the central valley, linking the key outposts of San Rafael, San Miguel, Hincha, Caobas, Neyba, and Azua.[169]

As the deliberations progressed, the troubling realization must have settled upon the Cuban and other veteran regiments that they would become subordinate to García and the men in his immediate circle.[170] Instead of keeping the regional units together under the command of leaders with whom they had served for decades, the companies from neighboring Caribbean cities were split up and integrated into units of the army of Santo Domingo. To make matters worse, after the orgy of reward and promotion following the capitulation of Juana Méndez, many veterans discovered that they owed their obedience to men who possessed far less experience in actual combat and now outranked them. Not surprisingly, the infighting began almost immediately.[171] After the capture of Bayajá, Aristizábal had named Joaquin Sasso, colonel of the regiment of Puerto Rico, as commander of the forces in the west, but García overruled Aristizábal and appointed Casasola to the same post. The decision prompted an angry exchange between the two men, and before long, the news of the dispute among the leaders in Santo Domingo came to the notice of the king's counselors in Spain. In April 1794, an exasperated secretary of state, Antonio Valdéz, informed García that the king had approved Casasola's appointment, but at the same time, he reprimanded the captain general for his high-handed tactics, warning him that above all, Charles IV desired harmonious relations between the two men.[172] An equally confrontational relationship existed among the leaders in the outposts along the southern frontier, pitting comrades-in-arms Armona and Lleonart, who were in charge of garrisons at either end of the central valley, against García's favorite, Cabrera, who commanded the forces at San Miguel.[173]

Yet no issue was more contentious than García's support of the black auxiliaries. García had accepted unquestioningly the vows of loyalty and subordination from Juan Francisco and his captains, and he naively relied upon the assurances of the mulatto parish priest of Dajabón, José Vázquez, who served as the liaison between the Spanish forces and the black troops in the north. The captain general was less trusting of Toussaint, especially when the commander of the forts in the central valley, Lleonart, wrote several reports warning García that the black leader showed none of the experience or military discipline that Lleonart had come to rely upon in the veteran units of color in Havana and Santiago de Cuba.[174] Adding to the problem, the responsibility to feed, clothe, and house the black auxiliaries fell on the treasury of Santo Domingo. Civilians and regular military alike watched helplessly as the auxiliary troops "lived in opulence" while they suffered from the effects of deprivation and starvation.[175] The black auxiliaries also reveled in the orgy of reward when García distributed fifteen silver and gold commendation medals to Juan Francisco, Biassou, Toussaint, and their captains.[176] Among the recipients of the royal largesse was the priest of Dajabón, José Vázquez, who first became the chaplain of the auxiliary troops and later was selected to be the treasurer of the church of Santo Domingo.[177]

"They Fear Hunger More Than They Fear the Enemy"

By spring 1794, the Spanish army was poised to strike against the republican forces in Saint Domingue, while the supply networks intended to support the military campaigns barely functioned. Throughout 1793, the threat of famine haunted the Spanish Caribbean cities, in spite of Charles IV's royal order permitting the Spanish consuls in Philadelphia and New York to issue licenses for American ships to sail to Spain's Caribbean ports. The supply of flour and other food was reduced further by the number of enemy ships and corsairs that roamed the waters around Hispaniola ready to prey on merchant vessels that attempted to make the run to the island. As a consequence, by December 1793, the flour that made it to the war theater from North America was of such poor quality that the quartermaster had to send agents to Curaçao to purchase fresh supplies.[178] A month later, nothing had changed. As deliberations began over the wisdom of carrying the campaign to the enemy in Saint Domingue, the military units still had an "abundance of munitions but a lack of provisions."[179] The importance of keeping food and medicine coming to the troops prompted García to name a colleague to supervise the effort in Santo Domingo City while he went to the frontier. Juan Antonio de Urízzar, the chief treasury official in the capital, was appointed to purchase and to allocate flour and other comestibles, and Juan Sánchez was charged with overseeing the receipt and distribution of provisions to the fighting forces.[180] In spite of their best efforts, pirate attacks deterred ships from northern cities from attempting to sail to Hispaniola, creating an "untenable scarcity," and as shipments out of the capital became more and more sporadic, complaints from the garrisons in the interior became more strident. On occasion, some smaller vessels made it through the pirate blockade. In June, a ship from La Guaira arrived in Santo Domingo with 200 sacks of hardtack, much of which went to supply the garrisons along the southern and western cordon.[181] At the same time, a few intrepid captains from North America evaded the pirates and made it to Bayajá with flour, rice, and dried beef.[182] Royal administrators in other Caribbean cities who had men fighting in Santo Domingo scrambled to gather supplies to send to their forces on the frontier, sometimes relying upon North American captains to brave the pirates to get provisions to their forces.[183]

Scarcity exacerbated the disease environment that decimated the troop numbers. En route to Hispaniola from Caracas, Aristizábal attempted to return to Puerto Cabello, but some of his boats were blown far off course by a winter storm and ultimately made safe harbor far to the east in Puerto Rico.[184] The men onboard were afflicted by an epidemic that sickened over 400 soldiers, and their recovery was slowed since they did not have any hospital facilities.[185] On land, the fevers continued to claim many victims, taxing physician Pedro Pablo Irigoyen of the regiment of Cantabria to the limit of his abilities. When his unit was assigned to Dajabón, Irigoyen worked under terrible conditions to try to save his comrades in arms. His task was doubly difficult because the regiment did not have a pharmacist trained to prepare accurate formulas; nonetheless, Irigoyen "faithfully followed the receipts of Masdeval," the pioneer in prescribing quinine to treat malaria and other fever-inducing aliments (for example, dengue and dysentery).[186] Impassable roads and rising rivers on Hispaniola impeded the distribution of provisions and medicines to the garrisons along the southern and western cordon, prompting Urízzar to question the wisdom of trying to send supplies overland in the "worst of the rainy season."[187] His frustration was almost palpable in his missives to Madrid as he described his efforts to send quinine to the pharmacies in Bayajá, Dajabón, and Montecristi by sea in a small chest, but, in spite of his best efforts, the number of sick soldiers just kept growing.[188] Bad weather also plagued Cuba, making the roads in Havana province virtually impassable and limiting the quantity of domestically produced provisions that could be brought to market.[189]

In March 1794, García recklessly began preparations for the offensive against Guarico. His prudent comrades in the *junta de guerra* urged caution, especially when faced with the adverse conditions, which seemed to worsen daily.[190] Undaunted, García requested two more regiments from Puerto Rico, 400 additional troops from Havana, and the transfer of the remainder of the men from the Cantabrian unit still in the capital city.[191] He also implored Urízzar to make every effort to send more food because the troops had only forty barrels of flour and a small amount of dried beef. With few men fit for duty and faced with anarchy among the free coloreds and blacks, the Spanish forces "feared hunger more than they feared the enemy."[192] Nonetheless, on 8 May, García ordered the invasion of republican French Saint Domingue to begin. It took only six days for the Spanish army to be defeated at Yaguesi and chased back to Bayajá.[193] At the same time, reports from the commanders along the cordon in the central valley cautioned that Toussaint's insubordination had increased. Lleonart was suspicious of the number of fires that had been set in the central valley, and he warned García that the possibility that Toussaint or his men would commit a "vile act" had become even greater.[194]

As the defeated army retreated to Bayajá to recover and regroup, conditions on the frontier worsened as the death toll from fevers rose. After the campaign against Guarico, Casasola renounced his position as commander of the troops and requested permission to retire to recover from his illness.[195] Shortly thereafter, García also succumbed to the fevers and retreated to Hato de la Gorra along the north coast near Santiago.[196] The regiment of Cantabria had lost twenty officers, among them their physician, Irigoyen, who only abandoned caring for the men under him and retired to Montecristi when directly ordered to do so by Sasso, one of the few remaining senior officers still fit for duty.[197] By the end of May, physicians and pharmacists in the hospitals in Bayajá, Dajabón, and Montecristi struggled to care for over 600 patients who were sick with *calenturas*.[198] As a consequence of the attempted invasion of Saint Domingue, the army in Bayajá was a hollow shell, without leadership, with half its numbers dead or unfit for service and "reduced to eating any root and plant that they could find."[199]

As had happened thirty years previous, a hungry and disease-wracked Spanish army faced a deadly and determined enemy. Yet the combat conditions on the frontier of Santo Domingo in 1794, while resembling the disease environment in Cuba in 1762, were unlike any previous military campaign. In 1794, the French republic abolished slavery and sent Léger Félicité Sonthonax to pursue the military operations against Spain. A key element in Sonthonax's strategy was to lure the black auxiliaries away from their allegiance to Spain with the promise of freedom if they fought for the republic.[200] The failed attempt to conquer Guarico left the Spanish military vulnerable, and the possibility of freedom offered by the French republicans was irresistible. In spite of García's reward of thousands of pesos, provisions, and the privilege of selling slaves confiscated from royalist owners, the Spaniards could not count on the loyalty of Toussaint, Juan Francisco, or the men in their armies.[201]

As rumors of Toussaint's possible betrayal increased, García, still relying upon the assurances of Vázquez, sent Juan Francisco and his forces to the cordon to be able to repulse any attack. On the way to the central valley, Toussaint fired upon Juan Francisco's forces, but the latter made it to San Rafael, where he related the incident to Lleonart. Upon learning about the confrontation between Toussaint and Juan Francisco, Lleonart wrote Toussaint a letter requesting a conference so that each man could explain his actions. Toussaint delayed responding to the letter so that he could set his plans in action, and on the night of 3 July, he moved a large body of his troops toward Dondón. There he attacked Juan Francisco and his followers, causing numerous fatalities, taking many prisoners, and capturing horses and equipment. After the attack, Toussaint moved back across the frontier into republican territory, while Juan Francisco headed toward the north coast.[202]

In flight from Toussaint's surprise attack, Juan Francisco and his army appeared at the outskirts of Bayajá and demanded entry. According to the terms of the capitulation signed in January, the auxiliaries were forbidden to enter the city, and the commander on duty, Lieutenant Colonel Francisco de Montalvo of the third battalion of Cuba, refused their demand.[203] When Montalvo blocked the auxiliaries' entry, their chaplain, Vázquez, complained that García had forbidden the whites to try to order the blacks to obey against their will. None of the senior officers left in charge of Bayajá had the authority to countermand the orders of the captain general, and Montalvo was outranked and overruled.[204] Once inside, Juan Francisco's troops inexplicably slaughtered 700 French residents and refugees. Twenty residents of Bayajá managed to escape the massacre, fleeing on foot to an American boat under the protection of a mounted corps of guards under Montalvo's command. Upon witnessing the slaughter, while senior officers stood immobilized, Montalvo had mobilized his men in an attempt to save some of the city's doomed residents.[205]

With the armies in the north in full retreat and with the ability to reinforce and reprovision the cordon compromised, Lleonart's many warnings questioning Toussaint's loyalty came true. In October, Toussaint attacked the Spanish outposts in the central valley. In San Rafael, Lleonart and his men attempted to defend their posts, but after calling a *junta de guerra* to debate their alternatives, Lleonart and his officers were unanimous that there was no alternative but to retreat to save themselves from certain annihilation.[206] The Cuban forces fell back to the nearest outpost, San Miguel, where they joined the Cantabrian regiment, led by Cabrera. Again faced with a hopeless situation, the two regiments withdrew further to Hincha, under the command of Armona and garrisoned by the regiment of Havana. Finally, all three regiments abandoned the outposts and retreated to the well-fortified city of Dajabón, garrisoned by the Santo Domingo troops.[207] But the worst was yet to come. A vindictive García charged Lleonart and Armona in courts martial with abandoning their posts, and in an outrageous violation of military protocol, the two officers were physically jailed in Santo Domingo.[208] Cabrera, García's favorite, escaped even a word of censure.[209] García's actions widened the schism within the forces and further undermined the mission, and the men from units from the other cities in the Caribbean lost all confidence in their superior officers.

As summer passed into fall 1794, the ill winds of misfortune continued to assault the Hispanic Caribbean when a catastrophic hurricane struck western Cuba on 27 August. The storm made landfall around midnight, and its fury was felt in the countryside until 8:00 in the evening the following day.[210] Like the storm of 1768, it entered the island from the south and exited to the north, and its path of devastation extended from Matanzas to Havana. The stark meteorological data published a week later in the *Papel Periódico* barely convey the extent of the destruction. Most of the ships at anchor were destroyed or sustained considerable damage, including twelve boats belonging to the admiralty, sixty-four private vessels, and an "infinite number" of launches and dinghies. The list of casualties included forty-two schooners and eight sloops, the primary types of vessels used in the intra-island trade to ferry provisions and supplies among the Caribbean cities. In addition, seven private frigates and seven brigantines were lost.[211] The storm surge flooded the barrios outside the city and deposited several inches of sand in the streets and garden plots of the barrio of San Lázaro.[212] After causing significant damage in western Cuba, the hurricane continued northwestward through the Gulf of Mexico, where it made landfall in Louisiana on 31 August.[213] There the Mississippi River overran its banks, causing serious flooding, which destroyed the subsistence crops of rice and corn, along with the commercial products of indigo and cotton.[214]

The hurricane of 1794 could not have come at a worse time for the Cuban regiments in Santo Domingo, who were dependent upon Havana for their sustenance. The subsistence crops in western Cuba were completely destroyed, and the second landfall in Louisiana put an additional strain on the already-compromised provisioning capability of Havana. The governor of Louisiana, the Baron de Carondelet, the brother-in-law of the Cuban captain general, requested emergency supplies of medicines and food, but when news came that Cuba had also been affected, he was forced to contract for 1,000 barrels of flour from Charleston, South Carolina.[215] In addition, the maritime lifeline now suffered a serious setback with so many vessels destroyed in the port, and for months thereafter, the Cuban forces in Santo Domingo could expect to receive no provisions from their home city.[216] Faced with impending starvation at home and at war, there was no alternative to accepting flour and other foodstuffs from the United States. The details of the transactions have not come to light, but less than a month later, the warehouse in Havana had sufficient provisions for the commissary to be able to send nearly 3,600 barrels of flour to Santo Domingo "for the army, the navy, the hospitals and [for the outposts] on the frontier."[217]

Eventually news of the multiple catastrophes made its way back to Madrid. In October, a letter from survivors of the massacre at Bayajá arrived at court via the secretary to the embassy in London, Carlos Martínez de Yrujo, the Marqués de Casa Yrujo. Its contents were read aloud in front of the Council of State, and the horrified men learned of the ineptitude in the expeditionary army.[218] Shortly thereafter, the king and his council received news of Toussaint's betrayal, and the Council of State was obliged to come up with solutions to the many problems in the theater of operations.[219] As a consequence of their deliberations, in fall 1794, several changes were made in the conduct of the war. The Marqués de Real Socorro was sent on a fact-finding mission as the monarch's personal representative to Santo Domingo.[220] Sebastián Calvo, the Marqués de Casa Calvo, was named governor of Bayajá and commander of the army of the operations against the republicans.[221] Aristizábal began hunting down the pirates so that provisions could make it through to the troops.[222] A separate distribution system was established in Bayajá, and Casa Calvo was given the authority to receive ships directly rather than having them come into the capital city, which avoided their cargoes having to be sent overland to the frontier.[223] In an attempt to resolve the number of deaths from *calenturas*, Charles IV ordered that all measures be taken to help the troops, and construction began on a new hospital in Santo Domingo.[224] The final thorny question was how to respond to the massacre at Bayajá and the attacks in the central valley. Believing that the attacks had more to do with the enmity between Juan Francisco and Toussaint than with Juan Francisco's treachery, direct orders came from the monarch to "do everything to win [his] friendship," while recognizing that Toussaint was now Spain's avowed enemy.[225]

The decisive actions by the Council of State gave a new direction to the Santo Domingo campaign, if only for a brief few months. García retreated to the capital, where he remained for the rest of the conflict, and in August, Casa Calvo, a veteran of the campaigns of the 1780s, took over the command of the army.[226] After the pillage by Juan Francisco and his troops, Montalvo and José de Horrutiner sought to recover some of the stolen articles, while Cabrera initiated actions against the outposts on the frontier and successfully dislodged Toussaint and his followers from the areas that they occupied.[227] Beginning in August, every shipment of provisions into Bayajá was personally inspected by Casa Calvo, and the commissioner in charge of distribution, Sánchez, supervised the food shipments to the frontier garrisons. The overwhelming majority of the shipments were brought by North American captains.[228]

The Bittersweet Homecoming of the "Six Skeletons"

The changes initiated after the disastrous summer of 1794 were too little, too late. Just when it appeared that the military campaign in Santo Domingo had turned a corner, the news arrived that in 1795, the ever-mercurial Charles IV, on the advice of his minister of war, Manuel de Godoy, had signed a peace treaty with republican France, the Treaty of Basel.[229] The toll on the Cuban regiments had been horrific. The Havana companies had lost more than half of their men, those of Santiago de Cuba had lost 573 men, and the Louisiana regiments had lost 344 men. All totaled, the campaign in Santo Domingo had been responsible for the deaths of over 1,800 Cuban soldiers. And just as many remained unfit for duty because fever still coursed through the ranks. Upon receiving the news of the multiple defeats, Joaquín Beltrán de Santa Cruz y Mopox, the Conde de Mopox, lamented: "The regiments that arrived in Santo Domingo were six battalions; now they are six skeletons."[230] With news that the war had ended, Casa Calvo surrendered his command to the French general, Estéban Laveaux, and prepared to bring Cuba's remaining soldiers home.

Safe in Santo Domingo, 240 miles (80 leagues) from the battlefront, García had other ideas, and he demanded that Casa Calvo transfer his regiment to the capital city. The order reignited the animosity within the ranks and prompted a sharp exchange between the two men. Casa Calvo refused to follow the order, maintaining that the transfer to the capital would take an additional toll on his soldiers. Fully aware that he would be guilty of insubordination, Casa Calvo wrote of "the pain of losing men in a futile campaign." Instead of marching his troops to the east, the commander of Bayajá disobeyed García's order and ordered his sick regiments to board one of the ships in Aristizábal's fleet.[231] At first the naval commander was reluctant to take the units home, citing the long voyage and the resistance to mixing sick soldiers with healthy men, but after his "mature reflection about the just causes [to transport the troops]," Aristizábal ultimately permitted the Cuban regiments to come onto his ships, and the six skeletons sailed for Cuba.[232]

Rarely is a community more vulnerable to political unrest than in the aftermath of disaster. Under the stress of coping with the loss of friends and family and the disappearance of the structures of everyday life, and in the face of starvation, any community will be taxed to the limits of its endurance. Such was the case in Havana in spring 1796 when the fighting forces of Cuba arrived home. In the months since the hurricane in 1794, conditions in Cuba had gotten worse, and the misery after the storm was compounded by the news coming from Santo Domingo. The families on the home front were immediately aware of the failure of the assault on Guarico, of the massacre at Bayajá, and of Toussaint's treason.[233] At the same time, the number of casualties among the men who had been called to active duty increased daily, and a growing number of military families received the news that their male relatives had perished in the futile and ill-conceived campaign. As more and more of Cuba's men were sent to Santo Domingo, 200 militia members were put on active duty to protect the island.[234] Cuba's military families were undoubtedly horrified when they learned of the arrest of two of their most senior leaders and their confinement in the capital of the neighboring island.[235] Finally, the news came that instead of punishing the black auxiliaries for the massacre and pillage at Bayajá, Juan Francisco and his officers would be indemnified and resettled somewhere in the Spanish empire.[236]

The community's disbelief at the events in the wake of the defeat in Saint Domingue added to their frustration with Las Casas's advocacy of the merchants in the Consulado.[237] By the winter of 1795–96, "everyone[,] including the military officials, the town council, the military corps, and all of Havana's inhabitants," spoke openly and critically of the captain general and his favorites. Instead of viewing the captain general as a benign and beloved governor like his predecessors, the Cuban people alternately saw him as a coward and as a hypocrite. Stories spread that after the hurricane, when a rumor of an uprising similar to the one in Bayajá swept through the city, Las Casas locked himself inside his house on the harbor front so that he could escape should the rebellion prove to be a reality. Worse still, he left Havana vulnerable when he sent the few troops he could spare to reinforce his brother-in-law, the Baron de Carondelet, in Louisiana. Las Casas's "indiscreet and ignorant" behavior brought the island dangerously close to the fate of neighboring Saint Domingue. A long list of subscribers, men and women, citizens of Havana, and military members alike, had signed their names to a petition criticizing his government, and others who refused to sign their names openly were ready to offer their money and influence anonymously. Although the Cuban people acknowledged that the hurricane had caused much destruction on the island, they celebrated that the storm had thwarted the schemes (*mañas*) of the governor and his friends. The most damning criticism compared Las Casas's behavior in the storm's aftermath to that of the men who served before him: "He [Las Casas] was a hard man.... He did not have the consideration of other governors who were very good."[238]

The bittersweet homecoming was made even more difficult when the veterans discovered that their commercially minded cousins in the Consulado had done everything possible to prevent North American provisions from arriving in Havana. For the better part of the war years, the Consulado had worked its hardest when trying to convince His Majesty to limit access to the Cuban market to the Americans. The battle over the flour trade came to a head in 1796 when, in January, the Consulado learned that permission for neutral ships to trade with the island had been rescinded. That jubilation was short-lived when late in the month, the *Bacchus* showed up in Havana harbor from Philadelphia with a license to import flour issued by the Spanish consul in the northern city. The *Bacchus*'s arrival and the subsequent stream of ships from North American cities gave impetus to a series of unending complaints from the Consulado, which the leading authority on the period has appropriately termed "*los llorones Cubanos*."[239] In March, the complaints were muted and respectful. The Consulado presented its arguments in measured fashion, requesting that the consuls in Philadelphia be advised that the licenses be stopped.[240] In April, Arango voiced his anger that the trade continued, while at the same time he complained that it had been impossible to implement the high tolls on the Puente Nuevo.[241] By May, the tone of the proceedings changed to the urgency of men whose complaints were going unanswered. Now they presented their requests as logic, asking for the privileges to limit North American commerce, permitted because of the war with France in 1793. Instead of allowing ships from Philadelphia or New York to capitalize on the lucrative trade, the Consulado suggested that flour from Spain, Mexico, or Buenos Aires be the privileged commodity.[242]

By mid-summer, from July through August, an air of hysteria tinged with fury pervaded its deliberations as the Consulado complained of the harm that the continued traffic in flour from the north did to the commerce of Havana. Indignantly, it reiterated the events of the previous months. In January, a royal order suspended the access that North American ships would have to the island. The Spanish consuls in Philadelphia and New York had ample time to announce the prohibition, but instead of limiting the number of licenses, merchant houses rushed to secure an even greater number. The net result was that by June, over 16,000 barrels of American flour competed with 13,000 *tercios* of Mexican flour and 700 barrels from Spain. Alarmed, the intendant of the army, Pedro Valiente, called for a meeting with the commandant of the navy, because "the honor of the nation will not excuse the continued admittance of foreign produce."[243] By 4 August, the Consulado's strategy reverted to logic, informing the Council of State of the benefits accruing from permission to import beef from Buenos Aires and suggesting that a similar concession be granted for flour.[244]

The answer to these entreaties came on 23 August, when the news reached Havana that Charles IV had granted the exclusive privilege to import flour from North America to the Conde de Mopox.[245] Citing the "extreme decadence" in the effective fighting forces, Mopox, resident in Madrid, approached his mentor at court, Manuel de Godoy, the minister of war, and offered to finance the costs of replacing the Cuban troops.[246] Already among the richest men on the island, Mopox requested only one special favor: that he be allowed to import sufficient flour from North America to make sure that in the future such defeats were not caused by his troops being too undernourished to fight. Charles IV responded with two concessions. The first granted Mopox the exclusive privilege to import flour from North America.[247] The second reward promoted him to brigadier and appointed him as inspector of the troops, making him the second-most-powerful man on the island.[248]

When the news made its way to Havana, a shocked Consulado met in special session (*junta extraordinaria*) to debate what to do. Two members, Pedro José Erice and José Antonio Arregui, were given the task of crafting a report to send to Spain outlining the damage that the concession would do to Cuba. The report, debated and approved over the following months, predicted the ruin of the island and the trade with Spain if the concession were allowed to stand.[249] Shortly thereafter, Antonio del Valle Hernández, one of Arango's closest associates, complained about the "scandalous concessions that the King has conceded to the Conde de Mopox for the importation of flour and other provisions from North America."[250]

By fall 1796, the consequences of the Santo Domingo campaign had taken their toll on Cuba. Within the Consulado, the air of defeat must have been palpable when the realization set in that the entire year dedicated to advancing its economic agenda had been for nothing. Ironically, it had gotten what it had wished for—the elimination of U.S. vessels from Cuba—but the outcome was even worse than the predicament. The lucrative trade in provisions had not gone to it but to its rival. Even more bad news made its way to Havana in December when the Consulado learned that its champion and beloved governor, Las Casas, had been recalled to Spain and would be replaced by the Conde de Santa Clara.[251] A far different attitude prevailed outside the Consulado's chambers, where the military forces and the ordinary people on the streets of Havana celebrated Las Casas's recall and looked forward to a return to the benevolent rule to which they had become accustomed.[252] With Las Casas's departure in November 1796, the island breathed a collective sign of relief when he was replaced by Juan Procopio de Bassecourt, the Conde de Santa Clara, and he, in turn, by Salvador Muro y Salazar, the Marqués de Someruelos.[253]

In time, the violations of military protocol and García's abuses were rectified, and after years of uncertainty, both Lleonart and Armona were exonerated.[254] In 1796, Casa Calvo received a reprimand from the minister of war and a temporary demotion, but by 1797, with Mopox as inspector general, his position among the leaders of the military in Cuba was restored. In the context of widespread political turmoil, hurricane season of 1796 was almost anticlimactic. In October, a weak hurricane struck western Cuba, causing minor damage to the plantain and other crops in Pinar de Río. Given the discontent that simmered just below the surface, the storm passed virtually unnoticed.[255] After four decades, the residents of the Hispanic Caribbean had come to rely upon the provisions that were so easily obtained from Philadelphia, New York, Charleston, New Haven, and other North American cities. The ability to secure flour and other provisions from the United States, a privilege transformed into a habit, had become a fact of life for the Cuban people.

CHAPTER SEVEN

So Contrary to Sound Policy and Reason

A T THE END OF THE eighteenth century, the warm climate anomaly subsided as suddenly as it began. By 1800, temperatures plunged to a level not experienced since the 1740s.[1] Hurricanes continued to make landfall in Cuba, including one in Oriente in 1799, but not until the 1840s would a series of three sequential storms (in 1842, 1844, and 1846) again devastate the island and lead to another critical juncture in the island's history; the cycle of inordinately warm temperatures would not be equaled until the early twentieth century.[2]

For the Havana Consulado, the years leading up to the turn of the century continued to be a time of disappointment. Not once in its campaign to eliminate North American influence and to wrest control of the commerce of Cuba away from its rivals did it succeed in convincing Charles IV that it could offer a better solution to the problems of supply. After the defeats of 1796, in January 1797, the Consulado received more bad news when neutral commerce was reinstated. Trade relations with the United States were maintained until January 1802, when the concession was rescinded. The jubilation in the Consulado's meeting on January 31 was almost tangible. Those at the meeting applauded the king's resolution and congratulated His Majesty for recognizing the evil that would befall the commerce of Cuba if such permissiveness were continued.[3] Exuberance was short-lived when three months later a fire in the extra-urban barrio of Guadalupe destroyed nearly all of the residences and again placed the city on an emergency footing. Much to its dismay, the Consulado was forced to reverse its own position and request that emergency provisions be permitted to enter the port of Havana.[4] Combined with the outbreak of hostilities in Europe with the Napoleonic Wars, trade relations with the United States had passed the point of no return. After 1803, licenses were routinely granted to individuals to import North American flour. With every new concession, it became the unpleasant task of the new captain general, the Marqués de Someruelos, to announce the news at the Consulado's regular meetings.[5] By 1805, international warfare and the high demand in Europe had pushed the price of flour up to twenty-two pesos per barrel; at the same time, planter interests suffered because drought had reduced the sugar harvest by 46,330 boxes.[6] To add insult to injury, by 1804, the North Americans had cornered the lumber market in Cuba, and even the boxes in which the planters shipped their sugar were being brought in from the northern nation. In 1804 alone, nearly 1 million board feet of lumber were imported into the island.[7]

The Havana Consulado was unsuccessful because for every argument it put forth, even more influential voices in Charles IV's government countered with their own arguments in favor of the gradual opening of trade, which had been occurring over the past thirty-five years. The Havana interests' antagonists were another powerful group of merchants with allies in Philadelphia, centering around consuls José María Ysnardi and Diego de Gardoqui, who pressured the monarch to continue to issue licenses to trade with the island. One such adviser was Carlos Martínez de Yrujo, Marqués de Casa Yrujo. Writing under the pseudonym "A Spaniard" in a pamphlet published in 1800, Martínez de Yrujo argued forcefully against the ill-advised policy of prohibiting neutral ships access to Spanish American ports. Citing the commerce of Havana specifically, he explained the benefits, especially the increased revenue, that free trade would bring to the empire, particularly in the context of the Spanish desire for a strong defense, which required a large military presence in the Caribbean. He focused on the need for flour in St. Augustine as an example. Prior to the new regulations, flour destined for the Florida city originated in Philadelphia, was shipped to Cádiz, then on to Havana, and finally to the garrison in St. Augustine. The costs of such a circuitous route meant that by the time it reached the forces in East Florida, a barrel of flour cost the royal treasury nearly thirty dollars. Martínez de Yrujo pointed out that the constant change in storage environments, from very dry to very damp, increased the possibility of spoilage, adding additional costs. This "absurd system" existed because antiquated regulations prohibited commerce with foreigners, even though the port of Savannah in Georgia was but 200 miles to the north, where flour could be purchased for five or six dollars a barrel. In reply to the Havana Consulado's desire to reinstate the previous regulations, Martínez de Yrujo criticized the old system as being "so contrary to sound policy and reason that posterity will hardly believe that it ever existed."[8]

At the time of Martínez de Yrujo's publication, a considerable number of elderly residents could still remember the absurd system that had created artificial scarcity and exorbitant prices for bread during the previous half century. Few, though, would doubt the positive changes that had occurred in their lives, even while the realities of living in the hurricane belt remained unchanged. Centuries of living with a perennial threat of disaster had prepared them for five decades of climatic stress, and the ability to survive previous disasters had given the people confidence that they could weather subsequent storms—knowledge that had been passed down to children and grandchildren over generations. At the same time, that certainty was tempered by a sense of survival, which transcended their obedience to metropolitan desires. In 1750, they suffered from irremediable food shortages because of an indifferent government that was far away in Havana or Madrid, and from an economic system that worked against their needs. When faced with crisis, time and again local residents took matters into their own hands and resorted to the contraband trade with regions that could provide life-saving provisions to ensure the community's survival.

Beginning at mid-century, when a warm anomaly occurred in worldwide temperatures, a series of El Niño/La Niña cycles struck the Spanish Caribbean. The accompanying environmental crisis that began in the 1750s generated tentative yet positive measures on the part of the Spanish government, which were implemented to cope with the unusual conditions. In 1762, weather-induced famine led to disease in the eastern portion of Cuba, along the southern coast of the island, and among the defenders in Havana, which made a quick and effective response to the British invasion of the island impossible. The shock of the occupation of Havana forced the Spanish crown to reevaluate every aspect of life in the empire to prevent such a disaster from happening again.

As a consequence, Charles III took an active role in promoting scientific advances by collecting information about his far-flung empire. This move toward explaining natural phenomena was notably secular, and scientific methodology replaced superstition in the decisions of royal administrators. In addition, recovery efforts became the responsibility of the state rather than the church, but the men sent to govern during Charles III's reign cooperated with religious authorities rather than working against them. While no one in the eighteenth century believed that the Divine Will could be thwarted, increasingly people looked toward the state for help in disaster's aftermath. In taking the initiative, for the first time the government shared the responsibility—and the popular gratitude—with the religious leaders.

Climatic stress continued at the end of the Seven Years' War, but an increasingly pragmatic and humanitarian attitude in Charles III's court meant that metropolitan responses would be geared toward mitigating the effects of disaster on colonial subjects. When another series of disasters struck in 1766, the major colonial powers responded to political and ecological crisis with what they believed were temporary measures to cope with shortages of provisions. Great Britain, France, and Holland created free-trade zones in their Caribbean colonies that essentially allowed any friendly nation to come and trade. Spain opposed opening her ports to foreigners, and the result was a hybrid mercantile system that embodied some degree of free trade while retaining vestiges of an antiquated monopoly system in the Compañía Gaditana de Negros or the Asiento.

The hurricane of Santa Teresa, on 15 October 1768, provided the first real test of the enlightened Spanish royal government. No area of royal response went unaffected, from measures to secure provisions to the humanitarian actions on the part of the captain general, Antonio María Bucareli, who personally visited villages and hamlets whose residents had suffered greatly from this very strong hurricane. Bucareli's actions unequivocally were mitigation efforts, and as a consequence, governmental response to crisis shifted from local and ecclesiastical authorities to the representatives of the modernizing state. In this, he set the precedent for subsequent captains general, and the degree to which each followed his example would determine the reaction on the part of the population. Ironically, once accustomed to looking to the state, the Cuban people held royal officials responsible and were outraged when future governments failed to behave in a manner that they had come to expect.

Beginning with the winter of 1771–72, drought alternating with excessive rainfall affected a wide area, extending from the Lesser Antilles to Central Mexico. The hurricane season of 1772 brought a record number of landfalls, which effectively destroyed Caribbean food production systems throughout the region. Cuba's provisioning needs could not be met by Mexico, which was suffering from an El Niño–triggered drought. Authorities on the island were forced to turn to foreign suppliers, primarily Philadelphia, for flour, which was transshipped via the Spanish cities of Cádiz, Ferrol, and La Coruña. The Spanish government intended that such reliance upon outsiders would be temporary, but as other areas recovered, bad weather and poor harvests continued in Cuba, in 1774, 1775, 1776, 1777, and 1778, and the island grew even more dependent upon provisions from the north. Demand came together with supply in Saint Domingue, where Spanish purchasers met openly with North American purveyors. The ability to market their agricultural products in the Spanish colonies was one reason why the American Patriots were willing to gamble on independence from Great Britain. By 1778, the Spanish declaration of free trade—the *Reglamento para el comercio libre*—was as much an acknowledgment of a trade that Spain did not want to stop as a move toward commercial freedom, while at the same time North American merchants recognized that they no longer needed Great Britain for their economic survival.

Besides its political consequences, the end of the war resulted in an unprecedented shift in trade patterns. For North Americans, the buoyant exhilaration of independence was accompanied by the hard realization that their wartime markets in the Caribbean had evaporated, when North American flour merchants were expelled from Havana in 1784. Once again, however, nature came to the United States' aid in the fourth consecutive decade of El Niño–generated environmental stress. For the most part, Cuba escaped the horrific consequences of catastrophe, but her satellite colonies, Florida and Louisiana, were not as fortunate. In characteristic domino effect, disaster in those areas affected Cuba and created an unprecedented drain on the island's resources. Worse still, her intended supplier, Mexico, suffered from an El Niño–triggered drought, which was even worse than that of the 1770s, causing a complete collapse of Mexican agricultural production. By 1785, Mexico could not supply her own residents, much less Cuba, and henceforth, the market on the island opened even wider to North Americans.

The 1790s was the fifth decade of ecological crisis to overlap with a period of political turmoil. Unlike previous decades of environmental crisis, however, by the 1790s the danger came from internal unrest in a population that was traumatized by forty years of unprecedented environmental stress rather than from the threat of foreign invasion. Political instability in the Caribbean was exacerbated by metropolitan ineptitude. In Spain, the death of Charles III in 1789 brought his unqualified son, Charles IV, and his venal group of advisers to power. That same year, the French Revolution devastated that country, and its effects were transferred to the Caribbean in the form of a slave revolt in Saint Domingue in 1791. Given the dire political situation throughout the Atlantic world, the onset of another severe period of El Niño/La Niña activity could not have come at a worse time.

The humane and compassionate reaction to crisis in the 1770s remained in the Cuban community's memory and served as a basis for its negative opinion of the royal representative, Captain General Luis de las Casas. Much of Las Casas's unpopularity stemmed from his intractable character, exacerbated by his inability to cope with another series of sequential hurricanes. By 1795, the island was plagued by widespread popular disturbances. Rebellion and resistance in Cuba spread to Louisiana and Florida, both suffering from the direct and collateral effects of environmental stress. Even more important, the inability of the administration in Havana to cope with both environmental crisis and political crisis seriously compromised the effectiveness of a Spanish expeditionary force sent to Saint Domingue to retake the colony from the republican army and rebellious slaves.

The Cuban regiments were ill-prepared to cope with the combination of poor judgment, a lack of provisions, and an environmental crisis that exacerbated the disease environment. The specter that this fatal combination played in the outcome of the campaigns of 1762, 1775, and 1781 hung over the army of operation. To no avail, the Cuban veterans warned their superior officers of the potential for defeat, especially since so many of them, such as Juan de Lleonart and Matías de Armona, had participated in the previous campaigns and had experienced firsthand the positive and negative consequences of the decisions of their leaders. Yet the political atmosphere that underpinned the war effort in 1794 was not 1781, or even 1762. An inexperienced and incompetent monarch sat on the Spanish throne without a strong body of advisers to provide him with guidance. Joaquín García was not Bernardo de Gálvez, whose clarity of judgment and fearless character had made bold action possible. Just as important, a group of commercial interests had come to challenge provisioning mechanisms for their own profit. Back home in Havana, rather than throwing open the port to North American vessels, the commercial interests sought to reduce the quantity of life-saving provisions that reached the troops. Worse still, the Cuban forces were forced to serve in an army that was at war with itself. From the beginning, the conduct of the Santo Domingo campaign was destined for disaster, and all of the Spanish troops in Santo Domingo suffered its fatal consequences.

Throughout this study, the influence of weather-induced crisis has occupied center stage in the analysis, all the while drawing comparisons with the existing historical narrative. Such comparisons could not have been accomplished without establishing the intersection of historical climatology and historical evidence. The science of historical climatology has grown by leaps and bounds—indeed, even within the decade required to produce this work. A growing number of scientific studies establish beyond a doubt that from 1750 through 1800 the northern hemisphere experienced warmer temperatures in relation to those of the previous centuries. Historical evidence—proxy data—from archives throughout the Atlantic basin demonstrate conclusively that severe weather events struck Caribbean communities in the region again and again. The evidence for these severe weather events may be compared and contrasted to studies of historical El Niño and La Niña cycles, which establish numerous instances of an El Niño and/or a La Niña event during the latter half of the eighteenth century. Proxy data for the Caribbean basin reveal a chronological correlation among the El Niño/La Niña cycles, drought in Mexico, hurricanes and drought in Cuba and its satellite cities, and the consequences in the Atlantic world. While other studies show the onset of an El Niño sequence in Chile and Mexico, the same signature hazards are undeniable in Cuba in 1766, in 1771–72, in the 1780s, and in the most powerful cycle beginning in 1791.

Interdisciplinary theories from a variety of other areas, such as political science, geography, sociology, environmental history, and disaster studies, bring depth and breadth to the analysis. In the vanguard are works in political science that demonstrate that disaster can be a force behind political change but that disasters do not necessarily have to become political. The authorities' behavior in the aftermath of disaster determines whether the population will react in a positive or a negative way. A complementary approach questions whether the disaster becomes the trigger that causes a "critical juncture" in political events. Clear-cut comparisons can be drawn between the measures enacted in the 1770s and the indifferent—even confrontational—response in the 1790s. Every captain general— Juan de Prado, Antonio María Bucareli, the Marqués de la Torre, Diego de Navarro, Bernardo de Gálvez, and Luis de las Casas—faced the effects of significant disasters. What makes their regimes different, one from another, is the way in which each leader dealt with the aftermath. Another post-disaster theoretical framework is how crisis in one area creates a domino or ripple effect in other areas. A portion of the community will be victims, but for others the crisis can provide unequaled opportunity. Certain sectors profited from shortages of food and labor, and slaves capitalized on the post-disaster confusion to escape bondage and form runaway communities. When the analysis is extended beyond the narrow, localized confines of Cuba to the Atlantic world, the domino effect of disaster can be used in conjunction with the principles of transnationality, an analytical tool that deemphasizes artificially created political boundaries and concentrates on forces (social movements, kinship networks, economic connections) that can cross arbitrarily created lines of demarcation. One obvious conclusion is that climate and catastrophe did not recognize national boundaries. Disaster in Cuba or in any of its subordinate colonies, in Florida or Louisiana, created a ripple effect that was felt everywhere in the interconnected Spanish imperial economy and even outside the Spanish empire into North America and other nations' islands of the Caribbean.

Sociological and anthropological research, especially that which examines the post-disaster confusion, also provides important theoretical frameworks. When the community suffered the effects of disaster's aftermath, the determinants of social ranking such as race or status became almost irrelevant, as rescue and recovery efforts took precedence over all else. Illegal activities that contributed to the stricken community's welfare were condoned by residents and authorities alike. Just as important, the culprits were often hailed as heroes rather than prosecuted by local officials.

When faced with environmental catastrophe in the 1770s, the need to cope with extremely dire circumstances as much as progressive thinking broke down the Spanish adherence to mercantilism. The movement toward free trade for the Spanish colonies reached its zenith at the same time that the Spanish crown debated entering the war against Great Britain. Also at the same time, the provisions trade from the north became jeopardized. Both events are too close to be coincidence. One of the reasons why the thirteen British colonies were willing to commit themselves to independence was that they had become accustomed to the lucrative market that trade with Cuba provided. This trade was possible because the island was forced to turn to foreign suppliers when the repeated El Niño cycles destroyed both Cuba's and Mexico's food production systems.

Two hundred years before academic inquiry sought to establish such linkages, Charles III and his representatives recognized the importance of effective disaster mitigation. Throughout his reign, royal officials responded quickly and decisively when crisis occurred. Illegal behavior such as smuggling and trading with off-limits areas that alleviated the consequences of disaster was tolerated, but behavior that offended community norms such as price gouging and speculation was severely punished. The hardship was shared by all sectors of society, from top to bottom, and Charles III set the example for his subjects by sharing the burden and by accepting reductions in his revenue in the form of concessions and tax relief. The generosity of the royal approach was not lost on the Cuban people, who expressed their appreciation clearly. As the island set out on the road to recovery in 1768, laudatory poetry dedicated to Captain General Bucareli praised the governor as "a magnificent and excellent leader."[9] Similar accolades accompanied de la Torre, whose departure in 1777 was "mourned by all who came under his gentle rule."[10] When Charles III died in 1788, Cubans praised their beloved monarch as "a beneficent sun that in his rotation transmitted his ardent kindness to his vassals. . . . Virtuous even into his old age[,] that same virtue made him worthy of the eminent place he had obtained on earth."[11] No such accolades accompanied Luis de las Casas's recall from Cuba in 1796; instead, his departure was celebrated by the people. Fortunately, the undeniable contrast in the metropolitan response to disaster in the Caribbean allows comparisons to be drawn between the policies of Charles III and those of his son.

Neither the social dynamic of the local community nor the larger, imperial structures provide an adequate explanatory framework for many events in late-eighteenth-century Caribbean and Atlantic world history. Similarly, studies of the history of disaster focus upon immediate casualties and rarely look beyond the immediate to the long-range consequences. Together, however, they offer a new conceptual framework to reinterpret historical processes. The undeniable existence of a climate shift, the political, economic, and social consequences of disaster, and a wider understanding of Cuba's place in the Atlantic world are the issues that most inform this research. This book has established the environmental reasons for why so many governors could exercise their autonomy when faced with a catastrophic situation. Disaster meant that royal officials were forced to respond in a positive manner or suffer the consequences. Positive metropolitan responses worked to Cuba's advantage. Whether individually or collectively, the effects of hurricanes—on commerce, on policy, and on military campaigns—remain understudied in the historical literature. From an Atlantic world perspective, with Cuba at its nexus, however, these consequences can no longer be ignored. Such a perspective provides a compelling challenge to existing historiography to uncover the reasons why this period was a critical juncture in history.

A Chronology of Alternating Periods of Drought and Hurricanes in Cuba and the Greater Caribbean, Juxtaposed with Major Historical "Events," 1749–1800

Fall 1749	Hurricane	Atlantic—NE of Cuba[1]
Aug. 1750	Hurricane	Atlantic—NE of Cuba[2]
Oct. 1751	Hurricane	Havana province[3]
Winter 1751–52	Drought	Oriente[4]
Fall 1752	Hurricanes (2)	Oriente[5]
Fall 1753	Storms	Oriente[6]
Spring 1754	Drought	Oriente[7]
Oct. 1754	Hurricane	Oriente[8]
Spring 1755	Drought	Oriente and Puerto Príncipe[9]
Aug.–Nov. 1755	Hurricane	Oriente[10]
Feb. 1756	Winter storm	South coast[11]
Oct. 1756	Hurricane	Havana province[12]
Nov. 1758	Drought	Puerto Príncipe[13]

January 1759: Charles III Ascends to the Throne of Spain

Sept. 1759	Hurricane	Puerto Príncipe[14]
Oct. 1760	Hurricane	Oriente[15]
Oct. 1761	Hurricane	Oriente (Bayamo)[16]

December 1761: Spain Enters the Seven Years' War

May–Aug. 1762	Storms	Islandwide[17]

August 1762–July 1763: Siege, Fall, and Occupation of Havana
July 1763: Return of Spanish Rule

Apr. 1765	Drought	Oriente[18]
June 1765	Drought	Havana province[19]

October 1765: First Declaration of Limited Comercio Libre
1765: Creation of the Real Compañía de Comercio de Cádiz (Asiento)

Aug. 1766	Hurricane	Puerto Rico[20]
Sept. 1766	Hurricane	Puerto Rico[21]
Sept. 1766	Hurricane	Louisiana[22]
Oct. 1766	Storms	Havana province[23]

Oct. 1766	Hurricane	Puerto Rico[24]
Oct. 1766	Hurricane	Oriente[25]
Oct. 1766	Hurricane	Louisiana[26]
Oct. 1768	Catastrophic hurricane	Havana province[27]
Aug. 1769	Drought	Oriente[28]
Feb.–Oct. 1770	Drought	Cuba, Caribbean littoral[29]
Jan. 1771	Winter storm	Central Gulf of Mexico[30]
Jan.–June 1772	Drought	Oriente to Havana[31]
June 1772	Hurricane	At sea N of Puerto Rico[32]
July 1772	Hurricane	Puerto Rico[33]
1–7 Aug. 1772	Hurricane	Hispaniola, Santiago de Cuba, Bayamo, Havana[34]
29 Aug.–4 Sept. 1772	Hurricane	Puerto Rico, Hispaniola, S coast of Cuba, Yucatán peninsula, Mobile, and New Orleans[35]
31 Aug.–3 Sept. 1772	Catastrophic hurricane	Leeward Islands, Puerto Rico, N coast of Cuba[36]
Jan. 1773	Winter storm	Oriente[37]

January 1773: Bankruptcy and First Reorganization of the Asiento

Nov. 1773	Drought	Matanzas[38]
Dec. 1773	Drought	Caribbean littoral[39]
Oct. 1774	Hurricane	Havana province[40]
Jan. 1775	Drought	Havana province[41]

May 1775: Second Reorganization of the Asiento

Aug. 1775	Catastrophic hurricane	Oriente[42]
June 1776	Hurricane	S coast of Cuba[43]
June 1776	Hurricane	New Orleans[44]
May–July 1776	Storms	N coast of Cuba, Hispaniola, and Jamaica[45]
Aug. 1776	Hurricane at sea	N coast of Cuba[46]

February 1776: José de Gálvez Authorizes Trade with Patriots
May 1776: Miguel Eduardo Leaves for Philadelphia
July 1776: American Declaration of Independence

Feb.–May 1777	Drought	Havana and Matanzas[47]
July–Aug. 1777	Storms	Matanzas[48]

August 1777: Charles III Issues Royal Order to Send a Representative to the United States

Oct. 1777	Storms	Trinidad[49]
Oct. 1777	Hurricane	Oriente–Saint Domingue[50]
Winter 1777–78	Drought	Puerto Príncipe and Cuatro Villas[51]

January 1778: Juan de Miralles Leaves for the United States
February 1778: Spanish Declaration of Comercio Libre for the Viceroyalty of Río de la Plata (Argentina)

Aug.–Sept. 1778	Hurricane	Havana province[52]

October 1778: Spanish Declaration of Comercio Libre for the Empire
October 1778: Miralles Forms Commercial Partnership with Robert Morris in Philadelphia

May 1779: Spain Enters the War against Great Britain

Aug. 1779	Hurricane	Havana province, Louisiana[53]
Feb. 1780	Winter storm	Havana province, Louisiana[54]
Apr.–May 1780	Drought	Puerto Príncipe[55]
July 1780	Storms	Havana province[56]
Aug. 1780	Storms	Havana province[57]
Oct. 1780	Hurricane	Puerto Príncipe[58]
10–16 Oct. 1780	Catastrophic hurricane I	Lesser Antilles[59]

October 1780: First Spanish Attempt against Pensacola

16 Oct. 1780	Catastrophic hurricane II	Gulf of Mexico[60]
25 Oct. 1780	Storms	Saint Domingue[61]

February 1781: Second Attempt against Pensacola
May 1781: Fall of Pensacola
October 1781: Cornwallis's Surrender at Yorktown
February 1782: Spanish Capture of New Providence (Nassau)

Feb. 1782	Winter storm	Matanzas[62]

April 1782: Spanish Attempt against Jamaica

July 1782	Hurricane	Havana province[63]

February 1783: Treaty of Paris Ends the War
July 1784: Florida Returned to Spanish Sovereignty

June 1784	Hurricane	Straits of Florida, Havana province[64]
July 1784	Storm	Florida[65]

December 1788: Death of Charles III
December 1788: Charles IV Ascends to the Spanish Throne

Feb.–June 1791	Drought	Islandwide[66]
June 1791	Hurricane	Havana province[67]
Jan.–May 1792	Drought	Santa Clara[68]
Oct. 1792	Hurricane	Havana province[69]
Nov. 1792	Storms	Oriente[70]
Jan.–Mar. 1793	Drought	Havana province[71]

January 1793: Spain Declares War against Republican France
February 1794: Cuban Forces Arrive in Santo Domingo

Winter 1793–94	Storms	Havana province[72]
Winter 1793–94	Storms	Santo Domingo[73]
Aug. 1794	Hurricane I	Louisiana[74]
Aug. 1794	Catastrophic hurricane	Havana province[75]
Aug. 1794	Hurricane II	Louisiana[76]
Nov. 1795–Aug. 1796	Drought	Havana province[77]

July 1795: Spain Signs Treaty of Basel Ending War with France
August 1796: Spain Declares War on Great Britain

Oct. 1796	Hurricane	Pinar del Río[78]
Oct. 1799	Hurricane	Oriente[79]

. . . .

Sources for the Maps

Sources for Map 3.1. Hurricane strikes in the Caribbean basin,
from August to October 1766

Strike 1: Martinique, August 1766. Millás, *Hurricanes of the Caribbean*, 219.

Strike 2: Puerto Rico, August 1766. Miner Solá, *Huracanes en Puerto Rico*, 30;
Millás, *Hurricanes of the Caribbean*, 219.

Strike 3: Jamaica, August 1766. Millás, *Hurricanes of the Caribbean*, 219.

Strike 4: Havana, August 1766. Bucareli to Cagigal, Havana, 2 September 1766,
expediente 39, legajo 19, CCG, ANC.

Strike 5: Louisiana, August 1766. Ludlum, *Early American Hurricanes*, 62–63;
Weddle, *Changing Tides*, 10–23.

Strike 6: Montserrat, September 1766. Millás, *Hurricanes of the Caribbean*, 219.

Strike 7: St. Christophers, September 1766. Millás, *Hurricanes of the Caribbean*,
219–20.

Strike 8: St. Eustatius, September 1766. Millás, *Hurricanes of the Caribbean*, 219–20.

Strike 9: Puerto Rico, September 1766. Miner Solá, *Huracanes en Puerto Rico*, 30.

Strike 10: Hispaniola, September 1766. Millás, *Hurricanes of the Caribbean*, 219–20.

Strike 11: Tortuga, September 1766. Millás, *Hurricanes of the Caribbean*, 219–20.

Strike 12: Guadaloupe (20-foot surge), October 1766. Millás, *Hurricanes of the
Caribbean*, 224; Miner Solá, *Huracanes en Puerto Rico*, 30.

Strike 13: Puerto Rico, October 1766. Millás, *Hurricanes of the Caribbean*, 224;
Miner Solá, *Huracanes en Puerto Rico*, 30.

Strike 14: Oriente, October 1766. Bartolomé de Morales to Governor of Cuba,
Santiago del Prado, 12 October 1766, expediente 62, legajo 24, CCG, ANC.

Strike 15: Louisiana, October 1766. Weddle, *Changing Tides*, 10–23; Ludlum,
Early American Hurricanes, 62–63.

Sources for Map 4.2. Hurricane strikes in the Caribbean basin,
from June to September 1772

Strike 1: At sea (*La Amable*), June 1772. Ayans de Ureta to de la Torre, Santiago de
Cuba, 25 June 1772, legajo 1141, PC, AGI.

Strike 2: North coast of Puerto Rico, July 1772. Millás, *Hurricanes of the Caribbean*,
229–30.

Strike 3: South of Hispaniola, August 1772, Millás, *Hurricanes of the Caribbean*, 230.

Strike 4: Oriente, August 1772. Josef Alvarado to de la Torre, Bayamo, 22 August 1772, legajo 1178; Ayans to de la Torre, Santiago de Cuba, 23 August, 4 September 1772, legajo 1141; de la Torre to Arriaga, Havana, 23 August 1772, legajo 1216, all PC, AGI.

Strike 5: Havana Province, August 1772. De la Torre to Arriaga, Havana, 23 August 1772, legajo 1216; Lleonart to de la Torre, Puerto Príncipe, 24 August 1772, legajo 1168; de la Torre to Lleonart, Havana, 11 September 1772, legajo 1172; de la Torre to lieutenant governor of Baracoa, Havana, 19 October 1772, legajo 1143, all PC, AGI.

Strike 6: Windward Islands, August 1772. Millás, *Hurricanes of the Caribbean*, 230; Miner Solá, *Huracanes en Puerto Rico*, 30.

Strike 7: Puerto Rico, August 1772. Joaquín Pover to the Council of the Indies, San Juan, 10? September 1772, legajo 2516, SD, AGI; Millás, *Hurricanes of the Caribbean*, 230.

Strike 8: At sea, September 1772. José Antonio Armona to the Council of the Indies, Havana, 30 March 1774, legajo 257-A, Correos, AGI; Millás, *Hurricanes of the Caribbean*, 230; Miner Solá, *Huracanes en Puerto Rico*, 30.

Strike 9: At sea between Cuba and Jamaica, August 1772. Armona to the Council of the Indies, Havana, 31 August, 19 October 1772, legajo 257-A, Correos; Ayans to Arriaga, Santiago de Cuba, 3 October 1772, legajo 1216; Ayans to de la Torre, Santiago de Cuba, 23 August, 4 September 1772, legajo 1141, PC, both AGI.

Strike 10: Western Cuba and at sea, August 1772. Armona to the Council of the Indies, Havana, 31 August, 19 October 1772, legajo 257-A, Correos, AGI.

Strike 11: Louisiana, August 1772. Unzaga to de la Torre, New Orleans, 9 September 1772, "Despaches of Spanish Governors," typescript and translation in Manuscripts Department, Tulane University Libraries; Ludlum, *Early American Hurricanes*, 63–64.

Strike 12: Windward Islands, late August 1772. Millás, *Hurricanes of the Caribbean*, 235–37; *Pennsylvania Gazette*, 23 September, 14 October 1772, LCP.

Strike 13: Virgin Islands, late August 1772. Millás, *Hurricanes of the Caribbean*, 235–37; *Pennsylvania Gazette*, 23 September, 14 October 1772, LCP.

Strike 14: Puerto Rico, early September 1772. Muesas to the Council of the Indies, San Juan, (day omitted) September 1772, legajo 2516, SD, AGI.

Strike 15: Bahamas, September 1772. Ayans de Ureta to de la Torre, Santiago de Cuba, 4 September 1772, legajo 1141, PC, AGI; Ludlum, *Early American Hurricanes*, 64.

Strike 16: North Carolina, September 1772. Ludlum, *Early American Hurricanes*, 64.

Sources for Map 5.1. Hurricane strikes in the Caribbean basin,
February 1780 and from June to October 1780

Strike 1: Havana, February 1780. Navarro to José de Gálvez, Havana, 25 February
  1780, legajo 2082; Diario formado por Estéban Miró, Havana, 1780, legajo 2543,
  both SD, AGI.
Strike 2: Between New Orleans and Mobile, February 1780. Martín de Navarro to
  Diego de Navarro, New Orleans, 18 April 1780, legajo 2609, SD, AGI.
Strike 3: Puerto Rico, June 1780. Rappaport and Fernández-Partagás, "The Deadliest
  Atlantic Tropical Cyclones."
Strike 4: Western Cuba, July–August 1780. Navarro to the Captain of Managua,
  Havana, 21 July1780; Various Residents of Arroyo Arenas to Navarro, Arroyo
  Arenas, 19 August 1780, legajo 1269, PC, AGI.
Strike 5: Jamaica and Puerto Príncipe, October 1780. Millás, *Hurricanes of the
  Caribbean*, 240; Ventura Díaz to Navarro, Puerto Príncipe, 11 October 1780; Juan
  Nepomuceno de Quesada to Navarro, Puerto Príncipe, 16 November 1780, both
  legajo 1256, PC, AGI.

The Great Hurricane of 1780
Strike 6: At sea east of Barbados. October 1780, Francisco de Saavedra, *Journal*, 9–15
  October 1780.
Strike 7: Barbados, October 1780. Millás, *Hurricanes of the Caribbean*, 241–49.
Strike 8: Martinique, October 1780. Millás, *Hurricanes of the Caribbean*, 241–49.
Strike 9: Guadaloupe, October 1780. Millás, *Hurricanes of the Caribbean*, 241–49.
Strike 10: Jamaica, October 1780. Millás, *Hurricanes of the Caribbean*, 241–49;
  Matthew Mulcahy, *Hurricanes and Society*, 108–11, 166–74.
Strike 11: At sea in the Gulf of Mexico, October 17–19. Ramon Lloret to Navarro,
  Campeche, 5 November 1780, legajo 1248, PC, AGI; Millás, *Hurricanes of the
  Caribbean*, 260–62.
Strike 12: Saint Domingue, October 1780. Governor of San Nicolás to Navarro,
  San Nicolás (Saint Domingue), 25 October 1780, legajo 1231, PC, AGI.

NOTES

Abbreviations Used in Notes

AGI	Archivo General de Indias
AGS	Archivo General de Simancas
AHM	Archivo Histórico Municipal
AHN	Archivo Histórico de la Nación
ANC	Archivo Nacional de Cuba
AP	Asuntos Políticos
BAN	*Boletín del Archivo Nacional de Cuba*
BNE	Biblioteca Nacional de España
BNJM	Biblioteca Nacional de Cuba José Martí
CCG	Correspondencia del Capitán General
EFP	East Florida Papers
Escoto Collection	Jose Escoto Collection
exp.	expediente (folder)
GM	Secretaría de Guerra Moderna
HSP	Historical Society of Pennsylvania
IG	Indiferente General
LC	Library of Congress
LCP	Library Company of Philadelphia
leg.	legajo (bundle)
PC	Papeles Procedentes de Cuba
PKY	P. K. Yonge Library Special Collections
PZ	*Pennsylvania Gazette*
SD	Audiencia de Santo Domingo
TC	Tribunal de Cuentas
Topping Collection	Aileen Moore Topping Collection

Chapter One

1. Hobsbawm, *Age of Revolution*.

2. Gaspar and Geggus, *A Turbulent Time*.

3. Caviedes, *El Niño in History*, 167, 206.

4. Schwartz, "The Hurricane of San Ciriaco"; Pérez, *Winds of Change*; Mulcahy, *Hurricanes and Society*; Steinberg, *Acts of God*; Johnson, "Rise and Fall"; Johnson, "El Niño, Environmental Crisis"; Johnson, "St. Augustine Hurricane of 1811"; Post, *Last Great Subsistence Crisis*.

5. Mann et al., "Proxy-Based Reconstructions," 132–52; Jones and Mann, "Climate over Past Millennia," 1–42. The number of studies of the northern and southern hemispheres

grows daily and may be accessed in the National Climate Data Center ⟨www.ncdc.noaa
.gov/paleo/html⟩.

6. Geographers and climatologists date the Little Ice Age as lasting from about 1450
through about 1850. Quinn and Neal, "Historical Record of El Niño Events"; Caviedes,
El Niño in History; Fagan, *Little Ice Age*; Pfister, Brazdil, and Glaser, *Climatic Variability in
Sixteenth-Century Europe*. See also NOAA's website on historic hurricanes, ⟨http://www
.nhc.noaa.gov⟩.

7. Climatologists have verified the minute upward spike in the earth's temperature in
the mid-eighteenth century in scientific sources such as deep-core samples of glaciers, soil
samples, and dendrochronology. See S. Huang, "Integrated Northern Hemisphere Surface Temperature Reconstruction"; and Jones and Mann, "Climate over Past Millennia."
The authors warn that the term "Little Ice Age" may be oversimplistic and suggest specific
studies to determine variations within the overall cooler phase, which lasted nearly 500
years.

8. The National Hurricane Center in Miami, ⟨www.nhc.nooa.gov⟩, cites proxy data in
Lloyd's List. Extant issues from 1741–84 and 1790–97 from Gregg International Publishers
Limited (1969), Westmead, Farnborough, Hants., England, as evidence of an increased
number of ships lost at sea, are one source to point to severe weather patterns in the late
eighteenth century.

9. For example, see the abstracts in Ortlieb and Macharé, *Paleo-ENSO Records*; and
Caviedes, *El Niño in History*, 42.

10. Bridgman, Oliver, and Glantz, *Global Climate System*; Diaz and Markgraf, *El Niño
and the Southern Oscillation*; Glantz, *Currents of Change*; Glantz, *Drought and Hunger in
Africa*; Caviedes, *El Niño in History*.

11. Caviedes, *El Niño in History*, 11, 120.

12. Gergis and Fowler, "A History of ENSO Events," 369–70.

13. Ibid., 369.

14. Ibid., 361, 372.

15. Ibid. The authors' results are summarized in a comprehensive table, pp. 367–72, and
are supported by their bibliography citing 120 secondary sources published in the last
two decades. Historians will be familiar with their analytical framework, which employs
historians' "rule of three," that is, the practice of using three examples at a minimum to
come to a valid conclusion.

16. Watts, *West Indies*; Richardson, *Economy and Environment in the Caribbean*.

17. Claxton, "Record of Drought"; Claxton, "Climatic and Human History in Europe
and Latin America"; Claxton, "Climate and History."

18. Cronon, *Changes in the Land*; Cronon, "Modes of Prophecy"; Crosby, *Ecological
Imperialism*; Worster, *Wealth of Nature*; Worster, *Ends of the Earth*. For Latin America, see
Melville, *Plague of Sheep*; Dean, *With Broadax and Firebrand*; Funes Monzote, *From Rain-
forest to Cane Field*; Radding, *Landscapes of Power and Identity*; and Radding, *Wandering
Peoples*.

19. Cronon, "Modes of Prophecy," 1122–31; Worster, *Ends of the Earth*.

20. Juan and Ulloa, *Noticias secretas*; Juan and Ulloa, *Voyage to South America*. For modern examinations of the scientific revolution in the eighteenth century, see Barrera, *Expe-*

riencing Nature; Casado Arbonés, "Bajo el signo de la militarización"; Lucena Salmoral, "Las expediciones científicas," 49–63; Puig-Samper Mulero and Valero, *Historia del jardín botánico de la Habana*; Puig-Samper Mulero, "Las primeras instituciones científicas en Cuba," 9–33; Álvarez Cuartero, "Las sociedades económicas de Amigos del País en Cuba," 36–39; Misas Jiménez, "La real sociedad patriótica de la Habana," 75–77; Guirao de Vierna, "La comisión real de Guantánamo," 85–87; Weddle, *Changing Tides*, 10–23; Solano Pérez-Lila, *La pasión de reformar*, 57–138; and Cañizares-Esquerra, *Nature, Empire, and Nation*.

21. Pfister, "Learning from Nature-Induced Disasters," 17–20.

22. Olson and Gawronski, "Disasters as Critical Junctures?"; Olson, "Towards a Politics of Disaster"; Olson and Drury, "Un-Therapeutic Communities"; Drury and Olson, "Disasters and Political Unrest"; Lobdell, "Economic Consequences of Hurricanes in the Caribbean"; García Acosta, "Introduction," 1:15–37; Oliver-Smith, "Anthropological Research on Hazards and Disasters," 303–28; Oliver-Smith, "Theorizing Disasters," 23–47; Prieto, "The Paraná River Floods," 285–303.

23. Capoccia and Keleman, "The Study of Critical Junctures," 341–69; Olson and Gawronski, "Disasters as Crisis Triggers," 32–33.

24. Johnson, "El Niño, Environmental Crisis"; Johnson, "Rise and Fall."

25. Peacock, Morrow, and Gladwin, *Hurricane Andrew*; Kreps, *Social Structure and Disaster*; Kreps, "Sociological Inquiry and Disaster Research"; Provenzo and Provenzo, *In the Eye of Hurricane Andrew*; Oliver-Smith, "Anthropological Research on Hazards and Disasters," 303–28; Oliver-Smith, "Theorizing Disasters," 23–47.

26. Olson and Gawronski, "Disasters as Crisis Triggers," 10–11.

27. Pfister, "Learning from Nature-Induced Disasters," 17–20.

28. Schwartz, "Hurricanes and the Shaping of Circum-Caribbean Cultures." See also Olson and Gawronski, "Disasters as Crisis Triggers," 33.

29. Jones and Mann, "Climate over Past Millennia," 3–4; Pfister et al., "Documentary Evidence as Climate Proxies"; Selkirk, "The Last Word on Climate Change," 48–49.

30. Olson and Gawronski, "Disasters as Critical Junctures?" 5–35. The warnings expressed by scholars are well taken, and this study avoids the pitfalls of reading too much significance into what often are impressionistic accounts of suffering. Yet reaching conclusions about the impact of a particular disaster is not as difficult as it may seem. By the 1760s, and definitively after the hurricane in Cuba in 1768, reports from every town and village impacted by the storm allow geographic reconstruction of the hurricane's trajectory and extent of impact. Frequently, concrete measurements are available; for example, after the same storm, harbor pilots and royal engineers both reported that the water level in Havana rose by 6 varas, or 18 feet. By comparing such figures with modern measurements, it is possible to suggest that the 1768 hurricane was at least a Category 4 or Category 5 storm. Conversely, the absence of any mention of a storm surge during the hurricane of 1791 combined with innumerable reports of days of continuous rainfall suggests that if the tempest was a hurricane at all it was likely only a Category 1. In other instances, such as events in Oriente province in 1762 prior to the British attack on Havana and again in 1776 after months of terrible weather, the onset of epidemic diseases is too close in time

to be mere coincidence. Only when such verifiable historical data were recoverable were conclusions reached.

31. Arrate, *Llave del nuevo mundo*, 70; Torre, *Lo que fuimos*, 102–3; Wright, *Early History of Cuba*, 20–55; Guerra y Sánchez, *Manual de historia*, 22–32; Piño-Santos, *Historia de Cuba*, 21–30.

32. Valdés, *Historia de la isla de Cuba*, 261–63; Wright, *Early History of Cuba*, 20–33.

33. Arrate, *Llave del nuevo mundo*, 102; Valdés, *Historia de la isla de Cuba*, 297–304.

34. McNeill, *Atlantic Empires*, 38, 126–29; Marrero y Artiles, *Cuba*, 7:11–20; Guerra y Sánchez, *Manual de historia*, 151.

35. Cook, *Born to Die*, 15–59, is the definitive treatment of the topic.

36. De la Sagra, *Historia económica-política*; Knight, *Slavery and the Transformation of Society*, 16–17; Inglis, *Constructing a Tower*. For a view of the dimension of error in all censuses in general, see Johnson, *Social Transformation*, 185–86.

37. Hoffman, *Spanish Crown*; Lyon, *Enterprise of Florida*.

38. Guerra y Sánchez, *Manual de historia*, 79; Torres Ramírez, *La armada de barlovento*.

39. Allan J. Kuethe, "Havana in the Eighteenth Century," in Knight and Liss, *Atlantic Port Cities*, 24. Kuethe maintains that Mexican elites "complained that their silver disappeared into Havana's financial maze." See also Marichal and Souto Mantecón, "Silver and Situados," 604; Kuethe, "Guns, Subsidies, and Commercial Privilege," 130; TePaske, "La política española en el Caribe," 79–82; and Le Riverend Brusone, *Historia económica*, 143–44. From 1780 through 1784, fully three-quarters of the money remitted to Spain from the Mexican treasury (33,346,972 pesos of 46,666,505 remitted) stayed in Havana. "Relación de valores y distribución de la real Hacienda de Nueva España en quinquenio de 1780 a 1784," box 6, folder 4, Domingo Del Monte Collection, Manuscript Division, LC.

40. G. Douglas Inglis, "The Spanish Naval Shipyard at Havana in the Eighteenth Century," in Inglis, *New Aspects of Naval History*, 47–58; Ortega Pereyra, *La construcción naval en la Habana*; Valdés, *Historia de la isla de Cuba*, 281–89; Le Riverend Brusone, *Historia económica*, 65–68. Marrero y Artiles, *Cuba*, 7:134–35, 8:15–22, describes the shipyard as the "orgullo de la Habana" ("pride of Havana").

41. Rivero Muñíz, *Tabaco*, 2:1–10; McNeill, *Atlantic Empires*, 156–72; Marrero y Artiles, *Cuba*, 7:41–92; Le Riverend Brusone, *Historia económica*, 54, 94–97; Arrate, *Llave del nuevo mundo*, 150–51; Guerra y Sánchez, *Manual de historia*, 140.

42. Moreno Fraginals, *El ingenio*, 1:95–102; Le Riverend Brusone, *Historia económica*, 108–48; Guerra y Sánchez, *Manual de historia*, 145–46; Thomas, *Cuba*, 49–52. Recent revisionist interpretations include McNeill, *Atlantic Empires*, 156–72; Marrero y Artiles, *Cuba*, 7:1–23; Johnson, *Social Transformation*; and Pérez, *Winds of Change*.

43. Le Riverend Brusone, *Historia económica*, 60–64.

44. Ibid., 70–110.

45. McNeill, *Atlantic Empires*, 117–22; Marrero y Artiles, *Cuba*, 7:102–65.

46. Nelson, "Contraband Trade under the Asiento," 55–67.

47. Kuethe, *Cuba*, chap. 1.

48. "Ordenanza Municipial de la Havana," box 3, folder 1, Domingo del Monte Collection, Manuscript Division, LC; Actas del Ayuntamiento (Town Council), vol. 11, Santiago de Cuba, 25 June 1774, AHM.

49. Portuondo Zúñiga, "Introducción," in *Nicolás José de Ribera*, 46–47.

50. Olson's recent typology of the "blame game" identifies this approach as the loss of the mandate of heaven; see Olson, "Towards a Politics of Disaster," 267–68.

51. Casado Arbonés, "Bajo el signo de la militarización"; Lucena Salmoral, "Las expediciones científicas," 49–63. See also Favier and Granet-Abisset, "Society and Natural Risks in France."

52. Solano Pérez-Lila, *La pasión de reformar*, 105–79; Gomis Blanco, "Las ciencias naturales," 308–19; González-Ripoll Navarro, "Voces de gobierno," 149–62; Misas Jiménez, "La real sociedad patriótica de la Habana," 74–89; Weddle, *Changing Tides*; Sellés García, *Navegación astronómica*; Gutiérrez Escudero, *Ciencia, economía y política*.

53. Weddle, *Changing Tides*; Casado Arbonés, "Bajo el signo de la militarización," 25; Cañizares-Esquerra, *Nature, Empire, and Nation*; Barrera, *Experiencing Nature*.

54. Weddle, *Changing Tides*.

55. Indeed, it would not be until the nineteenth century that advances were made to become what we recognize as modern meteorology. See Viñes, *Investigaciones relativas a la circulación*.

56. José Antonio Armona, Report, Havana, 13 November 1771, leg. 257-A, Correos, AGI.

57. Ulloa's *Observations* overshadowed his political career as captain general of Peru and later of Louisiana. Even after his ignominious departure from New Orleans in 1768, Ulloa went on to serve Charles III as admiral in the war with Great Britain in 1779. Ulloa, like many of his colleagues, brought along his long experience with weather. He also exemplified the royal attitude toward gathering scientific information and passing it on to Spain. See Solano Pérez-Lila, *La pasión de reformar*.

58. Men who served as captain general of Cuba—such as Juan Güemes de Horcasitas, the Conde de Revillagigedo; Juan Manuel de Cagigal, the Marqués de Casa Cagigal; Miguel de Muesas; and the Madariaga brothers, Lorenzo and Juan Ignacio—are examples.

59. Report of José Antonio Armona, Havana, 17 October 1773, leg. 256-A, Correos, AGI.

60. *Instrucción que deben observar los Patrones-Pilotos de los Paquetbotes destinados al Correo mensual entre España, y las Indias-Occidentales*, 1764, leg. 1212, SD, AGI.

61. *Ampliación a los ynstrucciones antecedentes relativas a capitanes y pilotos de los correos maritimas*, San Lorenzo el Real, 24 October 1772, leg. 257-A; Junta de Pilotos de la Havana, 10 October 1775, leg. 257-B; both Correos, AGI. Miscellaneous Legal Instruments and Proceedings, 1784–1819, 19 September 1787, bundle 261N5, EFP.

62. Report of José Fuertes, 16 August 1791, 25 October 1790, leg. 260-A, Correos, AGI.

63. José de Gálvez, Circular Order, Correspondence with the Captain General, Havana, 11 July 1784, bundle 40, EFP.

64. *Havana, 29 de Agosto de 1794*, PKY.

65. Luis de las Casas to the Ayuntamiento de Santa Clara, Havana, 22 July 1791, Actas Capitulares del Ayuntamiento, Archivo Provincial de Santa Clara, Santa Clara, Cuba.

66. Lavedan, *Aforismos*, 155–60.

67. Ibid.

68. "Sobre la compra y pago de terrenos y solares extramuros de esta ciudad," 1773, foxas 224, libro 6, exp. 1334, TC, ANC, in *BAN* 10 (May–June 1911), 130–31.

69. *PZ*, 21 February 1771, LCP.

70. "Instrucciones del Conde de Ricla a los capitanes del partido," Havana, 9 October 1763, 20 October 1763, folder 8, box 8, Escoto Collection.

71. Arcos y Moreno to Cagigal, Santiago de Cuba, 13 December 1752, exp. 104, leg. 6; Cagigal to Arcos y Moreno, Havana, 10 March 1753, exp. number omitted, leg. 6; both CCG, ANC.

72. Gala, *Memorias de la colonia francesa de Santo Domingo*, 154–55.

73. Ibid., 156.

74. Lavedan, *Aforismos*, 167–68.

75. Johnson, "La guerra contra los habitantes," 181–209.

76. *Papel Periódico de la Havana*, 7 August 1791, Colección Cubana, BNJM; Millás, *Hurricanes of the Caribbean*, 284–86.

77. Armona to the Council of the Indies, Havana, 28 October 1768, leg. 256-B, Correos, AGI.

78. Navarro, "Bando sobre que se destechen las Casas de Guano." In Cuba, "guano" refers to palm leaves used for thatch. José de Rivera to Juan Ignacio de Urriza, 16 February 1786, BAN, 53–54 (1954–55), 278. In 1754, 14 percent of Havana's houses were constructed of guano, as compared to Santiago de Cuba with 29 percent and Matanzas with 97 percent. Of thirty-three urban centers, in fourteen every house was constructed of guano. Marrero y Artiles, *Cuba*, 8:224.

79. Knowles, "Description of the Havana," 27; Leandro S. Romero, archaeologist of the City of Havana, personal communication, 1993.

80. Lavedan, *Aforismos*, 160.

81. "Particulars of a Most Violent Hurricane, Which Happened in the Bay of Honduras, on the 2d day of September Last," 9 January 1788, *PZ*, LCP.

82. Anderson, "Cultural Adaptation to Threatened Disaster," 300–305; Louis A. Pérez Jr. argues that Cuban resilience in the nineteenth century was so powerful that it "insinuates into the calculus of nation." Pérez, *Winds of Change*, 140, 146, 155.

83. Gist and Lubin, *Psychological Aspects of Disaster*.

84. Lavedan, *Aforismos*, 78–86.

85. Duffy, *Sword of Pestilence*; Geggus, *Slavery, War, and Revolution*, 347–72; Earle, "A Grave for Europeans," 283–97; Curtin, *Death by Migration*.

86. Romay, *Disertación sobre la fiebre maligna*, n.p.

87. The most complete treatment of the connection between diet and disease in the Caribbean is Kiple, *Caribbean Slave*, 76–103. Although focused on the slave population, Kiple's insights are equally applicable to all residents of the region. See also Myllyntaus, "A Natural Hazard with Fatal Consequences," 87–90, which discusses the debate over mortality during famines. See also Post, *Food Shortage, Climatic Variability, and Epidemic Disease*; and Walter and Schofield, *Famine, Disease and the Social Order*.

88. Pringle, *Observations on the Diseases of the Army*; Sims, *Observations on Epidemic Disorders*; Dancer, *A Brief History of the Late Expedition*; Romay, *Disertación sobre la fiebre maligna*; Hillary, *Observations on the Changes of the Air*; King, *Medical World of the Eighteenth Century*. Recent studies include Curtin, *Death by Migration*.

89. Juan and Ulloa, *Voyage to South America*, 90.

90. Bucareli to Ayans, Havana, 18 May 1771, exp. 151, leg. 19; Casa Cagigal to Morales, Santiago de Cuba, 21 April 1771, exp. 64, leg. 24; both CCG, ANC.

91. Rapún, *Reglamento para el gobierno interior, político, y económico de los hospitales reales.*

92. Captain Davidson, Cádiz, 11 January 1775, in *PZ*, 8 February 1775, LCP.

Chapter Two

1. Juan and Ulloa, *Noticias secretas*; Juan and Ulloa, *Voyage to South America*. See also Barrera, *Experiencing Nature*; Casado Arbonés, "Bajo el signo de la militarización"; Lucena Salmoral, "Las expediciones científicas," 49–63; Puig-Samper Mulero and Valero, *Historia del jardín botánico de la Habana*; Puig-Samper Mulero, "Las primeras instituciones científicas en Cuba," 9–33; Álvarez Cuartero, "Las sociedades económicas de Amigos del País en Cuba," 36–39; Misas Jiménez, "La real sociedad patriótica de la Habana," 75–77; Guirao de Vierna, "La comisión real de Guantánamo," 85–87; Weddle, *Changing Tides*, 10–23; Solano Pérez-Lila, *La pasión de reformar*, 57–138; and Cañizares-Esquerra, *Nature, Empire, and Nation.*

2. Ignacio de Sola, Governor of Cartagena de Indias, to the Council of the Indies, Cartagena de Indias, 23 May 1752, exp. 17, leg. 5, CCG, ANC; Millás, *Hurricanes of the Caribbean*, 207–9. Millás acknowledges that there may have been as many as three hurricanes in Jamaica in 1751 but only verifies hurricane number 57, which hit Jamaica, Santo Domingo, and Haiti in late October.

3. Sola to Alonso de Arcos y Moreno, Governor of Santiago de Cuba, Cartagena, 23 May 1752, exp. 17, leg. 5, CCG, ANC.

4. *PZ*, 25 August 1773, referred to the drought and hurricanes in 1752; in LCP. Ludlum, *Early American Hurricanes*, 62.

5. Sims, *Observations on Epidemic Disorders*, 10–11.

6. Caviedes, *El Niño in History*, 146–50. Caviedes's chapter "Altered States: From El Niño to La Niña," 146–71, describes the correlation betwen El Niño and La Niña, originally termed an "anti-niño." Caviedes explains (150): "In consonance with the seesaw interactions between the tropical Pacific and the tropical Atlantic, which are especially active during the warm and low phases of ENSO [El Niño Southern Oscillation, another name for the El Niño/La Niña], La Niña episodes in the tropical Atlantic are characterized by ocean warming. This condition is favorable for the generation of cyclonic depressions, which are likely to develop into major hurricanes when they move into the western tropical Atlantic. This realization has led climatologists to expect increased hurricane activity in the Caribbean region, along the west coast of Central America, and in the Gulf region of North America during La Niña years." Recent scholarship on the impact of hurricanes includes Schwartz, "The Hurricane of San Ciriaco," 303–34; Pérez, *Winds of Change*; Steinberg, *Acts of God*; Mulcahy, *Hurricanes and Society*; Johnson, "El Niño, Environmental Crisis," 365–410; Johnson, "Rise and Fall," 54–75; Johnson, "Climate, Community, and Commerce," 455–82; and Johnson, "St. Augustine Hurricane of 1811," 28–56.

7. Pares, *War and Trade*, 109–14, 517–33; McNeill, *Atlantic Empires*, 85–92, 97–104; Kuethe, *Cuba*, 8–13.

8. Marqués de Gandara to Arcos y Moreno, Santo Domingo, 23 May 1750, exp. 43, leg. 5, CCG, ANC.

9. Arcos y Moreno to Francisco Antonio Cagigal de la Vega, Captain General of Cuba, Santiago de Cuba, 28 January 1751, exp. 228, leg. 7, CCG, ANC.

10. The definitive study is Lewis, *Spanish Convoy of 1750*, which not only examines the international consequences of the disaster but also establishes the consequences for the twenty-first century.

11. Arcos y Moreno to Sola, Santiago de Cuba, 1 February 1751, exp. 128, leg. 6, CCG, ANC.

12. Ibid.

13. Ibid.

14. Arcos y Moreno to Cagigal de la Vega, Santiago de Cuba, 24 January 1751, exp. 30, leg. 5, CCG, ANC.

15. José Pablo de Agüero to Arcos y Moreno, Santo Domingo, 27 January 1751, exp. 102, leg. 7; Arcos y Moreno to Cagigal de la Vega, Santiago de Cuba, 28 January 1751, exp. 228, leg. 7; Sola to Arcos y Moreno, Cartagena de Indias, September 1752, exp. 128, leg. 6; all CCG, ANC.

16. Arcos y Moreno to Cagigal de la Vega, Santiago de Cuba, (date illegible) September 1750, exp. 50, leg. 5, CCG, ANC.

17. Arcos y Moreno to Cagigal de la Vega, Santiago de Cuba, 27 March 1752, exp. 74, leg. 6, CCG, ANC.

18. *PZ*, 2 January 1753, LCP.

19. Cagigal to Arcos y Moreno, Havana, 12 February 1754, exp. 50, leg. 6, CCG, ANC.

20. Lorenzo de Madariaga, Governor of Santiago de Cuba, to Cagigal de la Vega, Santiago de Cuba, Summer (?) 1754, exp. 368, leg. 6, CCG, ANC.

21. Madariaga to Cagigal de la Vega, Santiago de Cuba, 17 October 1754, exp. 242, leg. 6, CCG, ANC, enclosing a letter written in French from the captain of the *Dama Maria*.

22. Madariaga to Martín Estéban de Aróstegui, Lieutenant Governor of Puerto Príncipe, Santiago de Cuba, 8 April 1755, exp. 284, leg. 7, CCG, ANC.

23. Apoderado de la Real Compañía to Madariaga, Santiago de Cuba, 1 August 1755, exp. 407, leg. 7, CCG, ANC.

24. Madariaga to Cagigal, Santiago de Cuba, 31 August 1755, exp. 232, leg. 7, CCG, ANC.

25. Miguel Palomino to Madariaga, Cabañas, 26 February 1756, exp. 12, leg. 7, CCG, ANC.

26. Millás, *Hurricanes of the Caribbean*, 212–13.

27. Martín Estéban de Aróstegui to Madariaga, Puerto Príncipe, 9 November 1758, exp. 161, leg. 7, CCG, ANC.

28. Millás, *Hurricanes of the Caribbean*, 214–15, for 1759. Joseph Paulino de Salgado to Madariaga, Juragua, (day?) October 1760, exp. 195, leg. 10; Francisco Tamayo to Madariaga, Bayamo, 19 October 1761, exp. 61, leg. 11; both CCG, ANC.

29. Cagigal to Arcos y Moreno, Havana, 28 September 1750, exp. 67, leg. 5, CCG, ANC.

30. Arcos y Moreno to Cagigal, Santiago de Cuba, 3 October 1751, exp. 38, leg. 5, CCG, ANC.

31. Joseph Sunyer de Bastero to the President of Santo Domingo, Santiago de Cuba, 3 September 1751, exp. 33, leg. 5, CCG, ANC. Sunyer de Bastero described the voyage as "un pasaje más miserable."

32. Arcos y Moreno to Cagigal, Santiago de Cuba, 3 October 1751, exp. 38, leg. 5, CCG, ANC.

33. Cagigal to Arcos y Moreno, Havana, 9 December 1753, exp. 61, leg. 6; Madariaga to Cagigal, Santiago de Cuba, 9 January 1754, exp. 316, leg. 6; both CCG, ANC.

34. Staab et al., "Acute Stress Disorder," 219–25; Peacock, Morrow, and Gladwin, Hurricane Andrew; Provenzo and Provenzo, In the Eye of Hurricane Andrew.

35. Arcos y Moreno to Cagigal, Santiago de Cuba, 3 October 1751, exp. 38, leg. 5, CCG, ANC.

36. Arcos y Moreno to Francisco Crespo de Ortíz, Santiago de Cuba, 8 October 1752, exp. 222, leg. 6, CCG, ANC.

37. Arcos y Moreno to Cagigal, Santiago de Cuba, 3 October 1751, exp. 38, leg. 5, CCG, ANC.

38. Cagigal de la Vega to Arcos y Moreno, Havana, 15 September 1750, exp. 64, leg. 5, CCG, ANC.

39. Arcos y Moreno to Cagigal de la Vega, Santiago de Cuba, (date illegible) September 1750, exp. 50, leg. 5, CCG, ANC.

40. Arcos y Moreno to Cagigal de la Vega, Santiago de Cuba, 27 March 1752, exp. 74, leg. 6, CCG, ANC. The total in the register (tazmia) for the year's tobacco harvest was just 93,148 manojos, down from an average of 150,000 manojos. Portuondo Zúñiga, "Introducción," in Nicolás José de Ribera, 7 n. 5.

41. Cagigal to Arcos y Moreno, Havana, 12 February 1754, exp. 50, leg. 6, CCG, ANC.

42. Madariaga to Cagigal de la Vega, Santiago de Cuba, Summer (?) 1754, exp. 368, leg. 6, CCG, ANC.

43. Madariaga to Martín de Aróstegui, Santiago de Cuba, 8 April 1755, exp. 284, leg. 7, CCG, ANC.

44. Apoderado de la Real Compañía to Madariaga, Santiago de Cuba, 1 August 1755, exp. 407, leg. 7, CCG, ANC.

45. Millás, Hurricanes of the Caribbean, 212–13.

46. Martín Estéban de Aróstegui to Madariaga, Puerto Príncipe, 9 November 1758, exp. 161, leg. 7, CCG, ANC.

47. Johnson, "La guerra contra los habitantes," 190; Johnson, Social Transformation, 71–72.

48. Lewis, "Anglo-American Entrepreneurs," 113–14; Pares, War and Trade, 406, 492, 532–33; Pares, Yankees and Creoles, 84; Marrero y Artiles, Cuba, 7:159–63.

49. Arcos y Moreno to Cagigal, Santiago de Cuba, undated 1752, exp. 108, leg. 6, CCG, ANC.

50. Arcos y Moreno to Cagigal, Santiago de Cuba, undated 1753, exp. 156, leg. 6, CCG, ANC.

51. McNeill, Atlantic Empires, 134.

52. Madariaga to Cagigal, Santiago de Cuba, 5 March 1755, exp. 319, leg. 7, CCG, ANC.

53. "Ordenanza Municipial de la Havana," box 3, folder 1, Domingo del Monte Collection, Manuscript Division, LC; Actas del Ayuntamiento, vol. 11, Santiago de Cuba, 25 June 1774, AHM.

54. Pedro Jiménes to Arcos y Moreno, Santiago de Cuba, 21 March 1750, exp. 41, leg. 5, CCG, ANC.

55. Sola to Arcos y Moreno, Cartagena de Indias, 19 April 1748, exp. 5, leg. 5, CCG, ANC.

56. Arcos y Moreno to Sola, Santiago de Cuba, undated 1751, exp. 67, leg. 5, CCG, ANC. A *fanega* is approximately 16 bushels and weighs approximately 25 pounds.

57. Actas del Ayuntamiento, vol. 6, Santiago de Cuba, 16 May, 14 August 1742, AHM. Spanish captains also traded with other colonies such as the Dutch entrepôt Saint Eustatius.

58. Arcos y Moreno to Cagigal, Santiago de Cuba, 1 April 1751, exp. 39, leg. 5, CCG, ANC.

59. Arcos y Moreno to Cagigal de la Vega, 30 August (?) 1751, Santiago de Cuba, exp. 197, leg. 5, CCG, ANC.

60. Cagigal de la Vega to Arcos y Moreno, Havana, 29 July 1751, exp. 21, leg. 5, CCG, ANC.

61. Arcos y Moreno to Cagigal, Santiago de Cuba, 3 February 1752, exp. 121, leg. 6, CCG, ANC.

62. Arcos y Moreno to Antonio de la Fuente, Santiago de Cuba, 11 June 1754, exp. 84, leg. 5, CCG, ANC.

63. Actas del Ayuntamiento, vol. 11, Santiago de Cuba, 25 June 1774, AHM.

64. Madariaga to Cagigal de la Vega, Santiago de Cuba, Summer (?) 1754, exp. 368, leg. 6, CCG, ANC.

65. Ibid.

66. Cagigal to Madariaga, Havana, 14 November 1754, exp. 174, leg. 6, CCG, ANC.

67. Aróstegui to Madariaga, Puerto Príncipe, 24 March 1755, exp. 359, leg. 7, CCG, ANC.

68. Madariaga to Aróstegui, Santiago de Cuba, 8 April 1755, exp. 284, leg. 7; Aróstegui to Madariaga, Puerto Príncipe, undated 1755, exp. 110, leg. 7; both CCG, ANC.

69. Madariaga to Cagigal de la Vega, Santiago de Cuba, Summer (?) 1754, exp. 368, leg. 6, CCG, ANC.

70. Madariaga to Cagigal, Santiago de Cuba, 4 March 1755, exp. 362, leg. 7, CCG, ANC.

71. Madariaga to Cagigal, Santiago de Cuba, 9 March 1755, exp. 233, leg. 7, CCG, ANC.

72. Francisco Xavier de Palacios to Madariaga, Havana, 6 November 1756; Caney, 15 December 1756, exp. 137, leg. 7; both CCG, ANC. The legal proceedings about Basabe's case have not been located.

73. Carlos Basabe to Madariaga, Havana, undated 1755, exp. 114, leg. 8, CCG, ANC.

74. Cagigal to Madariaga, Havana, 20 April 1755, exp. 114, leg. 7, CCG, ANC.

75. Portuondo Zúñiga, "Introducción," in *Nicolás José de Ribera*, 27.

76. Ibid., 47; Johnson, *Social Transformation*, 86, 100; Frederick, "Luis de Unzaga."

77. Lieutenant Governor of Bayamo to Francisco Cagigal de la Vega, Bayamo, 1 September 1747, exp. 1, leg. 5, CCG, ANC.

78. Arcos y Moreno to José Antonio de Silva, Santiago de Cuba, 24 August 1752, exp. 125, leg. 6, CCG, ANC.

79. Arcos y Moreno to Cagigal, Santiago de Cuba, 29 March 1750, exp. 57, leg. 5, CCG, ANC; Pares, *Yankees and Creoles*, 60.

80. McNeill, *Atlantic Empires*, 121–22.

81. Actas del Ayuntamiento, Santiago de Cuba, vol. 11, 10 February 1775, 20 March 1775, 21 June 1775, AHM; vol. 14, 22 June 1793, AHM, describes Bayamo's resistance to the authority of Santiago de Cuba from "time immemorial."

82. Marrero y Artiles, *Cuba*, 7:187.

83. Arcos y Moreno to Cagigal, Santiago de Cuba, 24 August 1750, exp. 209, leg. 5, CCG, ANC.

84. Arcos y Moreno to Cagigal, Santiago de Cuba, 29 March 1750, exp. 57, leg. 5, CCG, ANC.

85. Arcos y Moreno to Cagigal, Santiago de Cuba, 20 March 1750, exp. 210, leg. 5, CCG, ANC.

86. "Relación de los presos que han de cargo de Stte Don Joseph Ruíz," Santiago de Cuba, September 1750, exp. 217, leg. 5, CCG, ANC.

87. Cagigal to Arcos y Moreno, Havana, 10 February 1754, exp. 157, leg. 6, CCG, ANC.

88. Cagigal to Arcos y Moreno, Havana, 7 April 1753, exp. 53, leg. 6, CCG, ANC.

89. Arcos y Moreno to Cagigal, Santiago de Cuba, 10 November 1752, exp. 102, leg. 6, CCG, ANC.

90. Circular Letter, Arcos y Moreno, Santiago de Cuba, 7 September 1752, exp. 199, leg. 6, CCG, ANC.

91. Arcos y Moreno to Cagigal, Santiago de Cuba, 10 November 1752, exp. 102, leg. 6, CCG, ANC.

92. Arcos y Moreno to Cagigal, Santiago de Cuba, 9 May 1753, exp. 305, leg. 6; Cagigal de la Vega to Arcos y Moreno, Havana, 8 March 1753, exp. 159, leg. 6; both CCG, ANC.

93. Cagigal to Madariaga, Havana, 10 April 1755, exp. 117, leg. 7, CCG, ANC.

94. Portuondo Zúñiga, "Introducción," in *Nicolás José de Ribera*, 46–47.

95. Arcos y Moreno to Cagigal, Santiago de Cuba, 2 May 1750, exp. 309, leg. 5, CCG, ANC. See also Portuondo Zúñiga, "Introducción," in *Nicolás José de Ribera*, 47.

96. Portuondo Zúñiga, "Introducción," in *Nicolás José de Ribera*, 48.

97. Cagigal to Madariaga, Havana, undated 1755, exp. 278, leg. 7, CCG, ANC.

98. Madariaga to Cagigal, Santiago de Cuba, 7 June 1755, exp. 253, leg. 7, CCG, ANC.

99. Francisco José de Ortíz to Cagigal de la Vega, Verracos, 15 December 1755, exp. 409, leg. 7, CCG, ANC.

100. Juan Fernández de Parra to Madariaga, Tánamo (Puerto Príncipe), 27 September 1757, exp. 26, leg. 7, CCG, ANC.

101. Ortíz to Cagigal de la Vega, Verracos, 15 December 1755, exp. 409, leg. 7; Fernández de Parra to Madariaga, Tánamo (Puerto Príncipe), 27 September 1757, exp. 26, leg. 7; both CCG, ANC.

102. Pares, *War and Trade*, 114.

103. Ibid., 556–59.

104. Ibid., 559–79.

105. The earliest historians of the event never fail to establish it as the most traumatic defeat in the island's history. See Valdés, *Historia de la isla de Cuba*, who devotes nearly

half of his work to detailing this idea of the most traumatic event that until that time had taken place in Cuba's history. See also Torre, *Lo que fuimos*, 167–70; and Guiteras, *Historia de la conquista de la Habana*. By 1962, in celebration of the two-hundredth anniversary of the siege, the principal academic institutions on the island, the Archivo Nacional de Cuba, the Academia de Historia, and the Biblioteca Nacional de Cuba José Martí, had published collections of documents relevant to the conquest from Spanish and British sources. Archivo National de Cuba, *Papeles sobre la toma de la Habana por los ingleses en 1762*; Rodríguez, *Cinco diarios*.

106. Hart, *Siege of Havana*; Syrett, *Siege and Capture of Havana*.

107. Cepero Bonilla, *Azúcar y abolición*; Moreno Fraginals, *El ingenio*; Le Riverend Brusone, *Historia económica*; Guerra y Sánchez, *Manual de historia*.

108. The pioneering study of the military in the context of the Bourbon Reforms is McAlister, *"Fuero Militar" in New Spain*. McAlister, in turn, inspired several subsequent studies, including Campbell, *Military and Society in Colonial Peru*; Archer, *Army in Bourbon Mexico*; Kuethe, *Military Reform and Society in New Granada*; and Domínguez, *Insurrection or Loyalty*. The number of studies about the effect of the Bourbon Reforms in other areas seems to be unlimited. The seminal works include Aiton, "Spanish Colonial Reorganization"; Phelan, The *People and the King*; Burkholder and Chandler, *From Impotence to Authority*; Lynch, *Spanish American Revolutions*; Barbier, *Reform and Politics in Bourbon Chile*; John R. Fisher, *Government and Society*; John R. Fisher, "Imperial 'Free Trade,'" 21–56; Farriss, *Crown and Clergy*; Fisher, Kuethe, and McFarlane, *Reform and Insurrection*; and Johnson, *Social Transformation*.

109. Parcero Torre, *La pérdida*; González Hernández, "Más allá de una capitulación."

110. Delgado, "El Conde de Ricla," 41–138; Torres Ramírez, "Alejandro O'Reilly," 1357–88; Knight, *Slave Society in Cuba*; Knight, "Origins of Wealth," 231–53; Knight, *Slavery and the Transformation of Society*; Kuethe, *Cuba*; Kuethe, "Havana in the Eighteenth Century," 13–39; Kuethe and Inglis, "Absolutism and Enlightened Reform," 118–43; Johnson, *Social Transformation*.

111. Kuethe, *Cuba*, 11.

112. McNeill, *Atlantic Empires*, 103–4; McNeill, "Yellow Jack and Geopolitics," 355–57.

113. Parcero Torre, *La pérdida*.

114. González Hernández, "Más allá de una capitulación."

115. Rodriguez, *Cinco diarios*, 9–11. Although studies have concentrated on Havana, other cities, such as Santiago de Cuba, San Juan, and Cartagena, were also informed that war had been declared.

116. Parcero Torre, *La pérdida*, 39–79.

117. Prado to the Council of the Indies, Havana, 1 February 1762, acknowledging the Royal Order of October 1961; Prado to the Council of the Indies, Havana, 21 May 1762; both leg. 169, Ultramar, AGI.

118. Prado to the Council of the Indies, Havana, 14 May 1762, leg. 169, Ultramar, AGI.

119. Nicolás José de Ribera, *Descripción de la isla*, in Portuondo Zúñiga, *Nicolás José de Ribera*, 154.

120. González Hernández, "Más allá de una capitulación," 18–23.

121. Joseph Paulino de Salgado to Madariaga, Juraguá, (?) October 1760, exp. 195, leg. 10, CCG, ANC.

122. Joseph Joaquín Cisneros to Madariaga, Juraguá, 29 October 1761, exp. 19, leg. 13, CCG, ANC.

123. Francisco Tamayo to Madariaga, Bayamo, 13 February 1761, exp. 61, leg. 11, CCG, ANC.

124. Francisco Tamayo to Madariaga, Bayamo, 19 October 1761, exp. 61, leg. 11, CCG, ANC.

125. Juan Leandro de Landa to Madariaga, Bayamo, 13 November 1761, exp. 78, leg. 11, CCG, ANC.

126. Guiteras, *Historia de la conquista de la Habana*, 135–40; Prado, "Diario Militar," 67–69.

127. González Hernández, "Más allá de una capitulación," 33–37; Pezuela, *Historia de la isla de Cuba*, 2:469–70.

128. Manuel Benito de Erasun to Madariaga, Santiago de Cuba, 7 February 1762, exp. 12, leg. 13, CCG, ANC.

129. "Dictamen del Tte Coronel Dn. Miguel de Muesas," El Morro (Santiago de Cuba), 16 May 1762, leg. 1209, SD, AGI.

130. Ibid.

131. Ibid.

132. Miguel de Muesas to Madariaga, El Morro (Santiago de Cuba), 8 June 1762, exp. 94, leg. 11, CCG, ANC.

133. Antonio Marín to Madariaga, Juraguá, 3 August 1762, exp. 119, leg. 11, CCG, ANC.

134. Joseph Péres to Madariaga, Aguadores, 13 September 1762, exp. 115, leg. 11, CCG, ANC.

135. Hilario Remírez de Esteñoz to Madariaga, El Morro (Santiago de Cuba), 26 June 1762, exp. 133, leg. 11, CCG, ANC.

136. Nicolás José de Ribera, *Descripción de la isla*, in Portuondo Zúñiga, *Nicolás José de Ribera*, 137–38.

137. Joseph Péres to Madariaga, Aguadores, 29 July 1762, exp. 129, leg. 11, CCG, ANC.

138. Rafael Antonio de Sierra to Madariaga, Juraguá, 7 August 1762, exp. 17, leg. 13, CCG, ANC.

139. Joseph Plácido Fernández to Madariaga, Juraguá, 7 October 1762, exp. 99, leg. 13, CCG, ANC.

140. Muesas to Madariaga, El Morro (Santiago de Cuba), 1 September 1762, exp. 42, leg. 13, CCG, ANC.

141. Sebastián Julián Troconis, "Estado de viveres existente en los almahazenes de mi cargo y otros por rezivir con destino para repuestos para las urgencias de la presente Guerra," El Morro (Santiago de Cuba), 4 September 1762, leg. 1209, SD, AGI.

142. Muesas to Madariaga, El Morro (Santiago de Cuba), 11 September 1762, exp. 36, leg. 13, CCG, ANC.

143. Muesas to Madariaga, El Morro (Santiago de Cuba), 10 September 1762, exp. 39, leg. 13, CCG, ANC.

144. Muesas to Madariaga, El Morro (Santiago de Cuba), 11 September 1762, exp. 35, leg. 13, CCG, ANC.

145. Pedro Valiente to Madariaga, Cabañas (Santiago de Cuba), 16 July 1762, exp. 82, leg. 11, CCG, ANC.

146. Luis Trufa to Madariaga, Aguadores, 23 July 1762, exp. 3, leg. 13, CCG, ANC.

147. Joseph Péres to Madariaga, Aguadores, 25 August 1762, exp. 134, leg. 11, CCG, ANC.

148. Bernardo Ramírez to Madariaga, Aguadores, undated, exp. 117, leg. 13, CCG, ANC.

149. Francisco Casals to Madariaga, Juraguá, 21 July 1762, exp. 169, leg. 11, CCG, ANC.

150. Rodríguez, *Cinco diarios*, 13–15.

151. Ibid., 122.

152. Prado to Juan Ignacio de Madariaga, Havana, 10 June 1762, leg. 169, Ultramar, AGI; Parcero Torre, *La pérdida*, 130–38; Rodríguez, *Cinco diarios*, 248. Hereafter the notes will distinguish between the brothers by using their full names.

153. Prado to Juan Ignacio de Madariaga, Havana, 10 June 1762; Juan Ignacio de Madariaga to Julián de Arriaga, Campo de San Juan, 21 June 1762; both leg. 169, Ultramar, AGI.

154. Prado to Juan Ignacio de Madariaga, Havana, 20 June 1762, ibid.

155. "Resolucion de la Junta de Guerra," Havana, 23 June 1762, leg. 169, Ultramar, AGI; Prado, "Diario militar," 86–87.

156. Lorenzo de Madriaga to Arriaga, Santiago de Cuba, 30 August 1765, leg. 1209, SD, AGI. Madariaga's explanation is contained in his *residencia* taken from 1765 to 1769.

157. Muesas to Arriaga, El Morro (Santiago de Cuba), 15 February 1763, leg. 1209, SD, AGI.

158. "Lista de los voluntarios," Santiago de Cuba, 24 July 1762, leg. 169, Ultramar, AGI.

159. Juan Álvarez to Lorenzo de Madariaga, Cabañas, 19 August 1762, exp. 57, leg. 13, CCG, ANC.

160. Álvarez to Lorenzo de Madariaga, Cabañas, 13 September 1762, exp. 54, leg. 13, CCG, ANC.

161. Antonio Marín to Lorenzo de Madariaga, Juraguá, 3 August 1762, exp. 119, leg. 11, CCG, ANC.

162. Bernardo Ramírez to Lorenzo de Madariaga, Aguadores, undated, exp. 117, leg. 13, CCG, ANC.

163. Muesas to Lorenzo de Madariaga, El Morro (Santiago de Cuba), 11 September 1762, exp. 36, leg. 13, CCG, ANC.

164. Ibid.

165. Hilario Remírez de Esteñóz to Lorenzo de Madariaga, El Morro (Santiago de Cuba), 19 October 1762, exp. 135, leg. 13, CCG, ANC.

166. Manuel Hernández to Lorenzo de Madariaga, Juraguá, 2 September 1762, exp. 86, leg. 11, CCG, ANC.

167. Joseph Sixto Ramírez to Lorenzo de Madariaga, Juraguá, 13 November 1762, exp. 152, leg. 13, CCG, ANC.

168. Francisco de Torralbo to Lorenzo de Madariaga, location unidentified, 5 December 1762, exp. 105, leg. 13, CCG, ANC.

169. Prado, "Diario militar," 87–88.

170. "Correspondencia entre Havana y Cuba," June–August 1762, leg. 1209, SD, AGI.

171. Prado to Lorenzo de Madariaga, Havana, 14 July 1762, leg. 169, Ultramar, AGI.

172. Prado to Juan Ignacio de Madariaga, Havana, 18 July 1762, 24 July 1762, leg. 169, Ultramar, AGI.

173. "Diario de Manuscrito de Madrid," in Rodríguez, *Cinco diarios*, 27; "Diario del Capitán D. Juan de Castas," in ibid., 47.

174. "Diario del Capitán D. Juan de Castas," in ibid.

175. Ibid., 50.

176. Prado to Lorenzo de Madariaga, Havana, 14 July 1762, leg. 169, Ultramar, AGI.

177. Prado to Juan Ignacio de Madariaga, Havana, 24 July 1762, ibid.

178. Prado to Juan Ignacio de Madariaga, Havana, 25 July 1762, ibid.

179. Prado to Juan Ignacio de Madariaga, Havana, 24 July 1762, ibid.

180. "Diario del Capitán D. Juan de Castas," in Rodríguez, *Cinco diarios*, 50.

181. Ibid. On 20 June, Prado ordered that all of the dogs in the city be killed. No explanation was offered for the action, but it was noted in the context of safety and maintaining the food supply.

182. Ibid., 52, 54; "Diario de Manuscrito de Madrid," in ibid., 33.

183. "Diario de Manuscrito de Madrid," in ibid., 32.

184. "Diario del Capitán D. Juan de Castas," in ibid., 49.

185. "Diario de Manuscrito de Madrid," in ibid., 33.

186. Ibid.; Municipio de la Habana, *La dominación inglesa en la Habana*. It is unclear how the hospitals of Santa Clara and Belén were used, since most civilian women had left the city. Nonetheless, many female military dependents stayed in Havana during the bombardment to care for their menfolk employed in the city's defense. See Johnson, "Señoras en sus clases," 22–23.

187. Juan José de Urriza, Intendant of Havana, to José de Gálvez, Havana, 18 August 1781, leg. 1551, IG, AGI.

188. Ibid.

189. "Relación de los meritos y exercicios literarios del Dr. D. Leandro Josef de Tagle," 1793, leg. 2218, SD, AGI.

190. "Diario de Manuscrito de Madrid," in Rodríguez, *Cinco diarios*, 33.

191. Prado to Juan Ignacio de Madariaga, Havana, 10 July 1762, leg. 169, Ultramar, AGI.

192. Prado to Lorenzo de Madariaga, Havana, 14 July 1762, ibid.

193. "Diario de Manuscrito de Madrid," in Rodríguez, *Cinco diarios*, 34.

194. Lorenzo de Montalvo to Albemarle, 4 September 1762, in Archivo Nacional de Cuba, *Papeles sobre la toma de la Habana por los ingleses en 1762*, 18. The effective forces in Havana were approximately 3,000 men, so about one-third of the soldiers were unfit for duty.

195. Prado to Juan Ignacio de Madariaga, Havana, 20 June 1762, leg. 169, Ultramar, AGI.

196. Juan de Lleonart to Marqués de la Torre, Puerto Príncipe, 13 June 1772, leg. 1168, PC, AGI; Juan de Lleonart, "Empleos," Havana, 31 December 1788, cuaderno (equivalent to an expediente) 2, leg. 7259, GM, AGS.

197. Joseph de los Reyes, "Estado de la Tropa de los Reximtos de Aragón, Havana y Edimbourgo," onboard *El Arrogante* anchored in Jagua harbor, 20 July 1762, leg. 169, Ultramar, AGI.

198. Juan de Lleonart, "Empleos," Havana, 31 December 1788, exp. 2, leg. 7259, GM, AGS.

199. Isidro de Limonta to Eugenio de Llaguna y Arriola, Santiago de Cuba, 26 August 1794, leg. 2236, SD, AGI.

200. Simón José Rodríguez to Albemarle, Matanzas, 4 November 1762, in Archivo Nacional de Cuba, *Papeles sobre la toma de la Habana por los ingleses en 1762*, 24.

201. "Diario de Manuscrito de Madrid," in Rodríguez, *Cinco diarios*, 28.

202. Ibid.

203. Prado to Juan Ignacio de Madariaga, Havana, 11 August 1762, leg. 169, Ultramar, AGI; Juan Ignacio de Madariaga to Lorenzo de Madariaga, Campo de Miraflores, 14 August 1762, leg. 1209, SD, AGI.

204. Albemarle to Lord Egremont, 13 July 1762, in Archivo Nacional de Cuba, *Papeles sobre la toma de la Habana por los ingleses en 1762*, 80.

205. Albemarle to Secretary of State, 17 July 1762, in ibid., 85.

206. Ibid., 91.

207. Salvador Vásquez, "Morbimortalidad colérica en Cuba," in Hernández Palomo, *Enfermedad y muerte en América y Andalucía*, 283-302.

208. Kiple, *Caribbean Slave*, 145-46.

209. Shattuck, *Diseases of the Tropics*, 334-51.

210. Antonio Marín to Madariaga, Juraguá, 3 August 1762, exp. 119, leg. 11, CCG, ANC.

211. Salvador Vásquez, "Las quinas del norte de Nueva Granada," in Hernández Palomo, *Enfermedad y muerte en America y Andalucía*, 403-26; Frías Núñez, "El discurso médico," 215-33; Alegre Pérez, "Drogas americanas en la Real Botica," 216-33.

212. Shattuck, *Diseases of the Tropics*, 421-39. The quote is on p. 423.

213. Ibid.

214. Johnson, "Señoras en sus clases," 22-23; Municipio de la Habana, *Libro de Cabildos*.

215. The premier British slave merchant was one John Kennion, a Liverpool merchant, who accompanied Albemarle's expedition. Thomas, *Cuba*, 49-57.

216. Inglis, personal communication to author, July 1999; Cosner, "Neither Black nor White, Slave nor Free." Most slaves introduced after 1762 were absorbed by the fortification projects. See Johnson, *Social Transformation*, 39-75; and Jennings, "War as the 'Forcing House of Change,'" 411-40.

217. A thorough census of the countryside was produced as a consequence of the hurricane in 1768 and may be found in leg. 1097, PC, AGI.

218. Richard Waln to William Drury, Philadelphia, 31 October 1762, Richard Waln Letterbook 1762-66, HSP.

219. Lorenzo de Montalvo to Albemarle, Havana, 4 September 1762, in Archivo Nacional de Cuba, *Papeles sobre la toma de la Habana por los ingleses en 1762*, 18.

220. "Exports of Merchandise, the Produce of Pennsylvania & c from Philadelphia, Anno 1759 to 1763 inclusive," Customs House Book, November 1761-October 1764, HSP.

221. Pares, *War and Trade*, 491-94, especially his discussion of prices on p. 494. He argues that the price of provisions was highest in 1759-60 and declined thereafter.

222. "Account of the Species of Goods Imported from Foreign Plantations in America & the Duties Received on the Same from the 10th Day of October 1762 to the 5th Day of January Following Christmas Quarter 1762," Customs House Records, HSP.

223. "Sales of the Sugars on Board the Prize Ship on 26 November 1762," Customs

House Book, HSP. Purchasers included Scott and McMichael, James Nicholson, Usher and Mitchell, and Willing and Morris. For additional sales, see 1 December 1762, prize sugar onboard the ship *Britannia*: Captain Beale, 2 hogsheads to Cornelius Vanderwilde; Conyngham and Co., 6 hogsheads; 2 bbl sugar weighing 64 lbs; Thomas Clifford, 5 hogsheads weighing 50 lbs; Oswel and Eve, 4 hogsheads, 40 lbs; S. Shewel, 4 hogsheads, 40 lbs. 4 December 1762: Captain Thomas Clifford, aboard the ship *Tyger* from New Providence selling French prize sugar; and 13 December 1762: John Cunningham, purchasing 4 hogsheads of French prize sugar from New Providence. Ibid. The international dimension of such seizures is explained in Pares, *War and Trade*, 450–68.

224. Arrival of the sloop *Abigal*, master J. McPherson, Philadelphia, 10 August 1762, "Account of the Species of Goods Imported from Foreign Plantations in America & the Duties Recieved on the Same from the 10th Day of October 1762 to the 5th Day of January following Christmas Quarter 1762," Customs House Records, HSP.

225. Arrival of the sloop *Discrete*, master James Wilson, Philadelphia, 22 November 1762, ibid.

226. Arrival of the sloop *Tyger*, Philadelphia, 4 December 1762, ibid.

227. Arrival of the sloop *Lovely Peggy*, Philadelphia, 5 October 1762, and arrival of the brig *Albemarle*, Philadelphia, 19 November 1762, ibid.

228. Book of Customs Duties Paid, folio 182, Customs House Records, HSP.

229. Arrival of the schooner *Industry*, Philadelphia, 18 December 1762, "Account of the Species of Goods Imported from Foreign Plantations in America & the Duties Recieved on the Same from the 10th Day of October 1762 to the 5th Day of January following Christmas Quarter 1762," Customs House Records, HSP.

230. Book of Customs Duties Paid, folio 182, Customs House Records, HSP.

231. Arrival of the ship *Marquis de Granby*, Philadelphia, 17 December 1762, "Account of the Species of Goods Imported from Foreign Plantations in America & the Duties Recieved on the Same from the 10th Day of October 1762 to the 5th Day of January following Christmas Quarter 1762," Customs House Records, HSP.

232. Records of merchant ships leaving for Havana before December 1763 are not available. Possible reasons include that the military was in charge of collecting provisions, that the opening of the Havana market took some by surprise, and that it took time to collect sufficient provisions to send to Havana. Nonetheless, Philadelphia's exports of flour and hardtack (biscuit) remained relatively consistent throughout the war years. "Exports of Merchandise, the Produce of Pennsylvania & c from Philadelphia, Anno 1759 to 1763 Inclusive," Customs House Book, November 1761–October 1764, HSP. Barrels are the unit of measurement in the following list:

1759: Flour 161,233, Bread 70,279; Total: 231,512
1760: Flour 169,874, Bread 59,103; Total: 228,977
1761: Flour 176,035, Bread 46,858; Total: 222,893
1762: Flour 164,018, Bread 58,134; Total: 222,152
1763: Flour 137,685, Bread 36,990, Total: 174,675

233. On 3 January, the brig *Albemarle*, master John McClelland; on 15 February, the schooner *Batchellor*, master David Gregory; on 1 March, the brig *Lydia*, master John

Conyngham; and on 18 March, the sloop *Chance*, master James Craig, cleared. On 23 March, three ships cleared: the snow *Sally*, master Thomas Powell; the sloop *Dispatch*, master John Davidson; and the sloop *Lovely Peggy*, master James Russell. On 12 April, the sloop *Lark*, master John Kennedy; and the *Hope*, master John Dee, cleared. On 23 April, the *Hound*, master Edward York; on 28 April, the *Speedwell*, master John Dutten; on 29 April, the *Hannah*, master Samuel Lombard; on 13 May, the *Fanny*, master James Poniers; and the *Industry*, master Thomas Fisher, cleared. Customs House Book, HSP.

234. Johnson, *Social Transformation*.

Chapter Three

1. McNeill, *Atlantic Empires*.

2. Admiral George B. Rodney sarcastically described Louisiana as "another desert for her [Spain's] empire." Quoted in Rea, "A Distant Thunder," 178.

3. Johnson, "Casualties of Peace," 91–125; Corbitt, "Spanish Relief Policy," 67–82; Gold, "Settlement of East Florida Spaniards," 216–31.

4. Lynch, *Bourbon Spain*, 318–19; Kuethe, *Cuba*, 78–112; Guerra y Sánchez, *Manual de historia*, 175–76.

5. Kuethe, *Cuba*, 70–72; Valdés, *Historia de la isla de Cuba*, 275; Le Riverend Brusone, *Historia económica*, 141–43; Marrero y Artiles, *Cuba*, 8:1–25.

6. Kuethe, "Havana in the Eighteenth Century," 22–23.

7. Johnson, *Social Transformation*, 39–70.

8. Ibid., 84–87; Jennings, "War as the 'Forcing House of Change,'" 411–40.

9. Johnson, *Social Transformation*, 39–70. See the explanation in the text and in Huet to de la Torre, 30 July 1775, 27 August 1775, leg. 1223, SD, AGI. Photocopy in the Levi Marrero Collection, Special Collections, Florida International University, Miami.

10. Johnson, *Social Transformation*, 64–69; Royal Order to Pascual de Cisneros, captain general of Cuba, Madrid (?), 19 October 1765, leg. 1220, SD, AGI; quoted in Lewis, "Anglo-American Entrepreneurs," 210 n. 10.

11. Ortíz de la Tabla y Ducasse, *Comercio exterior de Veracruz*; Walker, *Spanish Politics and Imperial Trade*; Ringrose, *Spain, Europe, and the "Spanish Miracle"*; Lewis, "Nueva España y los esfuerzos," 501–26.

12. García-Baquero González, *Cádiz y el atlántico*, 208–15; Muñoz Pérez, "La publicación del reglamento," 638–43; John R. Fisher, "Imperial 'Free Trade,'" 21–23; Kuethe, "Early Reforms of Charles III," 21–29; Kuethe and Inglis, "Absolutism and Enlightened Reform," 118–43. Such commercial reforms were one facet of a wider program known as the Bourbon Reforms.

13. Muñoz Pérez, "La publicación del reglamento," 638–43; John R. Fisher, "Imperial 'Free Trade,'" 21–22.

14. Alejandro O'Reilly to Julián de Arriaga, 12 April 1764, leg. 1509, SD, AGI; photocopy in Levi Marrero Collection, Special Collections, Florida International University, Miami; printed in Marrero y Artiles, *Cuba*, 8:262–67; and in Delgado, "El Conde de Ricla," 117–21.

15. Delgado, "El Conde de Ricla," 124–25; Torres Ramírez, "Alejandro O'Reilly," 1367.

16. Johnson, *Social Transformation*, 37–70; Johnson, "La guerra contra los habitantes," 181–209.

17. McNeill, *Atlantic Empires*, 197–201; Marrero y Artiles, *Cuba*, 7:108–62. Gárate Ojanguren, *Comercio ultramarino e ilustración*, 170–72, concludes that "in 1765, the Real Compañía de La Habana had ceased to be a privileged company." Among the privileges that the Havana Company lost were the exclusive right to trade with Cuba, the responsibility to administer the tobacco monopoly, the privilege of running the Havana shipyard, the authority to issue permission to cut timber, and the privilege to contract with New York merchants to bring food into Spanish Florida.

18. Torres Ramírez, *La compañía gaditana*. Among the many names that were used colloquially to refer to the Asiento were the Real Compañía, the Factoría, and the Armazón de Negros.

19. Johnson, *Social Transformation*, 84–87.

20. José Osorio to George Paplay (in Jamaica), Havana, July 1764 to February 1765, leg. 1212, SD, AGI; Jennings, "War as the 'Forcing House of Change.'"

21. Torres Ramírez, *La compañía gaditana*, 19–41, 108.

22. Ibid., 19–41, 53–77; Tornero Tinajero, *Crecimiento económico*, 34–44. Marrero y Artiles, *Cuba*, 7:158–59, discusses how nineteenth-century liberal economic philosophy has distorted the historical perception of the Havana monopoly company.

23. Governor of Puerto Rico to the Council of the Indies, San Juan, 1771, leg. 2516, SD, AGI. Although the Ministry of the Indies governed Spain's American colonies, most correspondence went to the consultative body, the Council of the Indies.

24. Alonso Arcos de Moreno to Juan Cagigal de la Vega, Santiago de Cuba, 30 August 1752, exp. 197, leg. 5, CCG, ANC.

25. O'Reilly to Ricla, Puerto Rico, 18 May 1765, leg. 1212, SD, AGI. He wrote, "Nunca puede ser completa mi satisfaccion hasta dar a VE un abrazo en Madrid. Los calores me inquietan. [My satisfaction will not be complete until I can give Your Excellency a hug in Madrid. The heat bothers me so much.]"

26. "Memorial of the Under Written Merchants Concerned in the Trade at the Havana to the Earl of Halifax," London (?), 14 November 1763, Archivo Nacional de Cuba, *Papeles sobre la toma de la Habana por los ingleses en 1762*, 110.

27. A. Keppel onboard His Majesty's ship *Valiant*, Jamaica, 18 November 1763, leg. 1212, SD, AGI; Conde de Ricla to Captain Briggs, Havana, 9 March 1765, leg. 1213, SD, AGI; Miguel de Altarriba to Captain Briggs, Havana, 9 March 1765, ibid.

28. Lorenzo de Montalvo to Marqués de Cruillas, Havana, 7 July 1763, Archivo Nacional de Cuba, *Nuevos papeles sobre la toma de la Habana por los ingleses en 1762*, 214.

29. Marqués de Cruillas to Montalvo, Mexico, 12 August 1763, ibid., 204.

30. Marqués de Casa Cagigal to the Conde de Ricla, Captain General of Cuba, Santiago de Cuba, 24 February 1764, exp. 65, leg. 26, CCG, ANC.

31. Miguel de Altarriba, Intendant of Havana, Havana, 7 March 1765, and Conde de Ricla, governor and captain general of Cuba, Havana, 7 March 1765, leg. 1212, SD, AGI, both granting permission to a packetboat, *La María*, with a French captain, Francisco Salvator, and French registry to go to New York to get food. See also Intendant of Havana

(Altarriba), Havana, 25 May 1764, 25 May 1765, 30 May 1765, leg. 1212; 21 March 1765, leg. 1213; all SD, AGI.

32. Martin José de Alegría to the Marqués de Casa Cagigal, Havana, 14 November 1764, exp. 19, leg. 19, CCG, ANC.

33. Alegría to the Marqués de Casa Cagigal, Havana, 14 November 1764, exp. 19, leg. 19, CCG, ANC.

34. Ringrose, *Spain, Europe, and the "Spanish Miracle."*

35. Johnson, *Social Transformation.*

36. Casa Cagigal to the Council of the Indies, Santiago de Cuba, 3 August 1763, leg. 1134, SD, AGI. One ship had fifty-four mules and six horses, and the other carried fifteen mules and seven horses.

37. Ibid.

38. Casa Cagigal to the Council of the Indies, Santiago de Cuba, 8 July 1764, ibid.

39. Casa Cagigal to the Council of the Indies, Santiago de Cuba, 10 October 1764, 22 October 1764, 24 October 1764, ibid.

40. Alegría to Casa Cagigal, Havana, 12 February 1765, exp. 10, leg. 19, CCG, ANC.

41. "Lista de los negociantes yngleses a quienes escrivi de orden del Exmo Sr Conde de Ricla para entregarme ynventarios de los géneros existentes en su poder: Sres. Hodey y Fanning, Sres. Jaffay y Wimot; Sres. Sims y Talbot, Sres. Bell y Fogo, Don Cornelio Coppinger, Don Alexandro Munro; Don Alexandro Macculloch; Don Pedro Ritchie; Sres. Stalker y Pyott; Sr. Kern," Havana, 1765, leg. 1212, SD, AGI.

42. Royal Order, San Lorenzo, 3 July 1765, leg. 1212, SD, AGI.

43. "Lista de los negociantes yngleses a quienes escrivi de orden del Emo Sr Cdr para entregarme ynventarios de los generos existentes en su poder," Havana, 1765, ibid.

44. Alexander Monroe to Conde de Ricla, Havana, 1765; Fogo and Napleton to Conde de Ricla, Havana, 1765; both ibid.

45. Marqués de Casa Cagigal to Jerónimo de Grimaldi, Minister of State, Santiago de Cuba, 22 April 1765, exp. 94, leg. 23, CCG, ANC.

46. Ricla to Julián de Arriaga, Havana, June 1765, exp. 7, leg. 21, CCG, ANC.

47. Conde de Ricla to Francisco Salvatore, Havana, 9 March 1765, leg. 1213, SD, AGI; Miguel de Altarriba to Francisco Salvatore, Havana, 9 March 1765, ibid.

48. Governor of Havana (Pascual de Cisneros) to the Council of the Indies, Havana, 21 June 1766, leg. 1135, SD, AGI.

49. José Antonio Armona to the Council of the Indies, Havana, 9 October 1766, leg. 1135, SD, AGI.

50. Millás, *Hurricanes of the Caribbean,* 219–25; Ludlum, *Early American Hurricanes,* 62.

51. Millás, *Hurricanes of the Caribbean,* 219.

52. Ibid., 219–20.

53. Ibid.

54. Ibid., 224.

55. Ibid., 219–20.

56. Miner Solá, *Historia de los huracanes en Puerto Rico,* 29–30.

57. Ibid., 30; Millás, *Hurricanes of the Caribbean,* 224.

58. Bartolomé de Morales to Governor of Cuba, Santiago del Prado, 12 October 1766, exp. 62, leg. 24, CCG, ANC.

59. Ibid.

60. Morales to Cagigal, Santiago del Prado, 25 October 1766, ibid. See also Díaz, The Virgin, the King.

61. Juan de Lleonart to Cagigal, Puerto Príncipe, 8 May 1767, exp. 42, leg. 24, CCG, ANC.

62. Alegría and Bernardo de Goicoa to Cisneros, Havana, 3 July 1766, exp. 11, leg. 19, CCG, ANC; Bucareli to Jose de Cuevas, Havana, 11 August 1767, enclosing a copy of a decree by Julian de Arriaga, Madrid, 6 January 1767, requiring relief for the victims of the earthquake, exp. 7, leg. 19, CCG, ANC.

63. Bucareli to Cagigal, Havana, 3 October 1766, exp. 34, ibid.

64. Bucareli to Cagigal, Havana, 2 September 1766, exp. 39, ibid.

65. Bucareli to Cagigal, Havana, 3 October 1766, exp. 34, ibid.

66. Weddle, Changing Tides, 10–23; Ludlum, Early American Hurricanes, 62.

67. Ibid.

68. Solano Pérez-Lila, La pasión de reformar, 220; Weddle, Changing Tides, 10–23; Acosta, "Las bases económicas," 331–75; Moore, "Antonio de Ulloa," 189–218; Moore, "Revolt in Louisiana," 40–55; Din, Francisco Bouligny, 31–35; Holmes, "Some Economic Problems of Spanish Governors," 521–24. Although Spanish rule in Louisiana is routinely characterized as tenuous, an alternate interpretation is offered by Hall, Africans in Colonial Louisiana, 276, who argues: "Spain actually ruled Louisiana between 1769 and 1803, a little over three decades. Spanish rule was a vast improvement over French rule."

69. Solano Pérez-Lila, La pasión de reformar.

70. Ulloa to Bucareli, Balisa, 12 December 1766, "Despatches of the Spanish Governors of Louisiana," Manuscripts Department, Howard-Tilton Memorial Library, Tulane University, New Orleans.

71. Ulloa to Bucareli, Balisa, 23 January 1767, 10 February 1767, 3 March 1767, ibid.

72. Ulloa to Bucareli, Balisa, 15 March 1767, ibid.

73. On Canada's agricultural production, see Pares, War and Trade, 390–93; Parry, Sherlock, and Maingot, Short History, 93–106; and McNeill, Atlantic Empires, 107–14, 137–54.

74. Millás, Hurricanes of the Caribbean, 220.

75. PZ, 27 November 1766 in Jamaica; 26 February 1767 in Grenada; 26 May 1768 in Montserrat; 20 April 1769 in Jamaica; and 16 May 1769 in Curaçao; all in LCP.

76. PZ, 27 November 1766, quoting letters from Jamaica, LCP.

77. PZ, 26 February 1767, quoting a report of 28 December 1766, LCP.

78. Ulloa to Bucareli, Balisa, 25 March 1767, "Despatches of the Spanish Governors of Louisiana," Manuscripts Department, Howard-Tilton Memorial Library, Tulane University, New Orleans; Moore, "Antonio de Ulloa," 189–218; Moore, "Revolt in Louisiana," 40–55; Din, Francisco Bouligny, 31–35.

79. Gurr, Handbook of Political Conflict.

80. Olson, "Towards a Politics of Disaster," 283.

81. PZ, 27 November 1766, quoting letters from Jamaica, LCP.

82. *PZ*, 26 February 1767, quoting a report of 28 December 1766, LCP.

83. Ulloa to Bucareli, Balisa, 25 March 1767, "Despatches of the Spanish Governors of Louisiana," Manuscripts Department, Howard-Tilton Memorial Library, Tulane University, New Orleans.

84. Pares, *War and Trade*, 540–55, 602–3; Parry, Sherlock, and Maingot, *Short History*, 133–34; Goebel, "British Trade to the Spanish Colonies," 289–91. See also *PZ*, 29 January 1767, 15 September 1768, and 12 October 1769, for notices of the opening of Caribbean free ports, LCP.

85. Parry, Sherlock, and Maingot, *Short History*, 133–34; Goebel, "British Trade to the Spanish Colonies," 290–91.

86. Goebel, "The 'New England Trade' and the French West Indies," 352 n. 57. The trade liberalization in the French colonies was similar to concessions in the British colonies. Goebel, "British Trade to the Spanish Colonies," 290–91; Pares, *War and Trade*, 540–55, 602–3.

87. Weddle, *Changing Tides*, 10–23.

88. Goebel, "The 'New England Trade' and the French West Indies," 352.

89. Nichols, "Trade Relations," 293; Le Riverend Brusone, *Historia económica*, 103.

90. Governor of Santiago de Cuba to Captain General of Cuba, 1 April 1751, Santiago de Cuba, exp. 39, leg. 5, CCG, ANC; Governor of Santiago de Cuba to Martín Estéban de Aróstegui (lieutenant governor of Puerto Príncipe), 8 April 1755, Santiago de Cuba, exp. 284, leg. 7, CCG, ANC.

91. Bucareli to Cagigal, Havana, 2 April 1767, exp. 66, leg. 19, CCG, ANC.

92. Parry, Sherlock, and Maingot, *Short History*, 133–34.

93. García-Baquero González, *Cádiz y el atlántico*, 210–15. See also Ortiz de la Tabla, *Comercio exterior*; Walker, *Spanish Politics and Imperial Trade*; John R. Fisher, "Imperial 'Free Trade'"; Ringrose, *Spain, Europe, and the "Spanish Miracle"*; and Stein and Stein, *Apogee of Empire*.

94. Olson, "Towards a Politics of Disaster," 283.

95. Goebel, "British Trade to the Spanish Colonies," 290–91. See also *PZ*, 29 January 1767, 15 September 1768, and 12 October 1769, LCP.

96. Juan de Ávalos to the Council of the Indies, Havana, 30 July 1766, leg. 2515, SD, AGI. Ávalos, like many, believed that espionage conducted by crew members on British ships visiting prior to 1760 contributed to success of the siege and capture of Havana during the Seven Years' War. See Knowles, "Description of the Havana."

97. Documentary evidence about the changing nature of contraband abounds in Cuban archives: in Governor of Santiago de Cuba to the Captain General of Havana, Santiago de Cuba, 11 September 1747, exp. 1, and 22 November 1750, exp. 44, both leg. 5; in the ongoing case begun in Governor of Santiago de Cuba to the Captain General of Havana, Santiago de Cuba, 7 April 1753, exp. 53, leg. 6; and in Governor of Santiago de Cuba to the Captain General of Havana, Santiago de Cuba, 27 September 1757, exp. 26, leg. 7; all CCG, ANC. The documentation continues in Spain: in Captain General of Cuba to the Council of the Indies, Havana, 10 September 1764, 22 May 1765, 6 September 1765, 6 December 1765, leg. 1134; and 5 February 1766, 19 April 1766, 5 June 1766, 6 June 1766, leg. 1135; all SD, AGI.

98. *PZ*, 29 January 1767, LCP; Goebel, "The 'New England Trade' and the French West Indies," 352.

99. Pares, *Yankees and Creoles*, 84.

100. Customs House Records Book, 12 April 1763, HSP.

101. Marcos de Vergara to the Council of the Indies, San Juan, 15 September 1766, leg. 2515, SD, AGI; Vicente de Zavaleta, factor for the Asiento, to Vergara, San Juan, 15 September 1766, ibid.

102. Vergara to the Council of the Indies, San Juan, 15 September 1766, ibid.

103. Council of the Indies to Vergara, Madrid, 17 February 1767, ibid.

104. Vergara to the Council of the Indies, San Juan, 24 February 1767, 3 March 1767, ibid. Torres Ramírez, *La compañía gaditana*, relates an occasion when Kennedy lent Vergara's widow 823 pesos in a letter of credit drawn on Brickdale's company in Cádiz, but that payment was denied in 1770 when she presented it for payment. Torres opines: "The aforementioned factor, with all of the company's money, had fled to the Danish islands, never to be heard from again. [El citado factor, con todo el dinero que tenía la compañía, habia huido a las islas danesas, sin volverser a saber nunca más de él.]" Ibid., 108.

105. Juan José de Goicoa, director of the Asiento, to the Council of the Indies, Madrid, 11 September 1768, leg. 2515, SD, AGI. This folder contains several of the complaints filed by the Asiento.

106. Goicoa to governor of Puerto Rico, Madrid, 22 April 1768, ibid.

107. José Tentor, governor of Puerto Rico, to Julián de Arriaga, Minister of the Indies, San Juan, 25 January 1769, ibid.

108. "Petición de los factores Pover y Noboa," San Juan, 20 February 1770; "Testimonio de los autos formados por la introducción de viruelas en esta ciudad por Don Miguel Barrera," Cumaná, 1771; and Council of the Indies to governor of Cumaná, Aranjuéz, 20 April 1771; all ibid.

109. Pedro Antonio de Florencia, "Certification," Havana, 26 March 1768, exp. 170, leg. 19, CCG, ANC.

110. "Escritura por Pedro Antonio de Florencia, escribano mayor de Real Hacienda de Havana," Havana, 10 January 1768, exp. 13, ibid. Havana also shipped 35 *tercios* of *habas* (white beans), 15 of garbanzos, and 100 of frijoles (likely red or black beans) to Santiago de Cuba, along with 20 *arrobas* of flour to Trinidad.

111. Juan Garvey to Bucareli, Santiago de Cuba, 24 October 1767, exp. 87, leg. 23, CCG, ANC.

112. *PZ*, 12 November 1767, outbound, sloop *Hibernia*, J. McCarthy to Jamaica; outbound, ship *Charming Susanna*, T. Connor to Jamaica; in LCP. For the *Hibernia*, see McCusker registration 1730, from McCusker, "Ships Registered," HSP; the *Charming Susanna*'s registration has not yet been located.

113. Council of the Indies, Decreto, San Lorenzo, 3 March 1769, leg. 2515, SD, AGI.

114. Griffin, *Stephen Moylan*, 1–4.

115. Vergara to the Council of the Indies, San Juan, 23 December 1767, leg. 2515, SD, AGI.

116. *PZ*, 24 March 1768, inbound, sloop *Hibernia*, J. McCarthy from Puerto Rico; 31 March 1768, inbound, ship *Charming Susanna*, T. Connor from Puerto Rico; in LCP.

117. Altarriba to Cagigal, Havana, 9 February 1768, exp. 14, leg. 19, CCG, ANC. Carbo's port of origin is unknown. It does not appear that he sailed from Philadelphia. It is also not known if he had any connection to the two captains, McCarthy and Connor, who sailed to San Juan.

118. Altarriba to Cagigal, Havana, 26 March 1768, ibid.

119. Julián Marín receipt, in Juan Garvey to Bucareli, Santiago de Cuba, 24 October 1767, exp. 87, leg. 23, CCG, ANC.

120. 8 January 1768, 12 January 1768, Actas Capitulares del Ayuntamiento, Archivo Municipal de Santiago de Cuba, Santiago de Cuba, Cuba.

121. Ibid., 11 March 1768.

122. Ibid.

123. Francisco de Arrate, testimony, ibid., 18 April 1768.

124. Juan Garvey, testimony, ibid., 23 April 1768.

125. Altarriba to Cagigal (letter #1), Havana, 8 April 1768, exp. 14, leg. 19, CCG, ANC.

126. Altarriba to Cagigal (letter #2), Havana, 8 April 1768, ibid.

127. Bucareli to Cagigal, Havana, (?) April 1768, exp. 48, ibid.

128. Ulloa to Bucareli, Balisa, August 1767, "Despatches of the Spanish Governors of Louisiana," Manuscripts Department, Howard-Tilton Memorial Library, Tulane University, New Orleans.

129. Ulloa to Bucareli, Balisa, 29 December 1767, ibid.

130. Ulloa to Bucareli, Balisa, 2 February 1768, ibid.

131. Ulloa to Bucareli, Balisa, 22 June 1768, 20 July 1768, ibid.

132. Ulloa to Bucareli, New Orleans, 1 August 1768, ibid.

133. Jose Antonio de Armona, administrator of the royal mail system, to the Council of the Indies, Havana, 30 March 1774, leg. 257-A, Correos, AGI. When royal officials spoke of the storm six years later, they reported that it "still made people tremble."

134. *Estado que comprehende las desgracias que causó el huracán el día 15 de octubre en la ciudad de la Havana* (Cádiz, 1768); and *Estado que comprehende las desgracias que causó el huracán el día 15 de octubre en la ciudad de la Havana* (Madrid, 1769); both leg. 1594, SD, AGI. The accounts differ only in the number of houses that were destroyed in Guanabacoa.

135. Juan José Galán to Bucareli, Batabanó, 9 November 1768, 12 November 1768, leg. 1093, PC, AGI.

136. Jose Antonio de Colina to Arriaga, Havana, 24 October 1768, folder 13, box 7, Escoto Collection.

137. Armona to the Council of the Indies, Havana, 28 October 1768, leg. 256-B, Correos, AGI.

138. "Relación" Joseph López, Captain and Master of the brigantine *San Juan Nepomuceno*, Castillo de Jagua, 19 October 1768; Andrés Brito Betancourt to Bucareli, Castillo de Jagua, 20 October 1768; both leg. 1093, PC, AGI.

139. Galán to Bucareli, Batabanó, 12 November 1768, leg. 1093, PC, AGI; Carlos de Castro Palomino to Bucareli, Guadelupe, 31 October 1768, ibid. The official governmental dossier on the storm consists of a series of reports in leg. 1097, PC, AGI. See also *Estado que comprehende*, leg. 1594, SD, AGI.

140. See Juan Álvarez de Miranda, "Poética relación Christiana y Moral," Havana, October 1768, leg. 1097, PC, AGI, a sixty-four-stanza poem that invokes religion and Divine Justice throughout its text. For the religious dimension, see Franciscan friar Bernardo de los Santos, parish priest Estéban Conde, and sacristan Francisco Josef de Melo to José de San Martín, Guanabacoa, 21 October 1768, leg. 1463, SD, AGI.

141. Just months prior to the storm, Charles III had given concessions to coffee planters in the hope of stimulating that industry in Cuba, Puerto Rico, Santo Domingo, and Venezuela. Royal Order, Aranjuéz, 8 June 1768, exp. 27, leg. 19, CCG, ANC.

142. Bucareli to José de San Martín, 17 October 1768, in Villa of Guanabacoa to the Council of the Indies, Guanabacoa, 1772, leg. 1463, SD, AGI.

143. Armona to the Council of the Indies, Havana, 28 October 1768, leg. 256-B, Correos, AGI.

144. Armona to Arriaga, Havana, ibid.; Valdés, *Historia de la isla de Cuba*, 162–63.

145. José de la Cuesta to Bucareli, Castillo del Morro (Havana), 16 October 1768, leg. 1097, PC, AGI.

146. Armona to the Council of the Indies, Havana, 28 October 1768, leg. 256-B, Correos, AGI. The loss of fifty-three ships also hampered recovery efforts, especially the ability to go in search of provisions from unaffected areas.

147. Colina to Arriaga, Havana, 24 October 1768, folder 13, box 7, Escoto Collection.

148. Armona to the Council of the Indies, Havana, 28 October 1768, leg. 256-B, Correos, AGI; *Estado que comprende*, 1768, leg. 1594, SD, AGI.

149. Johnson, "La guerra contra los habitantes," 190.

150. "Sobre la compra y pago de terrenos y solares extramuros de esta ciudad," exp. 1334, libro 6, 1773, TC, ANC, in *BAN* 10 (May–June 1911): 130–31.

151. *Estado que comprende*, Havana, 9 November 1768, leg. 1594, SD, AGI.

152. Le Riverend Brusone, *Historia económica*, 6–8.

153. Risco Rodríguez, *Cuban Forests*, 19–24; Funes Monzote, *From Rainforest to Cane Field*.

154. In 1773, Charles III reiterated the prohibition on cutting the island's trees for sugar boxes. Royal Order, 25 February 1771, Aranjuéz, 31 July 1772, leg. 1201, PC, AGI.

155. Francisco García to Bucareli, Batabanó, 16 October 1768, leg. 1097, PC, AGI.

156. "Resumen de los estragos," Havana, 24 October 1768, ibid.

157. *Estado que comprende*, 9 November 1768, leg. 1594, SD, AGI.

158. "Relacion de los estragos que ha ocasionado el Uracan de el Dia 15 de el pasado en este Partido de Jesus del Monte de mi cargo Mandado reconoser Yndividualmente de orden de el Exmo Sor Governador y Cap Genl pr el oficio en que los Ynformes a su excelencia genel, a saber." Leg. 1097, PC, AGI.

159. "Noticia por (?razon) de los estragos y quebrantos que padecio el Partido N. S. de los Remedios de Managua en el dia 15 de Octubre de este año de 68 como a los dos y media de la tarde acacionados por un fuerte uracan." Ibid. The vagrant was termed a *guachinango*, a pejorative term used for people of Mexican origin.

160. Cristóval Flores to the Council of the Indies, Guanabacoa, 8 November 1769, leg. 1463, SD, AGI.

161. Juan de Diós Castro Palomino to Bucareli, Guadeloupe, 25 October 1768, leg. 1097, PC, AGI.

162. Leandro Luís Jospe to Antonio Pedro Charnum, Ingenio de Río Blanco, 16 October 1768, ibid.

163. *Estado que comprende*, Havana, 9 November 1768, leg. 1594, SD, AGI.

164. Ibid.

165. Castro Palomino to Bucareli, Guadeloupe, 25 October 1768, leg. 1097, PC, AGI. Escaping serious damage were 108 mortar-and-tile buildings, along with 228 wattle-and-daub houses.

166. Ventura Doral to Bucareli, Havana, 29 October 1768, ibid.

167. José de San Martín to the Council of the Indies, Guanabacoa, 17 October 1768, 8 November 1769, leg. 1463, SD, AGI.

168. Ibid.

169. Franciscan friar Bernardo de los Santos, parish priest Estéban Conde, and sacristan Francisco Josef de Melo to José de San Martín, Guanabacoa, 21 October 1768, ibid.

170. Antonio José Cardoso to Bucareli, San Miguel del Padrón, 27 October 1768, leg. 1097, ibid.

171. Francisco Blandino to Bucareli, Regla, 16 October 1768, ibid.

172. Leandro Luis Jospe to Antonio Pedro Charnum, Ingenio de Río Blanco, 16 October 1768, ibid.

173. Antonio Fernández de Medina to Bucareli, Bajurano, 23 October 1768, ibid. The description of *plátanos* as *"pan de los pobres"* comes from Francisco Xavier Enríquez to de la Torre, San Pedro, 14 November 1774, leg. 1192, PC, AGI.

174. Cardoso to Bucareli, San Miguel del Padrón, 27 October 1768, leg. 1097, PC, AGI.

175. Bucareli to Juan Miguel de Arozena, Havana, 20 December 1768, ibid.

176. Galán to Bucareli, Batabanó, 9 November 1768, 12 November 1768, 1 December 1768, leg. 1093, PC, AGI.

177. Bucareli to Cabildo of Guanabacoa, Havana, 17 October 1768, leg. 1463, SD, AGI.

178. Bucareli to Miguel Ibañez Cuevas, Havana, 20 December 1768, leg. 1077, PC, AGI; Bucareli to Castro Palomino, Havana, 12 November 1768, leg. 1093, PC, AGI.

179. Bucareli, Circular Letter to local constables, Havana, 4 February 1769, leg. 1071, PC, AGI.

180. Bucareli to Nicolás Duarte, Havana, 19 November 1768, leg. 1093, PC, AGI.

181. Ibid.

182. Antonio José Cardoso to Bucareli, San Miguel del Padron, 27 October 1768, leg. 1097, PC, AGI.

183. Rosendo López Silvero to Bucareli, Alvarez, 12 November 1768, leg. 1093, PC, AGI.

184. Bucareli to Ibáñez Cuevas, Havana, 28 October 1768, 20 December 1768; Bucareli to Arozena, Havana, 22 April 1769, 9 May 1769; both leg. 1077, PC, AGI. Bucareli to Castro Palomino, Havana, 12 November 1768, leg. 1093, PC, AGI.

185. Bucareli to the Conde de Aranda, Havana, 26 October 1768, 31 October 1768, leg. 1071, PC, AGI; García Rodríguez, *Misticísmo y capitales*, 57–158.

186. Bucareli to Castro Palomino, Havana, 31 October 1768, leg. 1093, PC, AGI.

187. Bucareli to Ibañez Cuevas, Havana, 28 October 1768, 20 December 1768; Bucareli to Arozena, Havana, 22 April 1769, 9 May 1769; both leg. 1077, PC, AGI; Bucareli to Castro Palomino, Havana, 12 November 1768, leg. 1093, PC, AGI.

188. Residents of Buenaventura, Aguas Verdes, Santo Cristo del Salud y Batabanó to de la Torre, 21 August 1775, leg. 1201, PC, AGI, citing the original decree by Bucareli.

189. Brito Betancourt to Bucareli, Jagua, 14 November 1768, leg. 1093, PC, AGI.

190. Bucareli to Aranda, Havana, 20 November 1768, leg. 1077, PC, AGI.

191. Brito Betancourt to Bucareli, Jagua, 14 November 1768, leg. 1093, PC, AGI.

192. Bucareli to Ibáñez Cuevas, Havana, 20 December 1768, leg. 1077, PC, AGI.

193. Jospe to Charnum, Ingenio de Río Blanco, 16 October 1768, leg. 1097, PC, AGI.

194. Bucareli to the Conde de Aranda, Havana, 26 October 1768, 31 December 1768, leg. 1071, PC, AGI.

195. Brito Betancourt to Bucareli, Jagua, 14 November 1768, leg. 1093, PC, AGI.

196. Juan Manuel Ochagavía to Bucareli, Jesús María y José, 6 November 1768, ibid.

197. Francisco Br[illegible] to Bucareli, San Miguel, 16 October 1768, leg. 1097, PC, AGI.

198. Juan José Galán to Bucareli, Batabanó, 9 November 1768, leg. 1093, PC, AGI.

199. Bucareli to Azorena, Havana, 20 December 1768, leg. 1077, PC, AGI.

200. Colina to Arriaga, Havana, 24 October 1768, folder 13, box 7, Escoto Collection.

201. Armona to the Council of the Indies, Havana, 28 October 1768, leg. 256-B, Correos, AGI.

202. Juan Álvarez de Miranda, "Poética relación Christiana y Moral," 1768, leg. 1097, PC, AGI.

203. Armona to the Council of the Indies, Havana, 11 December 1768, leg. 256-B, Correos, AGI.

204. Moore, "Antonio de Ulloa," 189–218; Moore, "Revolt in Louisiana," 40–55; Din, *Francisco Bouligny*, 31–35.

205. Torres Ramírez, *Alejandro O'Reilly*, 97–127.

206. Din, *Francisco Bouligny*, 29. Weddle, *Changing Tides*, 15–22, describes how weather and navigational difficulty plagued the O'Reilly expedition.

207. Olson, "Towards a Politics of Disaster," 283.

208. Acosta, "Las bases económicas," 364, 349.

209. Athanase de Mézières to Luis de Unzaga y Amézaga, 1 February 1770, leg. 110, PC, AGI; in Bolton, *Athanase de Mézières*, 147. The correspondent, Athanase de Mézières, reported that the wheat crop failed because of the extremely rainy season.

210. Acosta, "Las bases económicas," 363.

Chapter Four

1. "Junta celebrada del orden del admin. gral de la renta de correos del puerto de la Havana," Havana, 17 October 1773, leg. 257-A, Correos, AGI.

2. José Antonio de Armona, "Relato," 30 March 1774, ibid.

3. "Junta celebrada del orden del admin. gral de la renta de correos del puerto de la Havana," Havana, 17 October 1773, ibid.

4. Acosta, "Las bases económicas," 349, 364.

5. Council of the Indies, Decreto, San Lorenzo, 3 March 1769, leg. 2515, SD, AGI.

6. Governor of Havana and Intendant of Havana to the Council of the Indies, Havana, 3 April 1769, leg. 2666, SD, AGI.

7. Council of the Indies, Decreto, San Lorenzo, 3 March 1769, leg. 2515, SD, AGI.

8. Joaquín Pover, factor for the Asiento, San Juan, 1773, leg. 2516, SD, AGI. The cities included San Juan, Puerto Rico; Santo Domingo, in present-day Dominican Republic; Havana and Santiago de Cuba, Cuba; and La Guaira, Caracas, and Cumaná, in present-day Venezuela.

9. Goicoa to the governor and royal officials of Puerto Rico, Madrid, 1 December 1770, leg. 2515, SD, AGI; James Duff, Testament, 17 October 1795, leg. 5910; John Welsh, Testament, 3 July 1765; both leg. 3642, 679–83, Protocolos Notariales, Archivo Provincial Histórico de Cádiz, Cádiz, Spain.

10. Tentor to Arriaga, San Juan, 19 February 1769, leg. 2515, SD, AGI.

11. Stephen Moylan was born in 1737, in Cork, Ireland, one of four sons of John Moylan and the Countess of Limerick. Griffin, *Stephen Moylan*, 2.

12. (?) Sterling to Moylan, Aquakanoc, 8 October 1778, congratulating Moylan on his engagement to the daughter of Philip Van Horne, Stephen Moylan Letterbook, HSP.

13. Casa de Contratación to Gerónimo Enrile, Cádiz, 11 May 1769, leg. 5512, n. 3 r. 28, Contratación, AGI; Tentor to Arriaga, San Juan, 14 July 1769, leg. 2515, SD, AGI.

14. Casa de Contratación to Manuel Félix Riesch, Cádiz, 24 July 1769, leg. 5512, n. 3 r. 32, Contratación, AGI; Tentor to Arriaga, San Juan, 14 July 1769, leg. 2515, SD, AGI. See also Solano Pérez-Lila, *La pasión de reformar*, 203–4.

15. The reason that the flour was not sent to Cádiz remains unclear.

16. A variety of sources were used to reach this conclusion, including John J. McCusker, "Ships Registered at the Port of Philadelphia before 1776: A Computerized Listing," and Philadelphia Customs House Records, both HSP; and the PZ, LCP. In 1770, three Philadelphia ships cleared for Puerto Rico. The first to leave, the sloop *Nancy*, captained by W. Marshall and owned by G. Price and C. Alexander, cleared in March. Her capacity was listed as 25 tons. McCusker Registration 2022, HSP; PZ, 8 March 1770, LCP. In April, Willing and Morris sent their ship *Tyger* with a capacity of 180 tons. McCusker Registration 2011, HSP; PZ, 5 April 1770, LCP. In May, the *Nancy* cleared for her second voyage, once again with Captain Marshall at the helm; PZ, 10 May 1770, LCP. By September, however, Marshall had exchanged his small command for the larger sloop *Santa Maria*, carrying 70 tons, which cleared on 6 September; PZ, LCP. In addition, he was listed as co-owner, with Willing, Morris, Moylan, and Kennedy. McCusker Registration 2069, HSP. John J. McCusker's "The Pennsylvania Shipping Industry in the Eighteenth Century (1973)," unpublished study by John J. McCusker, and his companion data set, "Ships Registered at the Port of Philadelphia before 1776: A Computerized Listing," both HSP, provided the fundamental information to identify shipowners. These were combined with the Philadelphia Customs House Books, 1766–75, 3 vols., Cadwalader Collection, Series 3: Thomas Cadwalader Papers, HSP. Miscellaneous volumes are also HSP. Of course, none of these sources would have come to light without Doerflinger's *Vigorous Spirit*, 108–13, and especially his excellent discussion of his source material.

17. Goicoa to the Council of the Indies, San Ildefonso, 21 September 1770, leg. 2515, SD, AGI: "Ningún perjuicio puede adquirirse de que la compañía compre las arinas en las colonias porque como es notorio todas las arinas que hay en Cádiz para su embarque a yndias son extrañeras."

18. Council of the Indies to the Directors of the Asiento de Negros, San Ildefonso, 21 September 1770; Council of the Indies to the Marqués de San Juan, San Ildefonso, 22 November 1770; both ibid.

19. Bucareli to Muesas, Havana, 13 April 1769, exp. 121, leg. 19, CCG, ANC.

20. Kuethe, *Cuba*, 92–93.

21. Goicoa to the governor and royal officials of Puerto Rico, Madrid, 1 December 1770, leg. 2515, SD, AGI; and enclosing a copy of James Duff to the directors of the Asiento, Cádiz, 15 November 1770, 17 November 1770, ibid. Only three shipments of flour went to Puerto Rico in 1771, and all seem to have been carried by individual shippers. The most interesting was the first, the mysterious ship *Hercules*, listed in McCusker's registration as "unknown owners, unknown residence, unknown year built, and unknown tonnage." McCusker Registration 2096, HSP. The *PZ* reported that the *Hercules* cleared for Puerto Rico on 9 May 1771. The other two shipments went on the *Don Carlos* in July (*PZ*, 11 July 1771, McCusker Registration 2124, HSP) and the *Morris* in August (*PZ*, 15 August 1771, McCusker Registration 1912, HSP).

22. Felipe Fonsdeviela, Marqués de la Torre, Captain General of Cuba, "Carta Circular," Havana, 6 June 1772, leg. 1168, PC, AGI. The Royal Order issued on 14 January 1772 was straightforward: "No permite comercio y no se permite la entrada en ningún puerto de S.M. a buques estrangeros si sean de guerra o de comercio. [Commerce is prohibited and no foreign boats, whether warships or merchantmen, will be permitted to enter any of His Majesty's ports.]"

23. Bucareli to Muesas, Havana, 13 April 1769, exp. 125, leg. 19, CCG, ANC.

24. Clemente Pérez to Ramón Vuelta Flores, Puerto Guayaja, 11 December 1771; Vuelta Flores to de la Torre, Puerto Príncipe, 13 December 1771; both leg. 1168, PC, AGI.

25. Clemente Pérez to Vuelta Flores, 11 December 1768, ibid.

26. Juan Joseph Eligio de la Puente to de la Torre, Santiago de Cuba, 12 December 1771, leg. 1165, PC, AGI.

27. Consulta, Council of the Indies, Madrid, 21 December 1772, leg. 508, Ultramar, AGI.

28. Bucareli to Juan Ayans de Ureta, Havana, 16 May 1771, exp. 146, leg. 19, CCG, ANC.

29. Juan de Lleonart to de la Torre, Puerto Príncipe, 6 March 1772, leg. 1168, PC, AGI.

30. Lleonart to de la Torre, Puerto Príncipe, 6 June 1772, ibid.

31. Lleonart to de la Torre, Puerto Príncipe, 24 August 1772, 5 October 1772, ibid.

32. Muesas to Manuel Varela, Cabo Comandante del Cobre, Santiago de Cuba, 28 August 1769, exp. 87, leg. 26, CCG, ANC. A peso equals 8 reales; an average worker earned 2 reales daily.

33. Ayuntamiento de Santa Clara, Actas Capitulares, 26 October 1770, tomo 8, 1772–79, Archivo Provincial de Santa Clara, Cuba.

34. Bucareli to Ayans, Havana, 18 May 1771, exp. 151, leg. 19, CCG, ANC.

35. Casa Cagigal to Bartolomé de Morales, Santiago de Cuba, 21 April 1771, exp. 64, leg. 24, CCG, ANC.

36. Luis de Unzaga y Amézaga to Bucareli, New Orleans, 8 July 1770, 31 August 1770, 11 November 1770, "Despatches of the Spanish Governors of Louisiana," vol. II, typescript and translation in Manuscripts Department, Howard-Tilton Memorial Library, Tulane University, New Orleans; Florescano, *Precios de maíz*, 60, 72–75.

37. Governor of Cumaná to the Council of the Indies, Cumaná, 1 February 1770, leg. 2515, SD, AGI.

38. Armona to the Council of the Indies, Havana, 7 February 1771, leg. 257-A, Correos, AGI.

39. Muesas to the directors of the Asiento, San Juan, 1771; Pover to Muesas, San Juan, 1771; directors of the Asiento to the governor of Puerto Rico, Cádiz, 1771; all leg. 2516, SD, AGI.

40. William Smith to Thomas Clifford, St. Eustatius, 7 January 1771, Clifford Family Correspondence, HSP.

41. Ayans de Ureta to de la Torre, Santiago de Cuba, 2 May 1772, leg. 1141, PC, AGI.

42. De la Torre to Arriaga, Havana, 20 June 1772, leg. 1216, PC, AGI; de la Torre to Juan Bautista Bonet, Havana, 26 June 1772, leg. 1159, PC, AGI.

43. Julián de Arriaga to Miguel de Altarriba, Intendant of Havana, El Pardo, 20 February 1772, Cédulas and Orders, ANC, translation and transcription in Manuscripts Department, Howard-Tilton Memorial Library, Tulane University, New Orleans. The governor of Louisiana was exonerated, the vessel and its owner/captain, Bartolomé Beauregard, were released, and Beauregard was reimbursed for the damages incurred by the hasty seizure.

44. Goicoa to the Council of the Indies, Madrid, 6 July 1772, leg. 2516, SD, AGI.

45. Millás, *Hurricanes of the Caribbean*, 229–39; Caviedes, "Five Hundred Years of Hurricanes," 304. Millás believes that as many as nine storms made landfall, but it is probable that several of these individual storms were actually part of one hurricane. In addition, it is clear that at least two hurricanes that made landfall in North America (New Orleans and the outer banks of North Carolina) originated in the Caribbean. Ludlum, *Early American Hurricanes*, 63–64.

46. Ayans de Ureta to de la Torre, Santiago de Cuba, 25 June 1772, leg. 1141, PC, AGI.

47. Millás, *Hurricanes of the Caribbean*, 229–30.

48. Ibid., 230.

49. Josef Alvarado to de la Torre, Bayamo, 22 August 1772, leg. 1178; Ayans to de la Torre, Santiago de Cuba, 23 August 1772, 4 September 1772, leg. 1141; de la Torre to Arriaga, Havana, 23 August 1772, leg. 1216; all PC, AGI.

50. Millás, *Hurricanes of the Caribbean*, 230; Miner Solá, *Historia de los huracanes en Puerto Rico*, 30.

51. Armona to the Council of the Indies, Havana, 31 August 1772, 19 October 1772, leg. 257-A, Correos, AGI.

52. Unzaga to de la Torre, New Orleans, 9 September 1772, "Despatches of the Spanish Governors of Louisiana," Manuscripts Department, Howard-Tilton Memorial Library, Tulane University, New Orleans; Ludlum, *Early American Hurricanes*, 63–64.

53. Millás, *Hurricanes of the Caribbean*, 235–37, describes this "severe storm of 1772" as an event unprecedented in meteorological history. See also *PZ*, 23 September 1772, 14 October 1772, LCP.

54. Muesas to the Council of the Indies, San Juan, (day omitted) September 1772, leg. 2516, SD, AGI.

55. Ayans de Ureta to de la Torre, Santiago de Cuba, 4 September 1772, leg. 1141, PC, AGI; Ludlum, *Early American Hurricanes*, 64.

56. Armona to the Council of the Indies, Havana, 30 March 1774, leg. 257-A, Correos, AGI. Armona summarizes the progress made by the royal mail system in the wake of the terrible season in 1772.

57. Ayans to Arriaga, Santiago de Cuba, 3 October 1772, leg. 1216; Ayans to de la Torre, Santiago de Cuba, 23 August 1772, 4 September 1772, leg. 1141; both PC, AGI.

58. Lleonart to de la Torre, Puerto Príncipe, 24 August 1772, leg. 1168; de la Torre to Lleonart, Havana, 11 September 1772, leg. 1172; de la Torre to lieutenant governor of Baracoa, Havana, 19 October 1772, leg. 1143; all PC, AGI.

59. Ayuntamiento de Santa Clara, Actas Capitulares, 4 September 1772, Tomo 8, 1772–79, Archivo Provincial de Santa Clara, Cuba.

60. Altarriba to de la Torre, Havana, 21 October 1772, leg. 1154, PC, AGI.

61. Unzaga to de la Torre, New Orleans, 9 September 1772, "Despatches of the Spanish Governors of Louisiana," Manuscripts Department, Howard-Tilton Memorial Library, Tulane University, New Orleans.

62. Ayans de Ureta to de la Torre, Santiago de Cuba, 25 June 1772, leg. 1141, PC, AGI.

63. Muesas to the Council of the Indies, San Juan, (day omitted) September 1772; Pover to the Council of the Indies, San Juan, 10 September 1772; both leg. 2516 (1773), SD, AGI.

64. Ibid.

65. *PZ*, 30 September 1772, LCP.

66. Armona to the Council of the Indies, Havana, 30 March 1774, leg. 257-A, Correos, AGI.

67. The letter is reprinted in Millás, *Hurricanes of the Caribbean*, 238. See also Atherton, *A Few of Hamilton's Letters*.

68. Caviedes, *El Niño in History*, 104–5; Claxton, "Record of Drought," 216–22; Florescano, *Precios de maíz*, 60, 72–75.

69. *PZ*, 14 October 1772, LCP.

70. De la Torre to Ayans, Havana, 20 November 1772, leg. 1216, PC, AGI.

71. It is not clear how this contact was made. There is no record of any ships under Spanish colors entering the port of Philadelphia from September through December 1772. Possibly the news was relayed by one of a number of vessels that sailed from other islands northward.

72. Muesas to the Council of the Indies, San Juan, (day omitted) September 1772; Pover to the Council of the Indies, San Juan, 10 September 1772; both leg. 2516 (1773), SD, AGI.

73. De la Torre to Arriaga, Havana, 16 August 1772, leg. 1216, PC, AGI.

74. De la Torre to Arriaga [letter 264], Havana, 28 September 1772, ibid.

75. "Extracto relacionado del principio del Real Asiento de Negros," Council of the Indies, San Lorenzo, 6 July 1772, leg. 2516, SD, AGI.

76. Aguirre and Arístegui and Company to the Council of the Indies, Cádiz, 25 August 1772, leg. 2820A, IG, AGI. From 23 September 1765 through 25 August 1772 the company lost 928,916 pesos, and it claimed it was still owed 596,326 pesos for slaves and flour it had imported into Cuba.

77. Aguirre and Arístegui to the Council of the Indies, Cádiz, 13 November 1772, ibid.

78. Ibid., 28 June 1772.

79. "Junta formada del orden de V.M.," Madrid, 15 June 1773, leg. 1211, SD, AGI; photocopies in vol. 25, Levi Marrero Collection, Special Collections, Florida International University, Miami.

80. Chief Tobacco Factor to Ayans y Ureta, Havana, 2 October 1773, leg. 1227, PC. AGI. It is quite possible that many of the slaves that were employed in the construction projects in and around Havana became the property of the state simply because no other buyers came forward. See Jennings, "War as the 'Forcing House of Change.'"

81. Tomás Ortíz de Landázuri and Manuel Lanz de Casafonda to Julián de Arriaga, San Lorenzo, 13 July 1772, leg. 2820A, IG, AGI.

82. Ortíz de Landázuri and Lanz de Casafonda to Arriaga, San Lorenzo, 13 November 1772, ibid.

83. Casa de Contratación to Enrile, Cádiz, 17 May 1773, leg. 5518, n. 2, r. 11, Contratación, AGI.

84. "Instrucción para el nuevo Giro y Govierno que desde ahora ha de tener la continuación de Assiento de Negros concedidas . . . por SM," San Lorenzo, 13 November 1772, leg. 2820A, IG, AGI; de la Torre to the Governor of Puerto Rico, Havana, 9 September 1773, leg. 2516, SD, AGI.

85. Ortíz de Landázuri and Lanz de Casafonda to Arriaga, San Lorenzo, 13 July 1772, leg. 2820A, IG, AGI.

86. Ortíz de Landázuri and Lanz de Casafonda to Arriaga, San Lorenzo, 13 July 1772, 13 November 1772, ibid.

87. Ibid.

88. Chief Tobacco Factor to Ayans y Ureta, Havana, 2 October 1773, leg. 1227, PC, AGI; Royal Order, 1 September 1772, Madrid, leg. 2820A, IG, AGI; "Junta formada del orden de V.M.," Madrid, 15 June 1773, leg. 1211, SD, AGI.

89. Ibid.; Arriaga to de la Torre, Aranjuéz, 30 April 1773, leg. 1212, PC, AGI.

90. James Duff, Duff and Welsh, to Willing and Morris, Cádiz, 17 July 1776, Folder July–August 1776, Willing and Morris Correspondence, Robert Morris Papers, Levis Collection, HSP.

91. PZ, LCP; McCusker, "Pennsylvania Shipping Industry" and "Ships Registered," HSP; Philadelphia Customs House Books, Vol. C, 1772–75, HSP.

92. Kuethe, "El fin del monopolio," 35–66. The rivalry among Spanish port cities after the 1760s is well studied in Ringrose, Spain, Europe, and the "Spanish Miracle," 88–91, 109–19. A contemporary account is "A Spaniard," Observations on the Commerce of Spain with Her Colonies in Time of War. Although anonymous, the author is believed to be the Marqués de Casa Yrujo, the Spanish consul in Philadelphia, who wrote: "Eight or ten commercial houses of Cádiz were, in reality, masters of the trade of Spain from Florida to California and the shipments in Spain, sales in America, and returns home only displayed a ruinous chain of the scandalous monopoly." Ibid., 11–12, LCP.

93. Arriaga to de la Torre, Madrid, 14 December 1772; Arriaga to de la Torre, El Pardo, 1 February 1773; both leg. 1212, PC, AGI.

94. Ayans de Ureta to de la Torre, Santiago de Cuba, 12 November 1772, 16 November 1772, 4 December 1772, 8 December 1772, leg. 1141, PC, AGI.

95. De la Torre, "Carta Circular," Havana, 21 October 1772, leg. 1154, PC, AGI.

96. De la Torre to Arriaga [letter 263], Havana, 28 September 1772, leg. 1216, PC, AGI.

97. De la Torre, "Bando," Havana, 24 August 1773, leg. 1167, PC, AGI.

98. De la Torre to Ayans, Havana, 22 October 1772, leg. 1143, PC, AGI.

99. Ayans de Ureta to de la Torre, Santiago de Cuba, 12 November 1772, 16 November 1772, 4 December 1772, 8 December 1772, leg. 1141; de la Torre to Ayans y Ureta, Havana, 2 December 1772, 16 December 1772, leg. 1143; de la Torre to the Council of the Indies, Havana, 2 November 1772, 11 February 1773, leg. 1151; all PC, AGI.

100. De la Torre to the Council of the Indies, Havana, 11 February 1773, leg. 1151, PC, AGI.

101. Philadelphia continued to send ships to Puerto Rico over the summer of 1773. *PZ*, 21 July 1773, 29 September 1773, 29 December 1773, LCP. Two of the voyages, both by the brig *Repeal*, were financed by Willing and Morris. McCusker, "Ships Registered," and McCusker Registration 2324, HSP.

102. Lleonart to de la Torre, Puerto Príncipe, 12 November 1772, leg. 1168, PC, AGI.

103. Altarriba to de la Torre, Havana, 31 August 1772, leg. 1151, PC, AGI.

104. Lieutenant Governor of Baracoa to de la Torre, Baracoa, 19 October 1772, leg. 1143, PC, AGI.

105. Altarriba to de la Torre, Havana, 7 August 1772, 31 August 1772, leg. 1151, PC, AGI.

106. Altarriba to de la Torre, Havana, 7 August 1772, 31 August 1772, 3 September 1772, 29 October 1772, 10 November 1772, 20 February 1773, ibid.

107. Francisco Joseph Roxas Sotolongo to de la Torre, Jesús del Monte, 2 September 1772, leg. 1189; Esteban Rodríguez del Pino to de la Torre, Bejucal, 6 September 1772, leg. 1167; de la Torre to Rodríguez del Pino, Havana, 9 September 1772, 17 September 1772, leg. 1167; all PC, AGI.

108. Rodríguez del Pino to de la Torre, Bejucal, 6 September 1772; de la Torre to Rodríguez del Pino, Havana, 9 September 1772, 17 September 1772; both leg. 1167, PC, AGI.

109. De la Torre to Arriaga, Havana, 23 August 1772, leg. 1216, PC, AGI.

110. De la Torre to Arriaga, Havana, 3 September 1772, ibid.

111. De la Torre to Arriaga, Havana, 13 September 1772, ibid.

112. Arriaga to de la Torre, San Lorenzo, 25 November 1772, leg. 1229, PC, AGI.

113. De la Torre to Arriaga, Havana, 20 February 1773, leg. 1212, PC, AGI.

114. "Junta celebrada el día 23 de noviembre de 1772 sobre escoger los medios más oportunos para atajar los perjuicios que recibe la bahia de las arenas y tierras que introducen en ella." Leg. 1228, PC, AGI.

115. "Junta celebrada . . . ," Havana, 24 December 1772, ibid.

116. Lorenzo de Montalvo to de la Torre, Havana, 24 December 1772, leg. 1202, PC, AGI. The estimate included 29,841 pesos for materials and 16,600 pesos for labor. The total with additional costs was 49,503 pesos, 4 reales.

117. "Junta celebrada . . . ," Havana, 23 November 1772, leg. 1228, PC, AGI.

118. De la Torre to Urriza, Havana, 20 March 1776, leg. 1153, PC, AGI.

119. Arriaga to de la Torre, Madrid, 12 December 1772, leg. 1212, PC, AGI, replying to de la Torre's report of 13 September 1772.

120. De la Torre to Bonet, Havana, 11 May 1773, leg. 1159; Arriaga to de la Torre, El Pardo, 13 February 1773, leg. 1212; both PC, AGI. The phrase reads, "Todo ha sido muy del agrado de SM."

121. De la Torre to Real Agrado, Havana, 6 October 1772, leg. 1229, PC, AGI.

122. De la Torre to Real Agrado, Havana, 11 November 1772, ibid.

123. Miguel Núñez and Sebastián Rodríguez to de la Torre, Guanabacoa, 6 July 1774, ibid. The normal daily wage for a laborer was 2 *reales*, so the assessment was likely not unduly burdensome for most of the community.

124. De la Torre to Real Agrado, Havana, 11 November 1772, ibid.

125. Ignacio José Barboa to Real Agrado, Guanabacoa, 16 July 1773, ibid.

126. San Martín to Real Agrado, Guanabacoa, 23 March 1774, ibid.

127. Domingo Santaya to Real Agrado, Buena Vista, 29 August 1774; Real Agrado to de la Torre, Guanabacoa, 1 September 1774; both ibid.

128. Real Agrado to de la Torre, Havana, 4 September 1775, ibid.

129. De la Torre to Rodríguez del Pino, Havana, 9 September 1772, 17 September 1772, leg. 1167, PC, AGI.

130. Nicolás Duarte to de la Torre, Isla de Pinos, 4 November 1772, ibid.

131. He was likely related to the constable of Bejucal, Estéban Rodríguez del Pino.

132. De la Torre to Nicolás Rodríguez del Pino, Havana, 9 July 1772, leg. 1195, PC, AGI.

133. Felipe Nuñez de Villavicencio to de la Torre, San Miguel del Padrón, August 1773, leg. 1193; de la Torre to Pedro de la Cruz Guerra, Havana, 11 June 1775, leg. 1196; both PC, AGI.

134. Martín Estéban de Aróstegui to de la Torre, Managua, 2 May 1776, leg. 1201, PC, AGI.

135. Marrero y Artiles, *Cuba*, 7:43–56.

136. Roxas Sotolongo to de la Torre, Jesús del Monte, 2 September 1772; de la Torre to Roxas Sotolongo, Havana; both leg. 1189, PC, AGI.

137. De la Torre to Martín Navarro, Havana, 15 December 1772, ibid.

138. Ibid. Universal militia service was intended primarily as a defense measure, but its secondary purpose was internal police functions. All men between the ages of fifteen and forty-five were liable for militia service and were thus brought under the jurisdiction of military justice. Johnson, *Social Transformation*, 49.

139. Roxas Sotolongo to de la Torre, Jesús del Monte, 9 February 1773, leg. 1189, PC, AGI.

140. De la Torre, "Comisiones dadas para aprehensión de desertores, recolección de negros, etc.," Havana, 11 August 1773, leg. 1229, PC, AGI, citing Ricla's commission of 29 October 1763 and Bucareli's commission of 23 February 1768.

141. De la Torre to Josef Gil, Havana, 7 April 1772, ibid.

142. De la Torre to Patricio Enríquez, Havana, 22 May 1772, 11 December 1772, ibid. His commission reads that he is "un sujeto inteligente de notoria honradez y responsavilidad de buena conducta y demas circunstancias que se requieren."

143. De la Torre to Sebastián de Espinosa, Havana, 27 March 1776, leg. 1229, PC, AGI.

144. José Ignacio de Rapún, Intendant of Cuba, to de la Torre, Havana, 14 June 1772, 21 April 1773, leg. 1152, PC, AGI.

145. De la Torre to the Alcaldes de la Santa Hermandad, Havana, 5 August 1772, leg. 1229, PC, AGI.

146. María Basabe, Ingenio and hacienda de Divina Pastora, Guatao, 7 June 1773, ibid.

147. Marquesa de Cárdenas de Montehermosa to de la Torre, Bahia Honda, 24 March 1773, leg. 1167, PC, AGI.

148. Conde de Buena Vista, Félix de la Torre, Ignacio Loynaz, Félix de Acosta y Riana, and José de la Guardia to de la Torre, Managua, 21 May 1773; María Basabe to de la Torre, Guatao, 7 June 1773; both leg. 1229, PC, AGI.

149. Marquesa de Cárdenas de Montehermosa to de la Torre, Bahia Honda, 24 March 1773, leg. 1167, PC, AGI.

150. Joseph del Castillo to de la Torre, Matanzas, 11 December 1772; de la Torre to Estevan Fiallo, Havana, 20 December 1772; de la Torre to Antonio de Fuentes, Havana, 12 March 1773; Cabrera to de la Torre, Havana, 20 July 1773; de la Torre to Tomás Labrador, Havana, 1 September 1773; de la Torre to Joseph Quijano, Havana, 1 March 1774; all leg. 1229, PC, AGI.

151. Isidro José de Limonta to de la Torre, "Relación de las embarcaciones que han entrado y salido de este puerto," Santiago de Cuba, 12 December 1775, 15 December 1775, 9 April 1776, leg. 1142, PC, AGI.

152. Pedro Barral to Gerónimo Enrile, Kingston, 11 September 1773, leg. 2820A, IG, AGI.

153. Arriaga to de la Torre, Madrid, 9 March 1775, leg. 1165; Ayuntamiento de Havana to de la Torre, Havana, 11 May 1775, leg. 1229; both PC, AGI.

154. De la Torre to Vicente de Justíz, San Antonio, 10 January 1773, leg. 1184, PC, AGI.

155. Arriaga to de la Torre, El Pardo, 20 February 1773, leg. 1212, PC, AGI.

156. Bonet to de la Torre, Havana, 6 November 1773, leg. 1158, PC, AGI, relaying the news from Josef Días Amador in Matanzas.

157. Goicoa to the Council of the Indies, Madrid, 20 December 1773, leg. 2516, SD, AGI.

158. Nicolás Cárdenas Vela de Guevara to de la Torre, Jaruco, 2 November 1774, leg. 1165; Estéban Rodríguez del Pino to de la Torre, Bejucal, 12 November 1774, leg. 1167; Enríquez to de la Torre, San Pedro, 14 November 1774, leg. 1192; Sebastián de la Cruz to de la Torre, Güines, 17 November 1774, 19 November 1774, leg. 1195; all PC, AGI.

159. De la Torre to Bonet, Havana, 10 December 1774, leg. 1159, PC, AGI.

160. De la Torre to Luis de Toledo, Havana, 1 January 1775, ibid.

161. Justíz de Santa Ana to de la Torre, Matanzas, 26 December 1774, ibid.

162. Bartolomé de Montes to Bonet to de la Torre, Havana, 13 January 1775, leg. 1158, PC, AGI.

163. Ayans y Ureta to de la Torre, Santiago de Cuba, 21 September 1775, 7 December 1775, leg. 1142, PC, AGI.

164. Juan de Ayans y Ureta, "Noticia relativa de las que han dado los priores de campo en sus respectivos partidos de esta jurisdicción de Cuba sobre el estago hecho por el uracán de viento y agua los dias 28 y 29 de agosto proximo pasado." Santiago de Cuba, 1 November 1775, ibid.

165. Ayans de Ureta to de la Torre, Santiago de Cuba, 17 February 1776, leg. 1227, PC, AGI.

166. Ayans de Ureta to de la Torre, Santiago de Cuba, 1 September 1775, leg. 1142, PC, AGI.

167. Ayans de Ureta to de la Torre, Santiago de Cuba, 17 February 1776, leg. 1227, PC, AGI.

168. Ibid.

169. Ayans de Ureta to de la Torre, Santiago de Cuba, 17 February 1776, leg. 1142, PC, AGI. On the new military hospital, see de la Torre to Rebollar, Havana, 7 December 1774, leg. 1184, PC, AGI.

170. Estéban de Oloríz to de la Torre, Santiago de Cuba, 24 June 1776, leg. 1143, PC, AGI.

171. Oloríz to de la Torre, Santiago de Cuba, 23 August 1776, leg. 1143, PC, AGI; de la Torre to Conde de Ricla, Havana, 29 May 1776, leg. 2079, SD, AGI.

172. De la Torre to Conde de Ricla, Havana, 29 May 1776, leg. 2079, SD, AGI.

173. Ibid.

174. De la Torre to Mateo de Echavarría, Havana, 2 March 1776, leg. 1184, PC, AGI. His request was granted, with the stipulation that he leave a responsible person in charge of his duties.

175. Ayans de Ureta to de la Torre, Santiago de Cuba, 21 March 1776, leg. 1142, PC, AGI.

176. De la Torre to Juan Manuel de Rebollar, Havana, 3 April 1776, leg. 1184, PC, AGI.

177. Ayans de Ureta to de la Torre, Santiago de Cuba, 11 May 1776, leg. 1142, PC, AGI.

178. Isidro de Limonta, "Relacion de las embarcaciones que han entrado y salido de este puerto para otros continentes desde 1 de junio hasta el dia de la fecha, y son los siguientes." Santiago de Cuba, 2 June 1776, leg. 1143, PC, AGI.

179. Oloríz to de la Torre, Santiago de Cuba, 4 July 1776, ibid.

180. Oloríz to de la Torre, Santiago de Cuba, 18 October 1776, ibid.; de la Torre to Lleonart, Havana, 5 November 1776, leg. 1184, PC, AGI.

181. See Cook and Lovell's collection of essays, Secret Judgments of God, including chapters by Villamarín and Villamarín, "Epidemics in the Sábana de Bogotá, 1536–1810," 113–41; Lovell, "Disease in Early Colonial Guatemala," 49–83; and Casanueva, "Smallpox and War in Southern Chile in the Late Eighteenth Century," 183–212. See also Cook and Lovell's insightful summary, "Unravelling the Web of Disease," 213–42. See also Fenn, Pox Americana, 14–22. Cook blamed smallpox introduced into the Caribbean in 1518 for the "first pandemic" on the mainland, beginning in 1519. Cook, Born to Die, 60–85.

182. Cook and Lovell, "Unravelling the Web of Disease," in Cook and Lovell, Secret Judgments of God, 217–20; Kiple, The Caribbean Slave, 144–45; Fenn, Pox Americana, 14–22.

183. Fenn, Pox Americana, 21, argues that in the years leading up to Edward Jenner's discovery of the vaccine, the disease was increasing in its virulence.

184. Actas Capitulares, Ayuntamiento de Santiago de Cuba, vol. 12, Santiago de Cuba, 14 June 1776, AHM.

185. Ibid. Joaquín Pover, "Relación," San Juan, 20 February 1770; Miguel Barrera, "Testimonio," Cádiz, 1771; both leg. 2516, SD, AGI. Rigau-Pérez, "Smallpox Epidemics in Puerto Rico," 429–30, cites the Royal Order and establishes that this practice was carried out in

San Juan. In Santiago, these regulations were followed in precise detail. Using lime as a disinfecting agent was described in James Freeman Curtis, "Logbook of the Voyage of the US Schooner *Porpoise*," Curtis-Stevenson Family Papers, Massachusetts Historical Society, Boston.

186. Actas Capitulares, Ayuntamiento de Santiago de Cuba, vol. 12, Santiago de Cuba, 14 June 1776, AHM.

187. Ibid., 28 June 1776.

188. Estéban de Oloríz to de la Torre, Santiago de Cuba, 26 June 1776, 3 July 1776, leg. 1143, PC, AGI.

189. De la Torre to Limonta, Havana, 13 September 1776, leg. 1184, PC, AGI.

190. Juan José Eligio de la Puente to José de Gálvez, Santiago de Cuba, 28 June 1776, leg. 1216, SD, AGI.

191. Actas Capitulares, Ayuntamiento de Santiago de Cuba, vol. 12, Santiago de Cuba, 27 September 1776.

192. Ibid., 6 November 1775, 28 June 1776. Ayans de Ureta to de la Torre, Santiago de Cuba, 6 November 1775; Ayuntamiento de Santiago de Cuba to Ayans de Ureta, Santiago de Cuba, 6 November 1775; both leg. 1142, PC, AGI. For background on smallpox and the slave trade, see Miller, *Way of Death*; Stewart, "Edge of Utility," 54–70; and Brown, "African Connection," 2247–49.

193. Actas Capitulares, Ayuntamiento de Santiago de Cuba, vol. 12, Santiago de Cuba, 28 June 1776.

194. Conde de Ripalda to de la Torre, Trinidad, 28 May 1776, leg. 1171, PC, AGI.

195. "Memoriales pidiendo," Havana, 4 May 1776, leg. 1229, PC, AGI.

196. Ibid., 2 April 1776. Isidro Limonta, "Relacion de las embarcaciones que han entrado y salido de este puerto para otros continentes desde el dia 1 de este mes hasta el de la fecha." Santiago de Cuba, 19 March 1776, leg. 1142, PC, AGI.

197. Rigau-Pérez, "Smallpox Epidemics in Puerto Rico."

198. Limonta, "Relacion de las embarcaciones," Santiago de Cuba, 2 June 1776, leg. 1143, PC, AGI.

199. Rigau-Pérez, "Smallpox Epidemics in Puerto Rico," 429–30.

200. Fenn, *Pox Americana*, 57–78.

201. Ibid., 14–22. Fenn argues that this was a true pandemic since smallpox subsequently spread to many areas of the Americas, such as Mexico, Hudson Bay, and the northwest Pacific Coast.

202. Francisco Eligio de la Puente, "Relacion de los meritos de Francisco Eligio de la Puente," 1733, exp. 79, leg. 145, IG, AGI.

203. Corbitt, "Spanish Relief Policy," 67–82; Gold, "Settlement of East Florida Spaniards," 216–31.

204. De la Torre to Ayans y Ureta, Havana, 20 March 1776, leg. 1144, PC, AGI.

205. Actas Capitulares, Ayuntamiento de Santiago de Cuba, vol. 12, Santiago de Cuba, 25 October 1776.

206. Eligio de la Puente to José de Gálvez, Santiago de Cuba, 28 June 1776, leg. 1216, SD, AGI.

207. De la Torre to Eligio de la Puente, Havana, 6 May 1776, leg. 1182, PC, AGI.

Chapter Five

1. Ortíz de la Tabla y Ducasse, *Comercio exterior de Veracruz*, 12–16.

2. Rodríguez Casado, *La política marroquí*, 236; Ferrer del Río, *Historia del reinado de Carlos III*, 4:110; Saavedra, *Los decenios*, 82.

3. Lynch, *Bourbon Spain*, 253–54. See also Priestly, *José de Gálvez*, written nearly a century ago.

4. *PZ*, 12 October 1776, 16 October 1776, LCP.

5. Morales Padrón, *Journal of Don Francisco Saavedra*, xvi–xxiv.

6. James Duff to Robert Morris, Cádiz, February 1776, Folder January–August 1776, Willing and Morris Correspondence, Robert Morris Papers, Levis Collection, HSP; Johnson, *Social Transformation*, 39–60.

7. Antonio López de Toledo to Juan Ignacio de Urriza, Batabanó, 18 June 1776, leg. 1152, PC, AGI.

8. Urriza to Estévan del Pino, Havana, 18 June 1776, ibid.

9. François DePlessis to Morris, New Orleans, 16 August 1776, Willing and Morris Correspondence, Robert Morris Papers, Levis Collection, HSP.

10. Martín de Aróstegui to Diego de Navarro, Matanzas, 27 July 1777, leg. 1248, PC, AGI.

11. Joseph Alvarado to Navarro, Trinidad, 22 October 1777, leg. 1259, PC, AGI.

12. José Días Tejada to Navarro, Bayamo, 7 December 1777, leg. 1254, PC, AGI.

13. Navarro to Joseph Tentor, Havana, 9 January 1778, acknowledging Tentor's report of 9 November 1777, folder 1, box 4, Domingo del Monte Collection, Manuscript Division, LC.

14. José Melchor de Acosta to Navarro, Guarico, 22 December 1777, leg. 1245, PC, AGI.

15. Navarro to Tentor, Havana, 9 January 1778, folder 1, box 4, Domingo del Monte Collection, Manuscript Division, LC.

16. Aróstegui to Navarro, Matanzas, 5 August 1777, leg. 1248, PC, AGI.

17. Luis Huet to Navarro, Havana, 2 August 1778, 24 September 1778, leg. 1247, PC, AGI.

18. Villavicencio to Navarro, San Miguel del Padrón, 20 October 1778, leg. 1269, PC, AGI.

19. Claxton, "Record of Drought," 195–216.

20. Ibid., 65–78; Florescano, *Precios de maíz*, 61.

21. Claxton, "Record of Drought," 207.

22. Ibid., 216.

23. John Joachim Zubly, "John Adams's Notes of Debates," 12 October 1775, in Smith, *Letters of Delegates*, 2:165–66.

24. Chronology of Congress, September–December 1775, in ibid., 2:xiii–xv.

25. Secret Committee Minutes of Proceedings, 27 September 1775, in ibid., 2:75–76.

26. *PZ*, July 1774–July 1775, LCP; Bezanson, Gray, and Hussey, *Prices in Colonial Pennsylvania*, 49–51, explains why prices were so volatile in Philadelphia at the time.

27. Philadelphia Customs House Records and *PZ*, 1766–75, LCP; McCusker, "Ships Registered," and McCusker, "Pennsylvania Shipping Industry," 236–62, LCP.

28. Firms such as Duff and Welsh were British merchant houses already established in Cádiz that normally received finished goods from Britain for sale to Spanish firms. The

firms' receipt of flour from Philadelphia involved little additional effort on their part except to market the product quickly to avoid spoilage.

29. Duff and Welsh to Willing and Morris, Cádiz, 17 July 1776, Folder June–July 1776, Willing and Morris Correspondence; Noble and Harris to Willing and Morris, Cádiz, 16 March 1776, Folder Willing and Morris Business Accounts, 1775; all Robert Morris Papers, Levis Collection, HSP.

30. Etienne Cathalan to Willing and Morris, Marseilles, 7 January 1776, Folder January–May 1776, Willing and Morris Correspondence; Etienne Cathalan to Willing and Morris, Marseilles, 18 October 1774, Willing and Morris Business Accounts, Folder 1771–74; all Robert Morris Papers, Levis Collection, HSP.

31. Morris to John Bradford, Philadelphia, 23 January 1777, Correspondence 1776, reel 12, Papers of Robert Morris, microfilm copies in Manuscript Division, LC.

32. Kuethe, "El fin del monopolio," 59–66; Ringrose, Spain, Europe, and the "Spanish Miracle," 107–10; "A Spaniard," Observations on the Commerce of Spain with Her Colonies in Time of War.

33. Duff and Welsh to Willing and Morris, Cádiz, 17 July 1776, Folder June–July 1776, Willing and Morris Correspondence, Robert Morris Papers, Levis Collection, HSP.

34. Duff and Welsh to Willing and Morris, (day omitted) November 1776, Cádiz, Folder November–December 1776, ibid. Duff wrote that seventy-eight barrels of superfine flour were sent to Havana and that payment for them was still pending.

35. "Junta formada por orden de V.M.," Madrid, 11 February 1775, leg. 2516, SD, AGI. The misgivings of the four counselors were among the reasons that Eligio de la Puente was sent on his fact-finding mission to the major port cities of the Hispanic Caribbean.

36. Armona and Urriza to Directores de Rentas Generales, Havana, 9 October 1775, box 4, folder 1, Domingo del Monte Collection, Manuscript Division, LC.

37. "Junta formada por orden de V.M.," Madrid, 11 February 1775, leg. 2516, SD, AGI; Royal Order, Aranjuéz, 14 June 1775, copy in Actas del Ayuntamiento, vol. 11, Santiago de Cuba, 14 July 1775, AHM.

38. Enrile to the Council of the Indies, Cádiz, 11 February 1775, leg. 2516, SD, AGI. The original articles of incorporation have not survived. Documents prepared and stored in the notarial office of Antonio Ynarejos Moreno (Notary 19) from July 1774 through June 1775 have been lost. Protocolos Notariales, Archivo Histórico Provincial de Cádiz, Cádiz, Spain. Fortunately, a copy was sent to the governors and town councils of all the cities served by the Asiento, and a copy can be found in the Actas del Ayuntamiento, vol. 11, Santiago de Cuba, 14 July 1775, AHM. Beaumarchais is well known to scholars of the American Revolution. See Van Tyne, "French Aid," 37–40.

39. The preeminent historian of the topic, Yela Utrila, says Beaumarchais was chosen personally by Charles III.

40. John Hans Delap to Willing and Morris, Bordeaux, 30 March 1776, Folder January–May 1776, Willing and Morris Correspondence, Levis Collection, HSP.

41. Cathalan to Morris, Marseilles, 7 January 1776, Folder January–May 1776, ibid.

42. Robert Morris to William Bingham, Philadelphia, 3 June 1776, Morris to Bingham, Philadelphia, 27 September 1776, reel 12, Robert Morris Papers, Manuscript Division, LC.

43. Etienne Cathalan, "Invoice of Sundry Goods," Marseilles, 10 June 1777, Robert

Morris Business Papers, Levis Collection, HSP; Ferguson, "Business, Government, and Congressional Investigation," 293–318; Albert, *Golden Voyage*, 25–73.

44. Gálvez to de la Torre, San Lorenzo, 28 February 1776, leg. 1227, PC, AGI.

45. Joaquín de Escalona to Eligio de la Puente, Havana, 31 March 1778; Payment to Antonio Montenegro, Julián de Flores, and Joaquín de Escalona, Havana, 15 April 1778; Josef Rocío to Urriza, Havana, 6 April 1779; Luciano de Herrera to Urriza, St. Augustine, 22 April 1779; all leg. 1242, PC, AGI. See also Johnson, "Casualties of Peace," 91–125; Portel Vilá, *Historia de Cuba*, 1:78–91; Cummins, *Spanish Observers*, 43–44, 74–75, 89, 100–105; and Lawson, "Luciano de Herrera," 170–76.

46. "Relación de Lorenzo Rodríguez," in Eligio de la Punte to de la Torre, Havana, 27 August 1773, leg. 1165, PC, AGI; Portel Vilá, *Historia de Cuba*, 1:78–91; Cummins, *Spanish Observers*, 19–21, 101–5.

47. Boyd and Navarro Latorre, "Spanish Interest in British Florida," 95–96, argue that this contact was discouraged. However, this did not prevent *floridanos* from unofficially continuing the practice, such as the voyage of the sloop *San Vicente Ferrer*, which carried fourteen members of the Talapuche nation to Havana. Urriza to Juan Lendian, Havana, 25 March 1775, leg. 1155, PC, AGI.

48. "Causa de D. Pedro Truxillo, Trinidad, 26 July 1766, exp. 11, leg. 19, CCG, ANC.

49. De la Torre to José de Gálvez, Havana, 10 May 1776, leg. 1227; de la Torre to Trujillo, Havana, 3 March 1777, leg. 1229; both PC, AGI.

50. De la Torre to José de Gálvez, Havana, 6 September 1776; Joseph Carrandi, Receipt, Havana, 26 October 1776; both leg. 1227, PC, AGI.

51. Rafael de Limonta to de la Torre, Baracoa, 28 June 1776, leg. 1229, PC, AGI.

52. Portel Vilá, *Historia de Cuba*, 1:78–91; Cummins, *Spanish Observers*, 39–42, 44.

53. Enrile to Miguel Eduardo, 6 May 1776, Havana, leg. 1227, PC, AGI.

54. Enrile to de la Torre, Havana, 2 May 1776, leg. 1227, PC, AGI.

55. "Instrucción reservada a que ha de arreglar d. Miguel Antonio Eduardo," 24 April 1777, in de la Torre to Gálvez, Havana, 18 November 1776, ibid.

56. Enrile to Eduardo, Havana, 2 May 1776, ibid.

57. Copy of Miguel Eduardo's "Instructions," Cádiz, 18 November 1776, ibid.

58. De la Torre to Gálvez, Havana, 18 November 1776, ibid.; Miguel Antonio Eduardo, "Diario de todo lo que ha occurido," in ibid.; Cummins, *Spanish Observers*, 39–42.

59. John Adams to Isaac Smith, 1 June 1776, in Smith, *Letters of Delegates*, 4:113.

60. *PZ*, 16 October 1776, LCP.

61. Conde de Ripalda to de la Torre, Puerto Príncipe, 29 May 1776, leg. 1229, PC, AGI.

62. "Copy of the letter from the general administrators of the interests of the compañía del Rl Asiento de negros in Kingston," Kingston, 5 July 1776, in Ripalda to de la Torre, Puerto Príncipe, ibid. The signatories were Thomas Hibbert and Nephew; Ford and Delprat; Watt and Allandice; Benson and Carter; Allan, Sean, Phipps and Sane; Bright, Duncomb, and Saunders; Hercules Ross; Joseph and Eliph Fitch; Thomas Bagnold; Dennie and Duffus; Thomas and Vildman; Thomas Donnvand; Dick and Milligan; Juan Wert; Brown and Welsh; Mamby and Goosman; Juan Appel; Jaime Robertson; Thompson and Campbell; Juan Westmorland and Comp; and Juan Jaques.

63. "Memoriales pidiendo liciencia para desembarcar Armazón de negros," Manuel Phélix Reisch, factor for the Asiento, Havana, 24 May 1772, 6 September 1773, ibid.

64. "Memoriales pidiendo liciencia," Havana, 26 August 1776, ibid.

65. De la Torre to José de Gálvez, Havana, 10 June 1776, leg. 1227; "Memoriales pidiendo liciencia," Havana, 15 June 1776, leg. 1229; both PC, AGI.

66. Sheridan, "Crisis of Slave Subsistence," 615–41.

67. Oloríz to de la Torre, Santiago de Cuba, 12 September 1776, leg. 1143, PC, AGI. See also Sheridan, "Crisis of Slave Subsistence," 615–41; and Sheridan, "Jamaican Slave Insurrection Scare," 290–308.

68. De la Torre to Gálvez, Havana, 4 September 1776, leg. 1227, PC, AGI.

69. "Memoriales pidiendo liciencia," Havana, 1 November 1776, leg. 1229; de la Torre to José de Gálvez, Havana, 4 November 1776, leg. 1227; both PC, AGI.

70. Dupuy to Morris, Mole San Nicolás, 14 August 1776, Willing and Morris Correspondence, Levis Collection, HSP.

71. De la Torre to José de Gálvez, Havana, 4 November 1776, leg. 1227, PC, AGI.

72. De la Torre to José de Gálvez, Havana, 27 October 1776; de la Torre to José de Gálvez, Havana, 4 November 1776; the letter from Ceronio is enclosed in Enrile to Governor to Gálvez, 26 January 1777; all ibid.

73. Chávez, *Spain and the Independence of the United States*, 45–88.

74. José de Gálvez to de la Torre, Madrid, 26 January 1777, leg. 1227, PC, AGI; de la Torre to José de Gálvez, Havana, 31 January 1777, folio 203, folder 17,616, Manuscritos de América, BNE.

75. José de Gálvez to Enrile, Aranjuéz, 26 January 1777, leg. 1227, PC, AGI.

76. Antonio de Raffelin, "Empleos," 31 December 1786, exp. 18, leg. 7259, GM, AGS.

77. Raffelin to Ricla, Havana, 5 December 1763, leg. 1212, SD, AGI.

78. Antonio de Raffelin, "Empleos," 31 December 1786, exp. 18, leg. 7259, GM, AGS.

79. De la Torre to Solano, Havana, 21 February 1777, leg. 1227, PC, AGI. See also Fernández, "José de Solano."

80. Raffelin to Solano, (day illegible) March 1777, Cabo Francés, leg. 1227, PC, AGI.

81. De la Torre to Gálvez, Havana, 7 March 1777, ibid.

82. De la Torre to José de Gálvez, Havana, 3 April 1777, ibid.

83. Raffelin to Solano, onboard the *Vaillant*, 25 April 1777, ibid.

84. Raffelin to Solano, Cabo Francés, 15 April 1777; Ceronio to Raffelin, Cabo Francés, 16 April 1777; Raffelin to Ceronio, Havana, 30 April 1777; all ibid.

85. "Relación de las embacarcaciones," Havana, 30 June 1777, 31 July 1777, leg. 1212, SD, AGI; "Permisos concedidos al asiento de negros para despachar sus barcos, año de 1777, año de 1778," Havana, 1777–78, leg. 1273, PC, AGI.

86. De la Torre to Tentor, Havana, 31 March 1777; Tentor to Gálvez, Santiago de Cuba, 18 Abril 1777; both leg. 2516, SD, AGI.

87. DePlessis to Morris, New Orleans, 16 August 1776, Willing and Morris Correspondence, Levis Collection, HSP; Chávez, *Spain and the Independence of the United States*, 89–100; Portel Vilá, *Historia de Cuba*, 1:78–91; Cummins, *Spanish Observers*.

88. Bernardo de Gálvez to de la Torre, New Orleans, 21 March 1777, Correspondence of

the Spanish Governors, ANC; typescript in Foreign Copying Project, Manuscript Division, LC.

89. De la Torre to Eligio de la Puente, Havana, 6 May 1776, leg. 1184, PC, AGI.

90. Conde de Macuriges to Gálvez, Havana, 17 October 1777, folio 211, folder 17,616, Manuscritos de América, BNE.

91. Kuethe, "El fin del monopolio," 63–64; Muñoz Pérez, "La publicación del reglamento," 615–43.

92. Saavedra, *Los decenios*, 110–11. A recent analysis is Piqueras, "Los amigos de Arango," 155–60.

93. *Reglamento y aranceles para el comercio libre de España a Indias de 12 de octubre de 1778.*

94. The most relevant to this study and also providing excellent summaries of the vast literature are Kuethe and Inglis, "Absolutism and Enlightened Reform," 118–43; Muñoz Pérez, "La publicación del reglamento," 615–43; John R. Fisher, *Commercial Relations*; John R. Fisher, "Imperial 'Free Trade,'" 21–24; and Fisher, Kuethe, and McFarlane, *Reform and Insurrection*.

95. Ortíz de la Tabla y Ducasse, *Comercio exterior de Veracruz*, 12–16.

96. Torres Ramírez, *La compañía gaditana*, 13.

97. Sarah Logan Fisher, "Diary of Trifling Occurrences," 454, 458, 461; Arthur L. Jensen, "Inspection of Exports in Colonial Pennsylvania," 292.

98. Sarah Logan Fisher, "Diary of Trifling Occurrences," 454.

99. Ceronio to Robert Morris, Cap Français, 9 November 1778, Robert Morris Papers, Levis Collection, HSP.

100. Navarro to José de Gálvez, Havana, 11 November 1777, exp. 77, leg. 44, Diversos, AHN.

101. Juan de Miralles to Gálvez, Havana, 16 December 1777, folio 227, leg. 17,616, Manuscritos de América, BNE. See also Johnson, "Casualties of Peace."

102. The most comprehensive account of Miralles's activities in Philadelphia is Cummins, *Spanish Observers*, 105–67.

103. "Señor fiscal de Real Hacienda contra de Juan de Miralles," Madrid, 11 August 1753, Escribanía de Cámara, AGI.

104. Rodríguez, *Cinco diarios*, 249–50.

105. Torres Ramírez, *La compañía gaditana*, 80–81.

106. De la Torre to Eligio de la Puente, Havana, 6 May 1776, leg. 1184, PC, AGI.

107. Navarro to José de Gálvez, Havana, 11 November 1777, exp. 77, leg. 44, Diversos, AHN.

108. Matrimonios de Españoles, 31 August 1781, Libro 7, 1771–94, S.M.I. Catedral de la Habana, Archivos Parroquiales, Archdiocese of Havana, Havana.

109. O'Reilly to Julián de Arriaga, Havana, 21 September 1765, leg. 2515, SD, AGI.

110. Johnson, *Social Transformation*, 146–63.

111. Portel Vilá, *Historia de Cuba*, 1:78–91; Cummins, *Spanish Observers*, 105–10; Rodríguez Vicente, "El comercio cubano," 94–104.

112. Miralles to Navarro, Charleston, 24 March 1778, leg. 1281, SD, AGI, typescript and translation in Topping Collection; Portel Vilá, *Historia de Cuba*, 1:78–91; Cummins, *Spanish Observers*, 105–10; Rodríguez Vicente, "El comercio cubano," 94–104.

113. In June, Miralles formally requested his license to travel to Philadelphia, specifically requesting permission to import slaves and food. Miralles to José de Gálvez, Edenton, N.C., 6 June 1778, leg. 1281, SD, AGI, in Topping Collection.

114. Bertha McGeehan to Margaret Armstrong, 22 January 1942, McGeehan Collection, HSP.

115. Galloway, "Diary of Grace Growden Galloway."

116. Bertha McGeehan to Margaret Armstrong, 22 January 1942, McGeehan Collection, HSP.

117. *Pennsylvania Packet*, 5 December 1778, McGeehan Collection, HSP.

118. Sterling to Moylan, Aquakanoc, 8 October 1778, congratulating Moylan on his engagement to the daughter of Philip Van Horne, Stephen Moylan Letterbook, HSP; Griffin, *Stephen Moylan*, 2.

119. Saavedra, *Los decenios*, 4.

120. José de Gálvez to Manuel Antonio Flores, Madrid (?), May 1779, leg. 577-A, Audiencia de Santa Fe, AGI, quoted in Marchena Fernández, *Oficiales y soldados*, 195.

121. Miralles to Navarro, Philadelphia, 31 August 1778, leg. 1281, SD, AGI, in Topping Collection.

122. Miralles to the governor of Puerto Rico, Philadelphia, 31 August 1778, leg. 1606, IG, AGI, in Topping Collection.

123. Miralles to Antonio Ramón del Valle, Philadelphia, 30 September 1778, leg. 1283, SD, AGI, in Topping Collection.

124. Miralles to Navarro, Philadelphia, 18 November 1778, leg. 1281, SD, AGI, in Topping Collection.

125. "Letters from Cap François on the 29th of October," *PZ*, 13 December 1780, LCP.

126. Saavedra, *Los decenios*, 5; Chávez, *Spain and the Independence of the United States*, 126–36.

127. Navarro, "Estado de la fueza y su complejo de los cuerpos de infanteria y cavalleria asi veterana como milicias que existen en esta plaza y sus imediaciones," Havana, 11 August 1779, leg. 2082, SD, AGI; Kuethe, "Havana in the Eighteenth Century," 22–23; Marchena Fernández, *Oficiales y soldados*, 338.

128. Beerman, "Arturo O'Neill," 31.

129. Lewis, "Anglo-American Entrepreneurs," 116–26. A recent evaluation of the signal importance of Miralles's mission to Philadelphia is Chávez, *Spain and the Independence of the United States*, 173.

130. Lewis, "Anglo-American Entrepreneurs," 116. Lewis demonstrates that 10,152 *tercios* of flour from Mexico were imported into Havana, as were 10,128 barrels of flour from the United States. A *tercio* and a U.S. barrel were roughly equivalent, each weighing about 200 pounds. He also believes that the figures for Havana only represent a part of the North American flour that was sold. His figures are for imports to Havana and do not include sales to the navy or for the expeditionary force to Saint Domingue in 1782. Ibid., caption to Table 38. Imports into Louisiana are also not counted in these figures. A secondary— and as yet unexplored—avenue for flour shipments to the Spanish forces was sales to the French fleet.

131. José Antonio Morejón to Navarro, San Pedro, 2 September 1779, leg. 1269, PC, AGI.

132. Bartolomé Rodríguez to Navarro, Santa Clara, 14 September 1779, leg. 1260, PC, AGI.

133. Rodríguez to Navarro, Santa Clara, 3 October 1779, ibid.

134. Juan Moler to Navarro, Trinidad, 30 June 1779, ibid.

135. Chávez, *Spain and the Independence of the United States*, 170–72.

136. Starr, *Tories, Dons, and Rebels*, 142–60; Chávez, *Spain and the Independence of the United States*, 166–83.

137. Starr, *Tories, Dons, and Rebels*, 161–74; Mowat, *East Florida as a British Province*; Wright, *Florida in the American Revolution*.

138. Chávez, *Spain and the Independence of the United States*, 169–74.

139. Millás, *Hurricanes*, 241–59; Mulcahy, *Hurricanes and Society*, 108–11, 166–74.

140. Navarro to José de Gálvez, Havana, 25 February 1780, leg. 2082, SD, AGI.

141. Diario formado por Estéban Miró, Havana, 1780, leg. 2543, SD, AGI, 674.

142. Huet to Navarro, El Morro, 25 February 1780, leg. 1247; Bonet to Navarro, Havana, 24 February 1780, leg. 1245; both PC, AGI. See also Guerra y Sánchez, *Manual de historia*, 178; Torre, *Lo que fuimos*, 104, 108; and Valdés, *Historia de la isla de Cuba*, 155–56.

143. Juan José Bachoni to Navarro, Castillo de la Cabaña, 23 February 1780, leg. 1248, PC, AGI.

144. Luis Huet, "Plano que demuestra el acantonamiento en barracones del este," 1 June 1780, Mapas y Planos, SD, AGI, 462; Huet to Navarro, El Morro, 25 February 1780, leg. 2082, SD, AGI.

145. Juan Bautista Vaillant, "Inspección de los quarteles de Luyanó, San Miguel y Guatao," Havana, 20 September 1779, 2 December 1779, leg. 1248, PC, AGI.

146. Bonet to Navarro, Havana, 24 February 1780, leg. 1245, PC, AGI.

147. Navarro to Francisco de Albuquerque, Havana, 23 February 1780; Albuquerque to Navarro, El Morro, 24 February 1780; both leg. 1248, PC, AGI.

148. Ibid.

149. Bonet to Navarro, Havana, 29 February 1780, leg. 1245, PC, AGI.

150. Bonet to Navarro, Havana, 4 March 1780, ibid.

151. Chávez, *Spain and the Independence of the United States*, 174–76.

152. José de Gálvez to Bernardo de Gálvez, San Lorenzo, 10 January 1780, exp. 1, leg. 6912, GM, AGS.

153. Navarro to José de Gálvez, Havana, 23 January 1781, leg. 1300, PC, AGI.

154. Morales Padrón, *Journal of Don Francisco Saavedra*, Introduction.

155. Gerónimo Girón to Navarro, Havana, 23 May 1780, leg. 1248, PC, AGI.

156. Navarro, Circular Letter, Havana, 28 April 1780, leg. 1254, PC, AGI. The passage reads, "Que los prisioneros ingleses sean tratados con la umanidad y dulzura que permita su condición y la seguridad de esta isla."

157. Bernardo de Gálvez to Navarro, enclosed in Navarro to Gerónimo Girón, Havana, 23 May 1780, leg. 1248, PC, AGI.

158. Bond, Juan de Miralles, and Robert Morris to Francis Hopkinson, Philadelphia, 1 March 1780, McGeehan Collection, HSP.

159. Morris to Hudson and Co. Merchants, Philadelphia, 8 June 1780, ibid.

160. José de Gálvez to Urriza, Aranjuéz, 20 April 1781, exp. 8, leg. 3, AP, ANC; "Extract

of a letter from a gentleman on board the Letter of Marque Brig *Fox*," 17 August 1780, *PZ*, September 20, 1780, LCP; James Buchanan, captain of the brig *Fox*, Baltimore, 29 November 1780, *PZ*, LCP. Lewis, "Anglo-American Entrepreneurs," demonstrates that trade from North America declined in 1780, only to resume exponentially in 1781.

161. Martín de Navarro to Diego de Navarro, New Orleans, 18 April 1780, leg. 2609, SD, AGI.

162. Navarro to Ripalda, Havana, 30 April 1780, 22 May 1780, leg. 1256, PC, AGI.

163. Urriza to Ripalda, Havana, 19 June 1780; Ventura Díaz to Urriza, Puerto Príncipe, 22 June 1780, 27 June 1780; all ibid.

164. Navarro to the Captain of Managua, Havana, 21 July 1780, leg. 1269, PC, AGI.

165. Various Residents of Arroyo Arenas to Navarro, Arroyo Arenas, 19 August 1780, ibid.

166. Ventura Díaz to Navarro, Puerto Príncipe, 11 October 1780, leg. 1256, PC, AGI.

167. Juan Nepomuceno de Quesada to Navarro, Puerto Príncipe, 16 November 1780; Díaz to Navarro, Puerto Príncipe, 11 October 1780; both ibid.

168. Quesada to Navarro, Puerto Príncipe, 11 January 1781, ibid.

169. Ibid.

170. Mulcahy, *Hurricanes and Society*, 108–9.

171. Ibid., 108–11, 166–74; Millás, *Hurricanes of the Caribbean*, 241–59.

172. Morales Padrón, *Journal of Don Francisco Saavedra*, 9–15 October 1780.

173. Millás, *Hurricanes of the Caribbean*, 251–59.

174. Ibid., 260–62; Fernández, "José de Solano."

175. Ventura Díaz to Navarro, Puerto Príncipe, 11 October 1780, leg. 1256; Junta de Guerra, Havana, 2 November 1780, leg. 1264; both PC, AGI.

176. Junta de Guerra, Havana, 11 November 1780, leg. 1264, PC, AGI.

177. Ramon Lloret to Navarro, Campeche, 5 November 1780, leg. 1248, PC, AGI.

178. Lloret to Navarro, Campeche, 21 November 1780, ibid.

179. Junta de Guerra, Havana, 18 November 1780, 22 November 1780, leg. 1264, PC, AGI.

180. Junta de Guerra, Havana, 20 November 1780, ibid.

181. Urriza to the Junta de Guerra, El Morro, 30 November 1780, ibid.

182. Vicente Garciny to Navarro, El Morro, 12 November 1780, leg. 1247, PC, AGI.

183. Junta de Guerra, Havana, 2 December 1780, leg. 1264, PC, AGI.

184. Junta de Guerra, Havana, 20 November 1780, ibid.

185. Vicente Garciny to Navarro, El Morro, 13 November 1780, 15 November 1780, leg. 1247, PC, AGI.

186. Governor of Mole San Nicolás to Navarro, Mole San Nicolás (Saint Domingue), 25 October 1780, leg. 1231, PC, AGI.

187. Governor of Martinique, 1780, quoted in Fernández de Castro, *Huracanes en Cuba*, 47.

188. Navarro to Gálvez, Havana, 28 November 1780, leg. 1300, PC, AGI.

189. Junta de Guerra, Havana, 25 November 1780, leg. 1264, PC, AGI.

190. Navarro, Circular Order, Havana, 22 January 1781, leg. 1248, PC, AGI.

191. Huet, Circular Order to all Capitanes del Partido, Havana, 13 March 1781, ibid.

192. Navarro to the Militia Colonels, Havana, 16 January 1781; Juan Batista Vaillant to Navarro, Quartel de Luyanó, 17 January 1781; both ibid.

193. Marino Murguia to Navarro, Pinal del Río, 15 March 1781, leg. 1269, PC, AGI, acknowledging the circular order and confirming his jurisdiction's compliance.

194. In addition to Caughey, *Bernardo de Gálvez*, other important works are Woodward, *Tribute to Don Bernardo de Gálvez*, xvii–xxvii; Reparáz, *Yo Solo*; Gálvez, *"Yo Solo,"* Introduction; Coker and Rea, *Anglo-Spanish Confrontation*; Baker and Bissler Haas, "Bernardo de Gálvez's Combat Diary," 176–99; Haarman, "Spanish Conquest of British West Florida," 107–34; Haarman, "Siege of Pensacola," 193–99; and Murphy, "Irish Brigade," 216–25. A recent study of the influence of the Enlightenment and the expansion of the quest for scientific knowledge in service to military operations is Weddle, *Changing Tides*, 113, 286. Unlike most of the laudatory works, Weddle is critical of Bernardo de Gálvez and the nepotistic practices of the age.

195. Porras Muñóz, "El fracaso de Guarico," 569–609.

196. Corbitt, "Administrative System: Part I," 42.

197. Rojas y Rocha, *Poema épico*; Corbitt, "Administrative System, Part I," 43; Saavedra, *Los decenios*, 219–25.

198. Royal Order, Madrid, 5 January 1781, exp. 1, leg. 3, AP, ANC.

199. José de Gálvez to Matio de Gálvez, El Pardo, 5 March 1781, exp. 4, leg. 3, ibid.

200. José de Gálvez to Urriza, Aranjuéz, 20 April 1781, exp. 8, leg. 3, ibid.

201. José de Gálvez to Urriza, Aranjuéz, 11 June 1781, exp. 13; José de Gálvez to Urriza, Aranjuéz, 11 June 1781, exp. 15; both leg. 3, ibid.

202. Joseph Grafton, "Journal of a Voyage from Salem to the Havanna on the Brigantine Romulus, Joseph Waters Commander, 1781–1784," microfilm copies in the Peabody Essex Museum, Salem, Mass.

203. Ibid.

204. Diego de Belmonte to Cagigal, Matanzas, 3 February 1782; Murguia to Cagigal, Nueva Filipinas, 28 July 1782; both leg. 1317, PC, AGI.

205. Lewis, "Anglo-American Entrepreneurs," 122–23.

206. Grafton, "Journal of a Voyage from Salem to the Havanna on the Brigantine *Romulus*."

Chapter Six

1. "Relación de las Embarcaciones Españolas y Estrangeras que en el próximo pasado mes de _____ han salido de este puerto para las Colónias Estrangeras en solicitud de Negros; y de las que en el propio mes han entrado en este puerto con cargamento de ellos con distinción de su número, clases, y sexos por el orden siguiente." 1 July 1791, leg. 2207, SD, AGI. *Papel Periódico de la Havana*, 7 August 1791; Census Returns, St. Augustine, 1785, bundle 323A, EFP.

2. Lewis, "Anglo-American Entrepreneurs," 112–26.

3. *Papel Periódico de la Havana*, 7 August 1791; Rappaport and Fernández-Partagás, "Deadliest Atlantic Tropical Cyclones," 2.

4. Miscellaneous Legal Instruments and Proceedings, St. Augustine, 25 September 1791, bundle 261n5, EFP.

5. Rappaport and Fernández-Partagás, "Deadliest Atlantic Tropical Cyclones," 2.

6. Lynch, *Bourbon Spain*, 376–81; Ringrose, *Spain, Europe, and the "Spanish Miracle,"* 112–19.

7. Johnson, *Social Transformation*, 121–80.

8. John R. Fisher, *Commercial Relations*, 49, 65; Ringrose, *Spain, Europe, and the "Spanish Miracle,"* 106–13; Barbier and Klein, "Wars and Public Finances," 315–39; Ortíz de la Tabla y Ducasse, *Comercio exterior de Veracruz*, 167–223.

9. Las Casas's faction included his nephew, Pedro Pablo O'Reilly (the son of Field Marshal Alejandro O'Reilly), Pedro Calvo de la Puerta (Condé de Buena Vista), Francisco de Arango y Parreño, Pablo de Estévez, and José de Ilincheta. Johnson, *Social Transformation*, 121–63.

10. José de Gálvez to Vicente Manuel de Zéspedes, 31 October 1783, leg. 10, Fondo de las Floridas, ANC, in Lockey, *East Florida*, 174.

11. José de Gálvez to Thomas Hassett, 25 November 1783, bundle 39, EFP, in Lockey, *East Florida*, 176–77; Curley, *Church and State in the Spanish Floridas*, 73–86.

12. Zéspedes to Bernardo de Gálvez, St. Augustine, 16 July 1784, leg. 2660, SD, AGI, in Lockey, *East Florida*, 223–24.

13. Bernardo de Gálvez to Zéspedes, Havana, 4 July 1784, bundle 40, EFP.

14. Bernardo de Gálvez to José de Gálvez, Havana, 15 June 1784, leg. 1344, PC, AGI.

15. Zéspedes to Pedro Vásquez, Havana, 11 July 1784, leg. 2660, SD, AGI, in Lockey, *East Florida*, 228.

16. Zéspedes to Bernardo de Gálvez, 16 July 1784, ibid.

17. Ibid.

18. Zéspedes to Bernardo de Gálvez, St. Augustine, 13 October 1784, bundle 40, EFP.

19. Ibid., 4 July 1784.

20. Ibid., 8 August 1784, 13 October 1784.

21. The troublemakers in East Florida were termed *banditti* by contemporaries. Lockey describes them as "refugees and vagrants." Introduction, in Lockey, *East Florida*, 14–19.

22. Zéspedes to Bernardo de Gálvez, St. Augustine, 8 August 1784, bundle 40, EFP. The regiment of Asturias left for Spain on the *Sacra Familia* and the *San Pedro y San Antonio*.

23. Zéspedes to Bernardo de Gálvez, 16 July 1784, leg. 2660, SD, AGI, in Lockey, *East Florida*, 228.

24. Ibid., 227.

25. Zéspedes to Bernardo de Gálvez, St. Augustine, 8 August 1784, bundle 40, EFP.

26. Zéspedes to Bernardo de Gálvez, St. Augustine, 8 August 1786, bundle 41b4, EFP.

27. Zéspedes to Juan Ignacio de Urriza, St. Augustine, 20 September 1785, bundle 55, EFP, in Lockey, *East Florida*, 727–28; Zéspedes to Gálvez, St. Augustine, 1 October 1785, leg. 2660, SD, AGI, in ibid., 730–31.

28. Florescano, *Precios de maíz*, 61.

29. Zéspedes to Bernardo de Gálvez, St. Augustine, 25 August 1786, bundle 41b4, EFP.

30. Ibid., citing Zéspedes's original declaration in 1784.

31. Departure of Vessels, St. Augustine, July–December 1784, bundle 242H19; Arrival of Vessels, St. Augustine, July–December 1784, bundle 214F17; both EFP.

32. Bernardo de Gálvez to Zéspedes, Havana, 12 September 1784, 9 November 1784, 20 November 1784, leg. 1356, PC, AGI.

33. José de Gálvez to Zéspedes enclosing Royal Order, San Ildefonso (?), 4 November 1784, bundle 39, EFP, in Lockey, *East Florida*, 304.

34. Johnson, "Climate, Community, and Commerce," 455–82; Nichols, "Trade Relations," 289–313; Lewis, "Anglo-American Entrepreneurs," 112–26.

35. Juana de Torres to the Audiencia de Santo Domingo, Cádiz, 26 September 1791, leg. 1481, SD, AGI.

36. Escrituras, St. Augustine, 8 October 1791, 11 November 1791, bundle 367; Memorials, St. Augustine, 1 August 1792, bundle 182m14; Oaths of Allegiance, St. Augustine, 18 May 1791, bundle 350U4; all EFP.

37. Antonio de Raffelin to the Casa de Contratación, Havana, 12 July 1787, leg. 512, Ultramar, AGI.

38. Jose María Ysnardi, power of attorney to his father, Jose Ysnardi, Cádiz, 5 October 1796; Jose María Ysnardi, Testament, 25 October 1799; both Protocolos Notariales, Archivo Histórico Provincial de Cádiz, Cádiz, Spain.

39. Escrituras, St. Augustine, 6 May 1785, 8 October 1785, 20 December 1785, bundle 367, EFP; Moreno Fraginals, *El ingenio*, 1:100–101.

40. Census of 1785, St. Augustine, Census Returns, bundle 323A, EFP.

41. Arrival of Vessels and Cargoes, St. Augustine, 30 August 1784, bundle 214F17, EFP.

42. Ibid., 27 February 1785.

43. Ibid., 30 August 1784; Memorials, St. Augustine, 28 June 1787, 25 May 1789, bundle 180A14, EFP; Census of 1784, St. Augustine, Census Returns, bundle 323A, EFP; Moreno Fraginals, *El ingenio*, 1:108.

44. *PZ*, 25 May 1785, reported that Gálvez had reversed the expulsion and had set up tribunals to adjudicate the claims filed by American businessmen who had had their property seized in 1784; in LCP.

45. Salvucci, "Anglo American Merchants and Stratagems," 127–33; Salvucci, "Supply, Demand, and the Making of a Market," 13–39.

46. Grafton, "Journal of a Voyage."

47. Antonio Porlier to Las Casas, Aranjuéz, 4 April 1790, exp. 20, leg. 4, AP, ANC, reiterating the Royal Order issued on 12 October 1779.

48. *Reglamento para el comercio libre.*

49. "Vecinos pidiendo liciencia para establecer una sociedad económica en Santiago de Cuba," Santiago de Cuba, 2 November 1783, 13 March 1787, leg. 1476-B, SD, AGI; Álvarez Cuartero, "Las sociedades económicas," 36–37.

50. Álvarez Cuartero, "Las sociedades económicas," 37–41; Lampros, "Merchant-Planter Cooperation and Conflict." Although these were technically different organizations, their membership overlapped and the members exercised considerable political power by virtue of their election to the ayuntamiento.

51. Johnson, *Social Transformation*, 91–120.

52. González-Ripoll Navarro, "Voces de gobierno," 149–62; Thomas, *Cuba*, 72–73; Le Riverend Brusone, *Historia económica*, 265; Guerra y Sánchez, *Manual de historia*, 1:202; Aimes, *History of Slavery in Cuba*. Las Casas's detractors include Moreno Fraginals, *El ingenio*, 1:39–50; Tornero Tinajero, *Crecimiento económico*, 44–56; and Johnson, *Social Transformation*, 121–63. Kuethe, *Cuba*, is neutral.

53. The original documents, Arango's multiple submissions to the Council of the Indies (later Council of State), were used for this research and can be found in Arango y Parreño to the Councils of Indies and State, 1789–94, leg. 120, Ultramar, AGI. Printed versions usually use the last or most polished version submitted in 1792 and differ from the original in several aspects. See also Arango y Parreño, *Obras*; Pierson, "Francisco de Arango y Parreño," 451–78; Gay-Calbó, *Arango y Parreño*; Paquette, *Sugar Is Made with Blood*, 83–86; and Moreno Fraginals, *El ingenio*, 1:51–80, 100–33.

54. Urriza to José de Gálvez, Havana, 10 May 1777, leg. 257-B; Raymundo de Onís to José de Gálvez, Havana, 12 August 1779, leg. 258-A; both Correos, AGI.

55. Arango y Parreño to the Councils of Indies and State, 1789–94, leg. 120, Ultramar, AGI; Le Riverend Brusone, *Historia económica*, 125–29.

56. Marichal and Souto Mantecon, "Silver and Situados," 587–613; Johnson, *Social Transformation*, 37–96.

57. Arango y Parreño to the Council of State, Madrid, 10 May 1793, leg. 120, Ultramar, AGI.

58. Royal Order, 15 January 1780, leg. 2316, SD, AGI, cited in Torres Ramírez, *La compañía gaditana*, 13. At least five years before the Real Cédula of 1789, Miguel Antonio de Herrera was granted permission to import 560 slaves from St. Thomas; Havana, 12 September 1784. Francisco Sánchez brought in 52 slaves; Havana, 25 July 1783. And María Candelaria González del Alamo received permission to import three boatloads of blacks from foreign colonies on 6 June 1784. All leg. 1356, PC, AGI. Another reason Havana's planters did not agitate more forcefully to end the existing monopoly was that in the mid-1780s Cuba's sugar industry suffered the effects of an economic recession.

59. Johnson, "Rise and Fall," 52–75.

60. Grafton, "Journal of a Voyage"; Lewis, "Anglo-American Entrepreneurs," 123.

61. Gabriel Raymundo de Azcárate, Andrés de Loyzaga, and José Antonio Arregui to the Council of the Indies, Havana, 2 August 1788, leg. 2824, IG, AGI. See also Johnson, "Rise and Fall," 52–75; and Tornero Tinajero, *Crecimiento económico*, 43.

62. Modern works include Klein, *Middle Passage*, 209–27; Thomas, *Cuba*; Murray, *Odious Commerce*; Kiple, *Blacks in Colonial Cuba*; Johnson, "Rise and Fall," 52–75; Tornero Tinajero, *Crecimiento económico*; López-Valdés, "Hacia una periodización," 13–29; Lampros, "Merchant-Planter Cooperation and Conflict"; Knight, *Slave Society in Cuba*, 3–24; Knight, "Origins of Wealth," 231–53; and Moreno Fraginals, *El ingenio*, 1:39–71. Traditional works are also useful, such as Aimes, *History of Slavery in Cuba*; Ortiz, *Los negros esclavos*; Guerra y Sánchez, *Manual de historia*; and Le Riverend Brusone, *Historia económica*.

63. "Extract of a letter from Cadiz, dated 12th April, 1791, to a House in this city," *PZ*, 15 June 1791, LCP.

64. "Extract of a letter from Cadiz, dated April 1791, to a mercantile House in this city," *PZ*, 22 June 1791, LCP.

65. *Papel Periódico de la Havana*, 7 January 1802 and 1 July 1802, citing the decree of 1791.

66. Le Riverend Brusone, *Historia económica*, 211.

67. Andrés Moreno to the Ayuntamiento de Santa Clara, 11 February 1791, Actas Capitulares del Ayuntamiento, Tomo 8, 1787–91, Archivo Provincial de Santa Clara, Santa Clara, Cuba.

68. Lucas Álvares to Las Casas, Batabanó, 11 March 1791, leg. 1471, PC, AGI.

69. Miguel Núñez and José Pérez de Medina to Las Casas, Guanabacoa, 6 May 1791, leg. 1460, PC, AGI.

70. Millás, *Hurricanes of the Caribbean*, 284–86.

71. Barreto, *Contestación al impreso del Sr. Conde de O'Reilly*.

72. Tomás Borrego to Luis de las Casas, Wajay (Jubajay), 23 June 1791, leg. 1472, PC, AGI.

73. *Papel Periódico de la Havana*, 7 August 1791; Millás, *Hurricanes of the Caribbean*, 284–86. The National Weather Service ranks this storm as the fourteenth-most-destructive in history. Rappaport and Fernández-Partagás, "Deadliest Atlantic Tropical Cyclones," 2.

74. Luis Huet, "Plano de la Havana y sus contornos, para demostrar las baterías que se han de construir en tiempo de guerra, los puestos que se han de tomar y campos que se han de formar con un campo volante de tropas, en el caso de una imbación decidida contra el puerto, plaza y fuertes adyacentes." Havana, 15 June 1776, SD 418, Mapas y Planos, AGI.

75. Johnson, *Social Transformation*, 39–71; Johnson, "La guerra contra los habitantes," 181–209.

76. José Pérez to Las Casas, Calvario, 25 June 1791, leg. 1470, PC, AGI.

77. Félix Gonzáles to Las Casas, Jesús del Monte, 22 June 1791, 25 June 1791, leg. 1471, PC, AGI.

78. Vicente Soria to Las Casas, Santiago de las Vegas, 26 June 1791, leg. 1460, PC, AGI.

79. Thomas Borrego to Las Casas, Jubajay, 23 June 1791, leg. 1472, PC, AGI.

80. Millás, *Hurricanes of the Caribbean*, 285.

81. Vicente de Castilla to Las Casas, Prensa, 22 June 1791, leg. 1470, PC, AGI.

82. Castilla to Las Casas, Prensa, 22 June 1791, 23 June 1791, ibid.

83. Cristóval Pacheco to Las Casas, Quemados, 23 June 1791, leg. 1472, PC, AGI.

84. Pacheco to Las Casas, Quemados, 27 June 1791, 3 July 1791, ibid.

85. Pedro Monza to Las Casas, Guayabal, 12 July 1791, leg. 1471, PC, AGI.

86. Ayuntamiento de Guanabacoa to Las Casas, Guanabacoa, 25 June 1791, 9 November 1791, leg. 1460; Castilla to Las Casas, Prensa, 22 June 1791, leg. 1470; Antonio Blanco to Las Casas, Puentes Grandes, 22 June 1791, leg. 1470; Miguel Díaz to Las Casas, Luyanó, 5 July 1791, leg. 1472; Miguel Chávez to Las Casas, Horcón, 1 October 1791, leg. 1471; all PC, AGI.

87. José Fuertes to the Duque de Alcudía, Havana, 1 August 1791, leg. 260-A, Correos, AGI.

88. Fuertes to Alcudía, Havana, 16 August 1791, 21 November 1791, 23 November 1791, ibid.

89. Las Casas to Pacheco, Havana, 28 June 1791, leg. 1472, PC, AGI.

90. "Socorro para los pobres a la resulta de la inundación de 21 de junio," *Papel Periódico de la Havana*, 6 October 1791.

91. Only a few cases of resistance can be documented, three to the east of Guanabacoa and eight in Jesús del Monte. See chapter 4.

92. Navarro to the Captain of Managua, Havana, 21 July 1780; Various Residents of Arroyo Arenas to Navarro, Arroyo Arenas, 19 August 1780; both leg. 1269, PC, AGI.

93. Miguel Díaz to Las Casas, Luyanó, 5 July 1791; Las Casas to Cristóval Pacheco, Havana, 28 June 1791; both leg. 1472, PC, AGI.

94. "Plano, elevación, y vista en perspectiva rigorosa del puente Blanco de Ricabal arruniado en el año de 1791," number 561, Mapas y Planos, SD, AGI; "Plano, pérfiles y elevación de un Puente," number 562, Mapas y Planos, SD, AGI.

95. Ayuntamiento de Guanabacoa to Las Casas, Guanabacoa, 9 November 1791, leg. 1460, PC, AGI.

96. Pablo Interian to Las Casas, San José de las Lajas, 2 August 1791, leg. 1471, PC, AGI.

97. Gaspar Francisco de Archeta (?) to Las Casas, Arroyo Arenas, 28 April 1792, leg. 1470, PC, AGI.

98. Las Casas to Ayuntamiento de Guanabacoa, Havana, 18 February 1792, leg. 1460, PC, AGI.

99. Las Casas, Circular Letter, Havana, 7 May 1792, leg. 1471, PC, AGI.

100. Manuel Saldivar, Francisco Fernández, and Francisco Agustín to Las Casas, Govea, 29 May 1792; Las Casas to José Antonio Morejón, Havana, 11 June 1792; Morejón to Las Casas, Seiba de Agua, 20 June 1792; all leg. 1470, PC, AGI.

101. Felipe Núñez Villavicencio to Las Casas, San Pedro, 6 November 1792, leg. 1472, PC, AGI. Other recalcitrant villages included Güines and Seiba. Luis López Gavilán to Las Casas, Güines, 22 May 1791, leg. 1471, PC, AGI.

102. *Papel Periódico de la Havana*, 20 May 1792.

103. González-Ripoll Navarro, "Voces de gobierno," 156–59.

104. Díaz to Las Casas, Luyanó, 5 July 1791, leg. 1472, PC, AGI.

105. Gonzáles to Las Casas, Jesús del Monte, 1 October 1791, leg. 1471, PC, AGI.

106. Borrego to Las Casas, Jubajay, 23 June 1791, leg. 1472, PC, AGI.

107. Varios Vecinos to Ayuntamiento de Havana, Havana, 12 August 1791; Las Casas to Ayuntamiento de Havana, Havana, 3 September 1791; both leg. 1460, PC, AGI.

108. Las Casas to Ayuntamiento de Havana, Havana, 14 September 1791, ibid.

109. Las Casas to Ayuntamiento de Havana, Havana, 12 August 1791, ibid.

110. Soría to Las Casas, Santiago de las Vegas, 26 June 1791, leg. 1460; Félix Gonzáles to Las Casas, Jesús del Monte, 26 June 1791, leg. 1471; José López to Las Casas, Managua, 25 June 1791, leg. 1472; all PC, AGI.

111. Johnson, *Social Transformation*, 146–63.

112. "Acto de lectura del Informe del Tribunal del Protomedico sobre el casabe," 23 September 1791, in Ministerio de Salud Pública, *La medicina en La Habana*, 2:270–71.

113. "Dictamen del Protomedicato en favor del consume de casabe," 23 November 1791, ibid., 2:271–73.

114. Super and Wright, *Food, Politics, and Society*.

115. Hilario de León and Manuel de Ayala to the Ayuntamiento de Santa Clara, Santa Clara, 22 June 1792; Joaquín Moya, Juan Antonio Montenegro, and José Montenegro to the Ayuntamiento de Santa Clara, 14 September 1792; both Tomo 9, 1792–99, Actas Capitulares del Ayuntamiento, Archivo Provincial de Santa Clara, Santa Clara, Cuba.

116. Tomás Honorio Pérez de Morales to the Ayuntamiento de Santa Clara, Santa Clara, 10 May 1793, ibid.

117. Soría to Las Casas, Santiago de las Vegas, 8 November 1792, leg. 1460; Joseph Morejón to Las Casas, Seiba del Agua, 20 June 1792, leg. 1470; Miguel Díaz to Las Casas, Luyanó, 2 October 1792, leg. 1472; Francisco Naranjo to Las Casas, Río Blanco del Norte, 19 October 1792, leg. 1472; all PC, AGI.

118. Fuertes to Alcudía, Havana, 9 November 1792, leg. 260-A, Correos, AGI; Soría to Las Casas, Santiago de la Vegas, 2 November 1792, leg. 1460, PC, AGI; Borrego to Las Casas, Quemados, 26 June 1792, leg. 1472, PC, AGI.

119. Urriza to Alcudía, Havana, 8 May 1794, exp. 53, leg. 6856, GM, AGS.

120. Urriza to Alcudía, Havana, 24 July 1794, exp. 13, leg. 6852, GM, AGS, reporting on what had been accomplished since 1792.

121. Las Casas to Luis López Gavilán, Havana, 15 March 1793, leg. 1471; López Gavilán to Las Casas, Güines, 15 March 1793, leg. 1471; Las Casas to the Captain of Cano, 28 January 1793, leg. 1470; José López to Las Casas, Managua, 28 February 1793, leg. 1472; all PC, AGI.

122. Miguel de Sotolongo to Las Casas, Hanábana, 1 November 1792, leg. 1471, PC, AGI.

123. Ibid., 3 January 1793.

124. 6 November 1792, Actas Capitulares del Ayuntamiento, Tomo 16, Archivo Municipal de Santiago de Cuba, Santiago de Cuba, Cuba.

125. Since the 1760s, the state had been assuming many of the traditional post-disaster functions of recovery, but one area where the state under Charles III had never dared to intervene was in ecclesiastical prerogative in conduct of church ceremony.

126. Pedro Acuña, "A la muy noble y leal ciudad de la Havana," Aranjuéz, 9 February 1793, Correspondencia del Obispo de la Habana, Archivo del Arzobispo de la Habana, Havana, Cuba.

127. Síndico Procurador to the Obispo de la Habana, Havana, 10 February 1793, ibid.

128. Las Casas to Félix José Trespalacios, Havana, 11 February 1793, ibid.

129. Trespalacios to Las Casas, Havana, 13 February 1793, ibid.

130. Trespalacios to Las Casas, Havana, 15 February 1793, 16 February 1793, ibid.

131. "Edicto del D.D. Phelipe Jph de Trespalacios, a nuestros amados fieles vecinos y moradores estantes y habitantes en las ciudades, villas y lugares de nro obispado de qualquiera dignidad," Havana, 18 February 1793, ibid.

132. Ibid.

133. José López to Las Casas, Managua, 28 February 1793, leg. 1472, PC, AGI.

134. Las Casas to Felipe de Lima, Havana, 24 January 1793, leg. 1470, PC, AGI.

135. Naranjo to Las Casas, Río Blanco del Norte, 23 May 1793, leg. 1472, PC, AGI.

136. López Gavilán to Las Casas, Güines, 15 March 1793, leg. 1471; Manuel Benítez to Las Casas, Naranjal, 2 November 1793, leg. 1472; both PC, AGI.

137. López to Las Casas, Managua, 27 March 1793, leg. 1472, PC, AGI.

138. Antonio de la Torre to Las Casas, Gibacoa, 22 November 1792, leg. 1471, PC, AGI.

139. Las Casas to Ayuntamiento de Guanabacoa, Havana, 10 June 1793, leg. 1460, PC, AGI.

140. Gonzáles to Las Casas, Jesús del Monte, 22 June 1791, 25 June 1791, 30 October 1792, leg. 1471, AGI, PC.

141. Ibid., 20 December 1793, 12 January 1794.

142. Sevilla Soler, *Santo Domingo*, 377–408.

143. The rebellion in Saint Domingue that resulted in the creation of the independent nation of Haiti has generated a large body of scholarship. Monte Tejada, *Historia de Santo Domingo*, 3:138; "Particular Account of the Insurrection of the Negroes of St Domingo Begun in August 1791, Translated from the French: The Fourth Edition with Notes and an Appendix Extracted from Authentic Original Papers," *Gentleman's Magazine and Historical Chronicle for the year MDCCXCII*, LCP.

144. Monte Tejada, *Historia de Santo Domingo*, 3:143–45.

145. Ignacio Leyte Vidal to Juan Bautista Vaillant, Baracoa, 14 September 1791, exp. 33, leg. 4, AP, ANC. See also Sevilla Soler, *Santo Domingo*, 383.

146. "Yo, el Rey," Declaration of War against France, Aranjuéz, 30 March 1793, exp. 44, leg. 4, AP, ANC; Sevilla Soler, *Santo Domingo*, 388; Monte Tejada, *Historia de Santo Domingo*, 3:146.

147. Monte Tejada, *Historia de Santo Domingo*.

148. Ibid., 3:147.

149. Sevilla Soler, *Santo Domingo*, 377–94; Geggus, *Slavery, War, and Revolution*.

150. Urízzar, "Cuentas de viveres de bahiaja, isla de santo domingo, 1794–1796"; "Libro gral de entradas y consumos de viveres en los rles almacenes de la plaza de Bahiaxa en los años de 1794, 1795, 1796," Bahiaja, 10 February 1794; both leg. 1057, SD, AGI. For the dimension of the dispute among officials in Caracas, see Pedro Carbonell to Conde de Campo Alange, Caracas, 30 November 1793, exp. 36, leg. 7176; García to Campo Alange, Santo Domingo, 18 July 1796, exp. 30, leg. 7161; both GM, AGS.

151. De Laura [Pablo Estévez], *Parte tercera*.

152. Lleonart to Alcudia, Santo Domingo, 23 February 1795, exp. 9, leg. 18, Estado, AGI; Francisco de Montalvo, "Relación de la tropa," Havana, 20 November 1796, exp. 54, leg. 6876, GM, AGS.

153. Vaillant to the Ayuntamiento de Santiago de Cuba, Actas del Ayuntamiento, vol. 16, Santiago de Cuba, 24 June 1793, AHM.

154. Junta de Fomento, Havana, 20 July 1796, exp. 2757, leg. 71, Junta de Fomento, ANC, citing the Royal Order of 25 June 1793, ibid., 23 September 1793.

155. Pedro Aparicio to Intendant of Cuba, citing a Royal Order, Aranjuéz, 1 February 1793, exp. 41, leg. 4, AP, ANC.

156. Pedro Carbonell to Campo Alange, Caracas, 30 November 1793, exp. 36, leg. 7176, GM, AGS.

157. "Mapa de las isletas Grande y Pequeña, en el río Massacre o Dajabón, situados entre los pueblos de Dajabón (español) y Juana Méndez o Oüanaminthe (francés), y bahía de Manzanillo (isla de Santo Domingo)," number 870, Mapas y Planos, SD, AGI.

158. García to Campo Alange, Santo Domingo, 28 June 1793, exp. 27; Campo Alange to García, Madrid, 8 November 1793, exp. 38; both leg. 7158, GM, AGS.

159. Pedro LaFevillez, "Solicitud," Havana, 15 March 1799, exp. 5, leg. 7162, GM, AGS.

160. García to Campo Alange, Santo Domingo, 22 July 1793, exp. 38, leg. 7158, GM, AGS. San Miguel had 323 veteran troops; San Rafael had only 300 men from the regiment of Cantabria and the *fixo* of Santo Domingo; Hincha's defenses were its militia volunteers.

161. García to Campo Alange, Santo Domingo, 23 August 1793, exp. 38, leg. 7158, GM, AGS. For the negative effect of fever on the British troops, see Geggus, *Slavery, War, and Revolution*, 347–72.

162. García to Campo Alange, Santo Domingo, 22 July 1793, exp. 38; LeFevilliez to García, Dajabón, 27 June 1793, exp. 27; both leg. 7158, GM, AGS.

163. García to Campo Alange, Santo Domingo, 20 October 1793, exp. 17, leg. 7151, GM, AGS. In addition to the companies already in Santo Domingo, García requested 2,200 additional troops from Havana.

164. Carbonell to Campo Alange, Caracas, 30 November 1793, exp. 72, leg. 7175, GM, AGS.

165. García to Campo Alange, Santo Domingo, 19 August 1793, exp. 4, and 20 October 1793, exp. 17, leg. 7151, GM, AGS; García to Campo Alange, December 1793, exp. 35, leg. 7158, GM, AGS.

166. Aristizábal to García, onboard the *San Eugenio*, 29 January 1794; García to Campo Alange, Bayajá, 5 February 1794; Consejo de Indias to García, Aranjuéz, 28 March 1794; all exp. 27, leg. 7159, GM, AGS.

167. At the same time that Aristizábal blockaded the town, García wrote to the Council of State that it was not possible to attack Bayajá at that time because of his shortage of troops. He estimated that there were 6,000 to 8,000 enemy in Guarico and 1,000 in Bayajá. García to Campo Alange, Santo Domingo, 15 January 1794, exp. 5, leg. 7159, GM, AGS.

168. García to Campo Alange, Bayajá, 4 February 1794, exp. 6, leg. 7156, GM, AGS.

169. Juan de Salazar to Urízzar, Santo Domingo, 19 June 1794, exp. 58, leg. 1751, GM, AGS.

170. García to Campo Alange, Santo Domingo, 15 January 1794, exp. 5; García to Campo Alange, Bayajá, 17 May 1794, exp. 44; both leg. 7159, GM, AGS.

171. In December, García had informed Aristizábal about the best way to assault Bayajá. Fortunately, the naval commander had ignored García's advice. García to Campo Alange, Santo Domingo, 13 December 1793, exp. 35, leg. 7158, GM, AGS.

172. Antonio Valdéz to García, Aranjuéz, 6 April 1794, exp. 32, leg. 7159; "Sobre las buenas relaciones que deben observar el capitán general de la isla y el comandante general de la escuadra de América, Gabriel Aristizábal," Valdéz to García, Aranjuéz, 26 April 1794, exp. 30, leg. 7151; both GM, AGS.

173. Monte y Tejada, *Historia de Santo Domingo*, 3:152.

174. Lleonart to García, enclosed in García to Campo Alange, Bayajá, 17 May 1794, exp. 44, leg. 7159, GM, AGS.

175. Estimates of the numbers of men who served under Juan Francisco and Touissant fluctuated wildly, from as few as 500 to as many as 7,000. Correspondents often complained about the demands for food and money based upon exaggerated numbers of soldiers.

176. García to Alcudia, Bayajá, 18 February 1794, exp. 86, leg. 14, Estado, AGI.

177. Eugenio de Llaguno to García, Aranjuez, 14 June 1794, exp. 13, leg. 7156, GM, AGS. The chaplaincy came with the lucrative responsibility to oversee the church treasury. Simply put, it was a license to steal. Later, probably to cover up the disappearance of so many sacred objects from the churches in Santo Domingo, the Cuban regiments became the

convenient scapegoats for the archbishop of Santo Domingo to deflect suspicion away from his favorite.

178. García to Campo Alange, Santo Domingo, 13 December 1793, exp. 35, leg. 7158, GM, AGS.

179. Sasso to García, Bayajá, enclosed in García to Campo Alange, Santo Domingo, 15 January 1794, exp. 5; García to Campo Alange, Bayajá, 17 May 1794, exp. 44; both leg. 7159, GM, AGS.

180. García to Campo Alange, Santo Domingo, 22 January 1794, exp. 74, leg. 14, Estado, AGI; Urízzar, "Cuentas de viveres de bahiaja, isla de santo domingo, 1794–1796"; Juan Sánchez, "Libro gral de entradas y consumos de viveres en los rles almacenes de la plaza de Bahiaxa en los años de 1794, 1795, 1796," Bayajá, 10 February 1794–April 1796; both leg. 1057, SD, AGI.

181. Urízzar to Llaguno, Santo Domingo, 19 June 1794, exp. 58, leg. 7151, GM, AGS.

182. Sánchez, "Libro gral de entradas y consumos de viveres," Bayajá, 22 April 1794, leg. 1057, SD, AGI.

183. Urízzar to Llaguno, Santo Domingo, 20 July 1794, exp. 58, leg. 7151, GM, AGS.

184. García to Campo Alange, Santo Domingo, 13 December 1793, exp. 35, leg. 7158, GM, AGS.

185. Ibid.

186. Pedro Pablo Irigoyen, Solicitud, Montecristi, 22 June 1794, exp. 21, leg. 7156, GM, AGS.

187. Urízzar to Llaguno y Arriola, Santo Domingo, 19 June 1794, exp. 58, leg. 7151, GM, AGS.

188. Ibid.

189. Las Casas to Ayuntamiento de Havana, Havana, 27 February 1794, leg. 1460, PC, AGI.

190. García to Campo Alange, Bayajá, 19 March 1794, exp. 50, leg. 7159, GM, AGS.

191. Ibid.

192. Urízzar to Llaguno y Arriola, Santo Domingo, 25 April 1794, exp. 52, leg. 7151, GM, AGS.

193. García to Alcudia, Bayajá, 16 May 1794, exp. 91, leg. 14, Estado, AGI.

194. Lleonart to García, enclosed in García to Campo Alange, Bayajá, 17 May 1794, exp. 44, leg. 7159, GM, AGS.

195. García to Campo Alange, Hato de la Gorra, 5 July 1794, exp. 51, leg. 7159, GM, AGS.

196. García to Campo Alange, Quartel de Santiago, 6 August 1794, exp. 15, leg. 7161, GM, AGS.

197. Urízzar to Llaguno, Santo Domingo, 25 April 1794, exp. 52, leg. 7151; Irigoyen, Solicitud, Montecristi, 22 June 1794, exp. 21, leg. 7156; both GM, AGS.

198. García to Campo Alange, Bayajá, 22 May 1794, exp. 45, leg. 7151, GM, AGS.

199. Urízzar to Llaguno, Santo Domingo, 25 April 1794, exp. 52, leg. 7151, GM, AGS.

200. Monte y Tejada, *Historia de Santo Domingo*, 3:152.

201. García to Campo Alange, Bahiaja, 17 May 1794, exp. 44, leg. 7159, GM, AGS.

202. García to Campo Alange, Quartel de Santiago, 6 August 1794, exp. 15, leg. 7161, GM, AGS.

203. Aristizábal to García, onboard the *San Eugenio*, 29 January 1794; García to Campo Alange, Bayajá, 5 February 1794; Consejo de Indias to García, Aranjuéz, 28 March 1794; all exp. 27, leg. 7159, GM, AGS.

204. It is difficult to determine who gave the order to allow the auxiliaries into the city. With García's and Casasola's departures, the next officer of the regiment of Santo Domingo, Sasso, was in charge. García to Alcudia, Bayajá, 16 May 1794, exp. 91, leg. 14, Estado, AGI, identifies the commanders at Bayajá. The next senior officer was Brigadier Pedro Garibay of the regiment of New Spain, who had over 800 troops; the Cuban commander, Sebastián Calvo, commanded only 300 men. García to Campo Alange, Santo Domingo, 28 August 1796, exp. 8, leg. 7153, GM, AGS.

205. Council of State, Proceedings, San Ildefonso, 24 October 1794, exp. 69, leg. 7151, GM, AGS, quoting a letter sent by the refugees to Carlos Martínez de Yrujo from Mole San Nicolás on 24 July 1794. The writer states that "the blacks could have been contained with 100 seasoned and experienced troops, and they would have never entered the city if the orders of Francisco de Montalvo, lieutenant colonel of Havana, had been followed." Given the testimony put forth by eyewitnesses and by analyzing the composition of the forces in Bayajá, it seems that the opinion offered by Geggus, in *Haitian Revolutionary Studies*, 176, that the Cuban forces at Bayajá exhibited "predatory" behavior characterized by "criminality and cowardice" is unfounded. See also Ferrer, "Mundo cubano," 108–12.

206. Lleonart to Alcudia, Santo Domingo, 23 February 1795, exp. 9, leg. 18, Estado, AGI.

207. Ibid.

208. García to Council of State, Santo Domingo, 25 October 1794, exp. 13, leg. 7161, GM, AGS.

209. García to Real Socorro, Santo Domingo, 2 November 1794, GM, AGS.

210. George Josef Roxas to Las Casas, Batabanó, 30 August 1794, leg. 1470, PC, AGI. Antonio de la Torre to Las Casas, Gibacoa, 6 September 1794; López Gavilán to Las Casas, Güines, 15 September 1794; both leg. 1471, PC, AGI.

211. *Papel Periódico de la Havana*, 4 September 1794. Based upon the reports of the storm surge, it is possible to conclude that this was at least a Category 3 and more likely a Category 4 or Category 5 hurricane.

212. Agustín Ramos de Zayas to Las Casas, San Lázaro, 29 August 1794, leg. 1472, PC, AGI.

213. Gilbert Saint Maxent to Campo Alange, Placaminas, 31 August 1794, leg. 2563, SD, AGI. The hurricane scored a direct hit on the garrison town of Placaminas, and it followed a hurricane that had come ashore in Louisiana on 10 August 1794.

214. Ibid.

215. Ibid.

216. Johnson, "Rise and Fall." Guerra y Sánchez, *Manual de historia*, 206–7, labeled the subsequent economic cycle the "depression of 1796." Most sources attribute the depression to the British blockade, but the blockade was not instituted until after war was declared in August. See also Moreno Fraginals, *El ingenio*, 1:97.

217. Sánchez, "Libro gral de entradas y consumos de viveres," Bayajá, 30 September 1794, leg. 1057, SD, AGI.

218. Council of State, Proceedings, Madrid, 24 October 1794, exp. 69, leg. 7151, GM, AGS.

219. Jose de Anduaga to García, Madrid, 26 December 1794, exp. 77, leg. 7159, GM, AGS.

220. Ibid.; Fernández, "José de Solano."

221. Casa Calvo to Alcudia, Bayajá, 31 August 1794, exp. 63; 19 September 1794, exp. 64; both leg. 14, Estado, AGI.

222. Jose de Anduaga to García, San Ildefonso, 26 September 1794, exp. 52, leg. 7151, GM, AGS.

223. Sánchez, "Libro gral de entradas y consumos de viveres," Bayajá, September 1794–June 1796, leg. 1057, SD, AGI.

224. Campo Alange to García, San Ildefonso, 24 September 1794, exp. 45, leg. 7151; García to Campo Alange, Santo Domingo, 24 April 1795, exp. 22, leg. 7165; both GM, AGS.

225. Jose de Anduaga to García, Madrid, 26 December 1794, exp. 77, leg. 7159, GM, AGS.

226. Casa Calvo to Alcudia, Bayajá, 31 August 1794, exp. 63; 19 September 1794, exp. 64; both leg. 14, Estado, AGI.

227. Francisco de Herran, Report, Bayajá, 29 July 1794, exp. 46, leg. 6853; García to Campo Alange, Santo Domingo, 24 December 1794, exp. 15, leg. 7160; both GM, AGS.

228. Sánchez, "Libro gral de entradas y consumos de viveres," Bayajá, September 1794–June 1796, leg. 1057, SD, AGI.

229. Alcudia to García, San Ildefonso, 8 September 1795, exp. 4, leg. 17, Estado, AGI.

230. Joaquín Beltrán de Santa Cruz to Príncipe de Paz, in "Yo, el Rey," San Lorenzo, 6 November 1796, exp. 44, leg. 6857, GM, AGS. See also Johnson, *Social Transformation*, 166–68; Kuethe, *Cuba*, 152–54; and Marrero y Artiles, *Cuba*, 13:254–57.

231. Casa Calvo to García, Havana, 10 September 1796, exp. 30, leg. 7161, GM, AGS.

232. Aristizábal to Las Casas, onboard the *San Eugenio*, 23 June 1796, exp. 52, leg. 6857, GM, AGS.

233. González-Ripoll Navarro, *El rumor de Haiti*; Ferrer, "Mundo cubano," 107.

234. Príncipe de Paz to Las Casas, San Lorenzo, 6 November 1796, exp. 44, leg. 6857, GM, AGS.

235. Lleonart to Alcudia, Santo Domingo, 23 February 1795, exp. 9, leg. 18, Estado, AGI.

236. Las Casas to Príncipe de Paz, Havana, 20 December 1795, exp. 176, leg. 5B, Estado, AGI. The impact that Charles IV's decision to overlook Juan Francisco's treasonable behavior had on other regiments of color in Cuba and elsewhere has yet to be examined in detail.

237. Arango to the Junta de Gobierno del Consulado, Havana, 22 April 1796, vol. 82, Levi Marrero Collection, Special Collections, Florida International University, Miami. Arango complained that "the captain general has to work hard to extirpate any seeds of discord."

238. De Laura [Pablo Estévez], *Parte tercera*.

239. Kuethe, *Los llorones Cubanos*, 134–55.

240. Junta de Gobierno del Consulado, Havana, 30 March 1796, exp. 2755, leg. 71, Junta de Fomento, ANC.

241. Arango to the Junta de Gobierno del Consulado, Havana, 22 April 1796, vol. 82, Levi Marrero Collection, Special Collections, Florida International University, Miami.

242. Junta de Gobierno del Consulado, Havana, 25 May 1796, exp. 2755, leg. 71, Junta de Fomento, ANC.

243. "Sobre varias embarcaciones anglo americanos que se han presentado en este puerto con harinas después de la real orden que revoca este comercio," Junta de Gobierno del Consulado, Havana, 20 July 1796, exp. 2757, ibid.

244. Junta de Gobierno del Consulado, Havana, 4 August 1796, exp. 2759, ibid.

245. Junta de Gobierno del Consulado, Havana, 23 August 1796, exp. 2767, ibid. See also Johnson, Social Transformation, 168–69; Kuethe, Cuba, 152–54; and Marrero y Artiles, Cuba, 13:254–57. Moreno Fraginals, El ingenio, 1:58, 100–101, describes the concession as a "sucio negocio" (dirty business).

246. Beltrán de Santa Cruz to Príncipe de Paz, in "Yo, el Rey," San Lorenzo, 6 November 1796, exp. 44, leg. 6857, GM, AGS.

247. Junta de Gobierno del Consulado, Havana, 23 August 1796, exp. 2767, leg. 71, Junta de Fomento, ANC. See also Johnson, Social Transformation, 168–69; Kuethe, Cuba, 152–54; and Marrero y Artiles, Cuba, 13:254–57. Moreno Fraginals, El ingenio, 1:58, 100–101.

248. Kuethe, Cuba, 153; Marrero y Artiles, Cuba, 13:251.

249. Junta de Gobierno del Consulado, Havana, 31 August 1796, 16 November 1796, exp. 2762, leg. 71, Junta de Fomento, ANC.

250. Marrero y Artiles, Cuba, 13:262 n. x. The comment was appropriated by Jacobo de la Pezuela, who wrote of the "bad impression such monstrous prerogatives gave to the people." Historia de la isla de Cuba, 3:289–90.

251. 20 October 1796, leg. 16, numero 16, exp. 1, Estado, AGI.

252. De Laura [Pablo Estévez], Parte tercera.

253. Guerra y Sánchez, Manual de historia, 207–8.

254. Johnson, Social Transformation, 168–69.

255. Miguel Díaz to Luis de las Casas, Luyanó, 4 October 1796, leg. 1472, PC, AGI, reported only minimal damage. See also Millás, Hurricanes of the Caribbean, 292.

Chapter Seven

1. See Jones and Mann, "Climate over Past Millennia," especially the graph on p. 2; and Mann et al., "Proxy-Based Reconstructions."

2. Pérez, Winds of Change, establishes the destruction of the coffee industry and the subsequent advance of sugar cultivation. See also Jones and Mann, "Climate over Past Millennia."

3. Junta de Gobierno, Havana, 31 January 1802, exp. 2795, leg. 73, Junta de Fomento, ANC.

4. Junta de Gobierno, Havana, 10 July 1802, exp. 2795, ibid.

5. Junta de Gobierno, Havana, 31 August 1803, exp. 2803-A; Junta de Gobierno, Havana, 8 November 1803, exp. 2807-A; Junta de Gobierno, Havana, 16 August 1804, exp. 2817; all ibid.

6. Junta de Gobierno, Havana, 6 April 1805, exp. 2819, ibid.

7. Junta de Gobierno, Havana, 24 May 1804, exp. 2802, ibid.

8. "A Spaniard," Observations on the Commerce of Spain with Her Colonies in Time of War, 48–49.

9. Juan Álvarez de Miranda, "Poética relación Christiana y Moral," 1768, leg. 1097, PC, AGI.

10. Marqués de la Torre, "Relación de mi mando en Cuba, desde el dia 18 de noviembre de 1771 hasta que entrego a mi sucesor Diego José," 12 June 1777, box 3, folder 2, Domingo del Monte Collection, Manuscript Division, LC; Valdés, *Historia de la isla de Cuba*, 169.

11. *Elogio funebre que en las exequias que a Nuestro Catolico Monarca, el Senor Don Carlos III Hizo la muy noble y leal Ciudad de Sn Juan de Jaruco, sita en la isla de Cuba el 29 de Marzo de 1789 Dixo el P. Mro. F. Vicente Ferrer de Acosta*, San Juan de Jaruco, 29 March 1789, leg. 1609, IG, AGI.

Appendix 1

1. Marqués de Gandara to Alonso Arcos y Moreno, Santo Domingo, 23 May 1750, exp. 43, leg. 5, CCG, ANC.

2. José Pablo de Agüero to Arcos y Moreno, Santo Domingo, 27 January 1751, exp. 102, leg. 7; Arcos y Moreno to Francisco Cagigal de la Vega, Santiago de Cuba, 28 January 1751, exp. 228, leg. 7; Ignacio Sola to Arcos y Moreno, Cartagena de Indias, September (?) 1752, exp. 128, leg. 6; all CCG, ANC.

3. Millás, *Hurricanes of the Caribbean*, 205.

4. Arcos y Moreno to Cagigal de la Vega, Santiago de Cuba, 27 March 1752, exp. 74, leg. 6, CCG, ANC.

5. *PZ*, 2 January 1753.

6. Cagigal to Arcos y Moreno, Havana, 12 February 1754, exp. 50, leg. 6, CCG, ANC.

7. Lorenzo de Madariaga to Cagigal de la Vega, Santiago de Cuba, Summer 1754, exp. 368, leg. 6, CCG, ANC.

8. Madariaga to Cagigal de la Vega, Santiago de Cuba, 17 October 1754, exp. 242, leg. 6, CCG, ANC, enclosing a letter written in French from the Captain of the *Dama Maria*.

9. Madariaga to Martín Estéban de Aróstegui, Santiago de Cuba, 8 April 1755, exp. 284, leg. 7, CCG, ANC.

10. Apoderado de la Real Compañía to Madariaga, Santiago de Cuba, 1 August 1755, exp. 407, leg. 7, CCG, ANC. Millás, *Hurricanes of the Caribbean*, rejects this case, #59.

11. Miguel Palomino to Madariaga, Cabañas, 26 February 1756, exp. 12, leg. 7, CCG, ANC.

12. Millás, *Hurricanes of the Caribbean*, 210.

13. Aróstegui to Madariaga, Puerto Príncipe, 9 November 1758, exp. 161, leg. 7, CCG, ANC.

14. Millás, *Hurricanes of the Caribbean*, 214–15.

15. Joseph Paulino de Salgado to Madariaga, Juraguá, October (?) 1760, exp. 195, leg. 10, CCG, ANC.

16. Francisco Tamayo to Madariaga, Bayamo, 19 October 1761, exp. 61, leg. 11, CCG, ANC.

17. Muesas to Madariaga, El Morro (Santiago de Cuba), 8 June 1762, exp. 94; Antonio Marín to Madariaga, Juragua, 3 August 1762, exp. 119; Joseph Péres to Madariaga, Aguadores, 13 September 1762, exp. 115; all leg. 11, CCG, ANC.

18. Marqués de Casa Cagigal to Jerónimo de Grimaldi, Minister of State, Santiago de Cuba, 22 April 1765, exp. 94, leg. 23, CCG, ANC.

19. Conde de Ricla to Julián de Arriaga, Havana, June 1765, exp. 7, leg. 21, CCG, ANC.

20. Millás, *Hurricanes of the Caribbean*, 224.

21. Miner Solá, *Historia de los huracanes en Puerto Rico*, 30.

22. Ludlum, *Early American Hurricanes*, 62–63; Weddle, *Changing Tides*, 10–23.

23. Antonio María Bucareli to Cagigal, Havana, 2 September 1766, exp. 39, leg. 19, CCG, ANC.

24. Miner Solá, *Historia de los huracanes en Puerto Rico*, 30.

25. Bartolomé de Morales to Governor of Cuba, Santiago del Prado, 12 October 1766, exp. 62, leg. 24, CCG, ANC.

26. Ludlum, *Early American Hurricanes*, 62–63; Weddle, *Changing Tides*, 10–23.

27. *Estado que comprehende las desgracias que causó el huracán el día 15 de octubre en la ciudad de la Havana* (Cádiz, 1768); and *Estado que comprehende las desgracias que causó el huracán el día 15 de octubre en la ciudad de la Havana* (Madrid, 1769); both leg. 1594, SD, AGI.

28. Miguel de Muesas to Manuel Varela, Santiago de Cuba, 28 August 1769, exp. 87, leg. 26, CCG, ANC.

29. Ayuntamiento de Santa Clara, Actas Capitulares, 26 October 1770, Tomo 8, 1772–79, Provincial Archive of Villa Clara, Cuba.

30. José de Armona to the Council of the Indies, Havana, 7 February 1771, leg. 257-A, Correos, AGI.

31. Juan de Ayans de Ureta to Marqués de la Torre, Santiago de Cuba, 2 May 1772, leg. 1141, PC, AGI; de la Torre to Arriaga, Havana, 20 June 1772, leg. 1216, PC, AGI; de la Torre to Juan Bautista Bonet, Havana, 26 June 1772, leg. 1159, PC, AGI.

32. Ayans de Ureta to de la Torre, Santiago de Cuba, 25 June 1772, leg. 1141, PC, AGI.

33. Millás, *Hurricanes of the Caribbean*, 229–30.

34. Josef Alvarado to de la Torre, Bayamo, 22 August 1772, leg. 1178, PC, AGI; Ayans to de la Torre, Santiago de Cuba, 23 August 1772, 4 September 1772, leg. 1141, PC, AGI; de la Torre to Arriaga, Havana, 23 August 1772, leg. 1216, PC, AGI; Millás, *Hurricanes of the Caribbean*, 230.

35. Armona to the Council of the Indies, Havana, 31 August 1772, 19 October 1772, leg. 257-A, Correos, AGI; Luis de Unzaga to de la Torre, New Orleans, 9 September 1772, "Despatches of the Spanish Governors of Louisiana," Manuscripts Department, Howard-Tilton Memorial Library, Tulane University, New Orleans; Ludlum, *Early American Hurricanes*, 63–64; Millás, *Hurricanes of the Caribbean*, 230; Miner Solá, *Historia de los huracanes en Puerto Rico*, 30.

36. Muesas to the Council of the Indies, San Juan, (day omitted) September 1772, leg. 2516, SD, AGI; Ayans de Ureta to de la Torre, Santiago de Cuba, 4 September 1772, leg. 1141, PC, AGI; Ludlum, *Early American Hurricanes*, 64. Millás, *Hurricanes of the Caribbean*, 235–37, describes this "severe storm of 1772" as an event unprecedented in meteorological history. See also *PZ*, 23 September 1772, 14 October 1772, LCP.

37. De la Torre to Vicente de Justíz, San Antonio, 10 January 1773, leg. 1184; Arriaga to de la Torre, El Pardo, 20 February 1773, leg. 1212; both PC, AGI.

38. Bonet to de la Torre, Havana, 6 November 1773, leg. 1158, PC, AGI, relaying the news from Josef Días Amador in Matanzas.

39. Goicoa to the Council of the Indies, Madrid, 20 December 1773, leg. 2516, SD, AGI.

40. De la Torre to Bonet, Havana, 10 December 1774, leg. 1159; de la Torre to Luis de Toledo, Havana, 1 January 1775, leg. 1159; Nicolás Cárdenas Vela de Guevara to de la Torre, Jaruco, 2 November 1774, leg. 1165; Estéban Rodríguez del Pino to de la Torre, Bejucal, 12 November 1774, leg. 1167; Enríquez to de la Torre, San Pedro, 14 November 1774, leg. 1192; Sebastián de la Cruz to de la Torre, Güines, 17 November 1774, 19 November 1774, leg. 1195; all PC, AGI.

41. Bartolomé de Montes to Bonet to de la Torre, Havana, 13 January 1775, leg. 1158, PC, AGI.

42. Ayans y Ureta to de la Torre, Santiago de Cuba, 21 September 1775, 7 December 1775, leg. 1142, PC, AGI.

43. Antonio López de Toledo to Juan Ignacio de Urriza, Batabanó, 18 June 1776; Urriza to Estévan del Pino, Havana, 18 June 1776; both leg. 1152, PC, AGI.

44. François DePlessis to Robert Morris, New Orleans, 16 August 1776, Willing and Morris Correspondence, Levis Collection, HSP.

45. Isidro de Limonta, "Relacion de las embarcaciones que han entrado y salido de este puerto para otros continentes desde 1 de junio hasta el dia de la fecha, y son los siguientes," Santiago de Cuba, 2 June 1776; Oloriz to de la Torre, Santiago de Cuba, 4 July 1776; both leg. 1143, PC, AGI.

46. Oloriz to de la Torre, Santiago de Cuba, 18 October 1776, leg. 1143; de la Torre to Lleonart, Havana, 5 November 1776, leg. 1184, PC, AGI.

47. Urriza to Council of the Indies, Havana, 22 February 1777, 10 May 1777, leg. 257-B, Correos, AGI.

48. Martín de Aróstegui to Diego de Navarro, Matanzas, 27 July 1777, 5 August 1777, leg. 1248, PC, AGI.

49. Joseph Alvarado to Navarro, Trinidad, 22 October 1777, leg. 1259, PC, AGI.

50. José Días Tejada to Navarro, Bayamo, 7 December 1777; José Melchor de Acosta to Navarro, Guarico, 22 December 1777; both leg. 1245, PC, AGI; Navarro to Joseph Tentor, Havana, 9 January 1778, acknowledging Tentor's report of 9 November 1777, folder 1, container 4, Domingo del Monte Collection, Manuscript Division, LC.

51. Urriza to Council of the Indies, Havana, 25 February 1777, 10 May 1777, leg. 257-B, Correos, AGI.

52. Luis Huet to Navarro, Havana, 2 August 1778, 24 September 1778, leg. 1247, PC, AGI.

53. José Antonio Morejón to Navarro, San Pedro, 2 September 1779, leg. 1269, PC, AGI; Bernardo de Gálvez to Diego José Navarro, New Orleans, 2 December 1779, vol. 10, "Despatches of the Spanish Governors of Louisiana," Manuscripts Department, Howard-Tilton Memorial Library, Tulane University, New Orleans.

54. Navarro to José de Gálvez, Havana, 25 February 1780, leg. 2082; Martín de Navarro to Diego de Navarro, New Orleans, 18 April 1780, leg. 2609; both SD, AGI.

55. Navarro to Conde de Ripalda, Havana, 30 April 1780, 22 May 1780; Urriza to Ri-

palda, Havana, 19 June 1780; Ventura Díaz to Urriza, Puerto Príncipe, 22 June 1780, 27 June 1780; all leg. 1256, PC, AGI.

56. Navarro to the Captain of Managua, Havana, 21 July 1780, leg. 1269, PC, AGI.

57. Navarro to the Captain of Managua, Havana, 21 July 1780; Various Residents of Arroyo Arenas to Navarro, Arroyo Arenas, 19 August 1780; both leg. 1269, PC, AGI.

58. Ventura Díaz to Navarro, Puerto Príncipe, 11 October 1780, leg. 1256, PC, AGI.

59. Millás, *Hurricanes of the Caribbean*, 241–59; Mulcahy, *Hurricanes and Society*, 108–11, 166–74; Morales Padrón, *Journal of Don Francisco Saavedra*, 9–15 October 1780.

60. Ramon Lloret to Navarro, Campeche, 5 November 1780, leg. 1248, PC, AGI; Millás, *Hurricanes of the Caribbean*, 260–62.

61. Governor of San Nicolás to Navarro, San Nicolás (Saint Domingue), 25 October 1780, leg. 1231, PC, AGI.

62. Diego de Belmonte to Cagigal, Matanzas, 3 February 1782, leg. 1317, PC, AGI.

63. Mariano Murguia to Cagigal, Nueva Filipinas, 28 July 1782, leg. 1317, PC, AGI.

64. Bernardo de Gálvez to José de Gálvez, Havana, 15 June 1784, leg. 1344, PC, AGI.

65. Ibid., 8 August 1784, 13 October 1784.

66. Andrés Moreno to the Ayuntamiento de Santa Clara, 11 February 1791, Actas Caputulares, Tomo 8, 1787–91, Archivo Provincial de Santa Clara, Santa Clara, Cuba; Lucas Álvares to Luis de las Casas, Batabanó, 11 March 1791, leg. 1471, PC, AGI; Miguel Núñez and José Pérez de Medina to Las Casas, Guanabacoa, 6 May 1791, leg. 1460, PC, AGI.

67. José Fuertes to the Duque de Alcudía, Havana, 1 August 1791, leg. 260-A, Correos, AGI; *Papel Periódico de la Havana*, 7 August 1791; Rappaport and Fernández-Partagás, "Deadliest Atlantic Tropical Cyclones," 2.

68. Hilario de León and Manuel de Ayala to the Ayuntamiento de Santa Clara, Santa Clara, 22 June 1792; Joaquín Moya, Juan Antonio Montenegro, and José Montenegro to the Ayuntamiento de Santa Clara, 14 September 1792, Actas del Ayuntamiento de Santa Clara, Tomo 9, 1792–99, Archivo Provincial de Santa Clara, Santa Clara, Cuba.

69. José Fuertes to Duque de Alcudía, Havana, 9 November 1792, leg. 260-A, Correos, AGI; Vicente Soría to Las Casas, Santiago de la Vegas, 2 November 1792, leg. 1460, PC, AGI; Francisco Borrego to Las Casas, Quemados, 26 June 1792, leg. 1472, PC, AGI.

70. Actas del Ayuntamiento de Santiago de Cuba, Santiago de Cuba, 6 November 1792, Actas del Ayuntamiento, Tomo 16, Archivo Municipal de Santiago de Cuba.

71. Síndico Procurador to the Obispo de la Havana, Havana, 10 February 1793, "Edicto del D.D. Phelipe Jph de Trespalacios, a nuestros amados fieles vecinos y moradores estantes y habitantes en las ciudades, villas y lugares de nro obispado de qualquiera dignidad," 18 February 1793, Correspondencia del Obispo de la Habana, Archivo del Arzobispo de la Habana, Havana, Cuba.

72. Las Casas to Ayuntamiento de Havana, Havana, 27 February 1794, leg. 1460, PC, AGI.

73. Juan Antonio Urrizar to Campo Alange, Santo Domingo, 19 June 1794, exp. 59, leg. 7151, GM, AGS.

74. Gilbert Saint Maxent to Campo Alange, Placaminas, 31 August 1794, leg. 2563, SD, AGI. The hurricane scored a direct hit on the garrison town of Placaminas, and it followed a previous hurricane that came ashore in Louisiana on 10 August.

75. George Josef Roxas to Las Casas, Batabanó, 30 August 1794, leg. 1470, PC, AGI; *Papel Periódico de la Havana*, 4 September 1794. Based upon the reports of the storm surge, this was at least a Category 3 and more likely a Category 4 or Category 5 hurricane.

76. Saint Maxent to Campo Alange, Placaminas, 31 August 1794, leg. 2563, SD, AGI.

77. Consulado to Gardoqui, Havana, 8 October 1796, leg. 2191, SD, AGI, copies in Levi Marrero Collection.

78. Miguel Díaz to Las Casas, Luyanó, 4 October 1796, leg. 1472, PC, AGI, reported only minimal damage; Millás, *Hurricanes of the Caribbean*, 292.

79. Ayuntamiento de Santiago de Cuba to Junta de Fomento, Santiago de Cuba, 31 August 1799, exp. 2784; 12 September 1799, exp. 2785; both leg. 72, Junta de Fomento, ANC.

BIBLIOGRAPHY

Manuscript Sources

Boston, Mass.
  Massachusetts Historical Society
    Curtis-Stevenson Family Papers
Cádiz, Spain
  Archivo Provincial Histórico de Cádiz
    Protocolos Notariales
Cambridge, Mass.
  Harvard University, Houghton Library
    José Escoto Collection
Gainesville, Fla.
  University of Florida, P. K. Yonge Library Special Collections
    East Florida Papers (microfilm)
    Papeles de Cuba (microfilm)
    Rare Books and Manuscripts
    Rochambeau Collection (microfilm)
    Stetson Collection (microfilm)
    Aileen Moore Topping Collection
Havana, Cuba
  Archivo del Arzobispo de la Habana
    Correspondencia del Obispo de la Habana, 1788–98
  Archivo Nacional de Cuba
    Asuntos Políticos
    Audiencia de Santo Domingo
    Correspondencia del Capitán General
    Escribanía de Gobierno
    Escribanía de Guerra
    Gobierno General
    Junta de Fomento
    Realengos, Extramuros
  Archivos Parroquiales
    S. M. I. Catedral de la Habana
  Biblioteca Nacional de Cuba José Martí
    Colección Cubana
    Colección Vidal Morales y Morales
    Mapas y Planos

Jacksonville, Fla.
    Diocese of St. Augustine Catholic Center
        Cathedral Parish Records, 1784–1821
Madrid, Spain
    Archivo Histórico de la Nación
        Diversos
    Biblioteca Nacional de España
        Colección Pezuela
        Manuscritos de América
Miami, Fla.
    Florida International University, Green Library
        Levi Marrero Collection
    University of Miami, Otto G. Richter Library
        Cuban Collection
        José Fernández Partagás, manuscript, "Poey, Viñes y Millás: Contribuyentes
            de Cuba al conocimiento básico de la meteorología"
New Orleans, La.
    Tulane University, Howard-Tilton Memorial Library
        Manuscripts Department
Philadelphia, Pa.
    Historical Society of Pennsylvania
        Clifford Family Correspondence
        Levis Collection
        John J. McCusker, "The Pennsylvania Shipping Industry in the Eighteenth
            Century (1973)"
        John J. McCusker, "Ships Registered at the Port of Philadelphia before 1776:
            A Computerized Listing"
        McGeehan Collection
        Philadelphia Customs House Books, 1766–75, 3 vols., Cadwalader Collection,
            Series 3: Thomas Cadwalader Papers
        Stephen Moylan Letterbook
    Library Company of Philadelphia
        Pennsylvania Gazette
Pinar del Río, Cuba
    Archivo Provincial de Pinar del Río
        Protocolos Notariales
Salem, Mass.
    Peabody Essex Museum
Santa Clara, Cuba
    Archivo Provincial de Santa Clara
        Actas Capitulares del Ayuntamiento
Santiago de Cuba, Cuba
    Archivo Municipal de Santiago de Cuba
        Actas Capitulares del Ayuntamiento

Seville, Spain
    Archivo General de Indias
        Audiencia de Santo Domingo
        Contratacción
        Correos
        Escribanía de Cámara
        Estado
        Indiferente General
        Ingenios y Muestras
        Mapas y Planos
        Papeles Procedentes de Cuba
        Ultramar
Simancas, Spain
    Archivo General de Simancas
        Secretaría de Guerra Moderna
Washington, D.C.
    Library of Congress, Manuscript Division
        Cuba Miscellany
        Domingo del Monte Collection
        East Florida Papers
        Foreign Copying Project
        Hispanic Collection
        Map Collection
        Rare Books and Pamphlets

Online Archival Collections

Ministerio de Cultura, Portal de Archivos Espanoles, Madrid, Spain (PARES),
    ⟨http://pares.mcu.es⟩
University of Florida, Digital Resources Online, University of Florida Libraries,
    Gainesville, Fla., ⟨http://ufdc.uf.edu⟩

Printed Primary Sources

Archivo General de Indias. *Colección de documentos inéditos para la historia de hispano-
    américa.* 15 vols. Madrid: Compañía Ibero-Americana de Publicaciones, 1927–32.
Archivo Nacional de Cuba. *Nuevos papeles sobre la toma de la Habana por los ingleses en
    1762.* Havana: Archivo Nacional de Cuba, 1951.
———. *Papeles sobre la toma de la Habana por los ingleses en 1762.* Havana: Archivo
    Nacional de Cuba, 1948.
Armona, José Antonio de. *Noticias privadas de casa útiles para mis hijos (Recuerdos
    del Madrid de Carlos III).* Edited, with introduction and notes, by Joaquín Álvarez
    Barrientos, Emilio Palacios Fernández, and María del Carmen Sánchez García.
    Madrid: Ayuntamiento de Madrid, 1988.

Barreto, José Francisco. *Contestación al impreso del Sr. Conde de O'Reilly*. Havana: Imprenta de D. Pedro Nolasco Palmer, 1812.

Calendario manual y guía de forasteros de la Isla de Cuba para el año 1795. Havana: Imprenta de la Capitanía General, 1795. Facsimile ed., Miami: Ediciones Cubanas, 1990.

Dancer, Thomas. *A Brief History of the Late Expedition against Fort San Juan, So Far as It Relates to the Diseases of the Troops: Together with Some Observations on Climate, Infection, and Contagion; and Several of the Endemial Complaints of the West Indies*. Kingston: D. Douglas and W. Aikman, 1781.

Du Broca, Louis. *Vida de J. J. Dessalines, gefe de los negros de Santo Domingo, con notas muy circunstancias sobre el origen, caracter y atrocidades de los principales gefes de aquellos rebeldes desde el principio de la insurreccion en 1791*. Translated from French by D.M.G.C. Mexico: Oficina de D. Mariano de Zúñiga y Ontiveros, 1806.

Fernández de Castro, Manuel. *Estudio sobre los huracanes ocurridos en la isla de Cuba durante el més de octobre de 1870*. Madrid: Imprenta de J. M. Lapuente, 1871.

Funes de Villalpando, Ambrosio, Conde de Ricla to Miguel de Altarriba, 14 May 1765. *Boletín del Archivo Nacional de Cuba* 43 (1944): 120.

Gala, Ignacio. *Memorias de la colonia francesa de Santo Domingo, con algunas reflexiones relativas a la isla de Cuba, por un viagero español*. Madrid: H. Santos Alonso, 1787.

Gálvez, Bernardo de. *"Yo Solo": The Battle Journal of Bernardo de Gálvez during the American Revolution*. Introduction by Eric Beerman. New Orleans: Polyanthos, 1978.

Garciny, Vicente de, to Marques de la Torre, 16 May 1777. *Boletín del Archivo Nacional de Cuba* 16 (1915): 212.

Hillary, William. *Observations on the Changes of the Air and the Concomitant Epidemical Diseases in the Island of Barbadoes. To Which Is Added a Treatise on the Putrid Bilious Fever Commonly Called the Yellow Fever and Such Other Diseases as Are Indigenous or Endemial in the West India Islands in the Torrid Zone, with Notes by Benjamin Rush*. Philadelphia: B&T Kite, 1811.

Humboldt, Alexander von. *Ensayo política sobre la Isla de Cuba*. Preliminary note by José Quintero Rodríguez and introduction by Fernando Ortiz. Reprint, Havana: Archivo Nacional de Cuba, 1960.

Juan, Jorge, and Antonio de Ulloa. *Noticias secretas de America, sobre el estado naval, militar, y político de los reynos del Peru y provincias de Quito, costas de Nueva Granada y Chile*. London: R. Taylor, 1826.

———. *Voyage to South America*. Translated from the original Spanish. Dublin: William Masterson, 1758.

Knowles, Charles. "Description of the Havana, 1761–1762." *Boletín del Archivo Nacional de Cuba* 37–38 (1941): 26–27.

Laura, Miseno de [Pablo Estévez]. *Parte tercera de las revoluciones periódicas de la Havana escribíala Miseno de Laura*. Havana: Imprenta de la Capitanía General, 1796.

Lavedan, Antonio. *Aforismos de Boerhave para conocer y curar las calenturas dados a luz en Latin por Maximilian Stoll, traducidos libremente al castellano por Antonio Lavedan*. Madrid: Don Francisco de la Parte, 1817.

Navarro, Diego José. "Bando sobre que se destechen las Casas de Guano, o Yaguas qe. estén dentro de la Ciudad." *Boletín del Archivo Nacional de Cuba* 28 (1929): 83–84.

———. "Estracto del padrón general de havitantes de la isla de Cuba." *Revista de la Biblioteca Nacional José Martí* 29 (1987): 25.

———. "Padrón General de la isla de Cuba formado a consequencia de Real Orden de 1º. de noviembre de 1776." *Revista de la Biblioteca Nacional José Martí* 29 (1987): 17–24.

Poey, Andrés. "A Chronological Table, Comprising 400 Cyclonic Hurricanes Which Have Occurred in the West Indies and in the North Atlantic within 362 Years, from 1493 to 1855; with a Bibliographical List of 450 Authors, Books, &c., and Periodicals, Where Some Interesting Accounts May Be Found, Especially on the West and East Indian Hurricanes." *Journal of the Royal Geographic Society of London* 25 (1855): 291–328.

Prado, Juan de. "Diario Militar de las Operaciones executadas en la Ciudad, y Campo de la Habana por disposicion de su Governador Don Juan de Prado." In *Cinco diarios del sitio de la Habana*, edited by Amalia Rodríguez, 65–126. Havana: Biblioteca Nacional José Martí, 1963.

Rapún, Nicolás José. *Reglamento para el gobierno interior, político, y económico de los hospitales reales, eregidos en la isla de Cuba, con destino a la curación de tropas, forzados, y negros esclavos de S.M., según las circunstancias, temperamento, y costumbres del país.* Madrid: J. de San Martin, 1776.

Raynal, Abbé. *A Philosophical and Political History of the Settlements and Trade of the Europeans in the East and West Indies.* Translated from the French by J. Justamond, M.A. 3rd ed. London: N.p., 1777.

Reglamento y aranceles para el comercio libre de España a Indias de 12 de octubre de 1778. Madrid: Imprenta de Pedro Marín, 1778. Facsimile ed., Seville: Facultad de Filosofia y Letras, Universidad de Sevilla y Escuela de Estudios Hispanoamericanos, 1978.

Ribera, Nicolás José de. *Descripción de la Isla de Cuba y algunas consideraciones sobre su población y comercios.* [1767.] Havana: Ministerio del Cultura de Cuba, 1973.

Rodriguez, Rafael. *Plano topográfico, histórico, y estadístico de la ciudad y puerto de la Habana.* Havana: Real Sociedad Patriótica, 1841.

Rojas y Rocha, Francisco de. *Poema épico de la rendición de Panzacola y conquista de la Florida Oriental por el Excmo Señor Conde de Gálvez.* Mexico City: Oficina de D. Mariano de Zúñiga y Ontiveros, 1785.

Romay, Tomás. *Disertación sobre la fiebre maligna llamada vulgarmente vómito negro enfermedad epidémica en las Indias Occidentales.* Havana: Sociedad Patriótica de la Habana, 1797.

Saavedra, Francisco de. *Los decenios: (autobiografía de un sevillano de la ilustración).* Transcribed, with introduction and notes, by Francisco Morales Padrón. Seville: Ayuntamiento de Sevilla, 1995.

Sagra, Ramón de la. *Historia económico-política y estadística de la isla de Cuba.* Havana: Imprenta de las Viudas de Arazoza y Soler, 1841.

Sims, James. *Observations on Epidemic Disorders with Remarks on Nervous and Malignant Fevers.* London: J. Johnson and G. Robinson, 1773.

"A Spaniard [Marqués de Casa Yrujo]." *Observations on the Commerce of Spain with Her Colonies in Time of War, by a Spaniard in Philadelphia.* Translated from the original manuscript by another Spaniard. Philadelphia: James Carey, 1800.

Secondary Sources

Abney, F. Glenn, and Larry B. Hill. "Natural Disasters as a Political Variable: The Effect of a Hurricane on an Urban Election." *American Political Science Review* 60, no. 4 (December 1966): 974–81.

Academia de Ciencias de Cuba. *La esclavitud en Cuba.* Havana: Editorial Académica, 1986.

Acosta, Antonio. "Las bases económicas de los primeros años de la Luisiana española (1763–1778)." In *La influencia de España en el Caribe, la Florida y la Luisiana,* edited by Antonio Acosta and Juan Marchena Fernández, 331–75. Madrid: Instituto de Cooperación Iberoamericana, 1983.

Acosta, Antonio, and Juan Marchena Fernández, eds. *La influencia de España en el Caribe, la Florida, y la Luisiana, 1500–1800.* Madrid: Instituto de Cooperación Iberoamericana, 1983.

Aimes, Hubert H. S. *A History of Slavery in Cuba, 1511–1868.* New York: G. P. Putnam's Sons, 1907.

Aiton, Arthur S. "Spanish Colonial Reorganization under the Family Compact." *Hispanic American Historical Review* 12 (August 1932): 269–80.

Albert, Robert C. The *Golden Voyage: The Life and Times of William Bingham, 1752–1804.* Boston: Little, Brown, 1969.

Alegre Pérez, María Esther. "Drogas americanas en la Real Botica." In *La ciencia española en Ultramar: Actas de las I Jornadas sobre "España y las expediciones científicas en América y Filipinas,"* coordinated by Alejandro R. Díez Torre, Tomás Malló, Daniel Pacheco Fernández, and Angeles Alonzo Flecha, 216–33. Madrid: Doce Calles, 1991.

Alvarez Cuartero, Izaskun. "Las sociedades económicas de amigos del país en Cuba (1787–1832): Una aportación al pensamiento ilustrado." In *Cuba: La perla de las Antillas: Actas de las I jornadas sobre "Cuba y su historia,"* edited by Consuelo Naranjo Orovio and Tomás Mallo González, 34–43. Madrid: Editorial Doce Calles, 1994.

Anderson, Jon W. "Cultural Adaptation to Threatened Disaster." *Human Organization* 27 (Winter 1968): 300–305.

Arango y Parreño, Francisco de. *Obras.* 2 vol. Havana: Dirección de Cultura, Ministerio de Educación, 1952.

Archer, Christon. The *Army in Bourbon Mexico, 1760–1810.* Albuquerque: University of New Mexico Press, 1977.

Arrate, José Martín Félix de. *Llave del Nuevo Mundo: Antemural de las Indias Occidentales.* Havana: Comisión Nacional Cubana de UNESCO, 1964.

Atherton, Gertrude, ed. *A Few of Hamilton's Letters, Including His Description of the Great West Indian Hurricane of 1772.* New York: Macmillan, 1903.

Bacardí y Moreau, Emilio. *Crónicas de Santiago de Cuba.* 10 vols. Madrid: Gráficas Breogán, 1924.

Baker, Maury, and Margaret Bissler Haas, eds. "Bernardo de Gálvez's Combat Diary for the Battle of Pensacola 1781." *Florida Historical Quarterly* 56 (October 1977): 176–99.

Barbier, Jacques A. "The Culmination of the Bourbon Reforms, 1787–1792." *Hispanic American Historical Review* 57 (February 1977): 51–68.

———. *Reform and Politics in Bourbon Chile, 1755–1796*. Ottawa: University of Ottawa Press, 1980.

Barbier, Jacques A., and Herbert S. Klein. "Wars and Public Finances: The Madrid Treasury, 1784–1807." *Journal of Economic History* 41 (June 1977): 315–39.

Barbier, Jacques A., and Allan J. Kuethe, eds. *The North American Role in the Spanish Imperial Economy, 1764–1819*. Manchester: Manchester University Press, 1984.

Beerman, Eric. "Arturo O'Neill: First Governor of West Florida during the Second Spanish Period." *Florida Historical Quarterly* 60 (July 1981): 29–41.

———. "José de Ezpeleta." *Revista de Historia Militar* 21 (1977): 97–118.

———. "'Yo Solo' Not 'Solo': Juan Antonio de Riaño." *Florida Historical Quarterly* 58, no. 2 (October 1979): 174–84.

Bezanson, Anne, Robert D. Gray, and Miriam Hussey. *Prices in Colonial Pennsylvania*. Philadelphia: University of Pennsylvania Press, 1935.

Bolton, Herbert Eugene, ed. *Athanase de Mézières and the Louisiana-Texas Frontier, 1768–1780*. Cleveland: Arthur H. Clark, 1914.

Boyd, Mark F., and José Navarro Latorre. "Spanish Interest in British Florida and in the Progress of the American Revolution." *Florida Historical Quarterly* 32 (1953): 95–96.

Bradley, R. *Climate since AD 1500 Database* (1992). NOAA/NCDC Paleoclimatology Program, World Data Center for Paleoclimatology, Boulder, Colo., available on NOAA's Paleoclimatology website, ⟨www.ncdc.noaa.gov/paleo⟩.

Bridgman, Howard, John E. Oliver, and Michael Glantz, eds. *The Global Climate System: Patterns, Processes, and Teleconnections*. Cambridge: Cambridge University Press, 2006.

Brown, Thomas H. "The African Connection: Cotton Mather and the Boston Smallpox Epidemic of 1721–1722." *JAMA: The Journal of the American Medical Association* 260 (1988): 2247–49.

Burkholder, Mark A. "The Council of the Indies in the Late Eighteenth Century: A New Perspective." *Hispanic American Historical Review* 56, no. 3 (August 1976): 404–23.

Burkholder Mark A., and David S. Chandler. *From Impotence to Authority: The Spanish Crown and the American Audiencias, 1687–1810*. Columbia: University of Missouri Press, 1977.

Calcagno, Francisco de. *Diccionario biográfico cubano*. Facsimile ed. Miami: Ediciones Cubanas, 1996.

Campbell, Leon G. *The Military and Society in Colonial Peru, 1750–1810*. Philadelphia: American Philosophical Society, 1978.

Cañizares-Esquerra, Jorge. *Nature, Empire, and Nation: Explorations of the History of Science in the Iberian World*. Stanford: Stanford University Press, 2006.

Carr, Peter. *Censos, padrones, y matrículas de la población de Cuba, siglos 16, 17 & 18*. San Luis Obispo, Calif.: The Cuban Index, 1993.

Casado Arbonés, Manuel. "Bajo el signo de la militarización: Las primeras expediciones científicas ilustradas a América (1735–1761)." In *La ciencia española en ultramar: Actas de las I Jornadas sobre "España y las expediciones científicas en América y Filipinas,"* coordinated by Alejandro R. Díez Torre, Tomás Malló, Daniel Pacheco Fernández, and Angeles Alonzo Flecha, 19–47. Madrid: Doce Calles, 1991.

Caughey, John Walton. *Bernardo de Gálvez in Louisiana, 1776–1783*. Berkeley: University of California Press, 1934.

Caviedes, César N. "Five Hundred Years of Hurricanes in the Caribbean: Their Relationship with Global Climate Variations." *Geojournal* 23 (April 1991): 301–10.

———. *El Niño in History: Storming through the Ages*. Gainesville: University Press of Florida, 2001.

Cepero Bonilla, Raul. *Azúcar y abolición*. [1948.] Barcelona: Editorial Crítica, 1976.

Chávez, Thomas E. *Spain and the Independence of the United States: An Intrinsic Gift*. Albuquerque: University of New Mexico Press, 2002.

Claxton, Robert H. "Climate and History: From Speculation to Systematic Study." *The Historian* 45 (February 1983): 220–36.

———. "Climatic and Human History in Europe and Latin America: An Opportunity for Comparative Study." *Climatic Change* 1 (1978): 195–203.

———. "The Record of Drought and Its Impact in Colonial Spanish America." In *Themes in Rural History of the Western World*, edited by Richard Herr, 194–226. Ames: Iowa State University Press, 1993.

Coatsworth, John H. "American Trade with European Colonies in the Caribbean and South America, 1790–1812," *William and Mary Quarterly*, 3rd ser., 24 (April 1967): 243–66.

Coker, William S., and Robert R. Rea, eds. *Anglo-Spanish Confrontation on the Gulf Coast during the American Revolution*. Pensacola, Fla.: Gulf Coast History and Humanities Conference, 1982.

Colburn, Forrest, ed. *Everyday Forms of Peasant Resistance*. Armonk, N.Y.: M. E. Sharpe, 1989.

Cook, N. David. *Born to Die: Disease and New World Conquest*. Cambridge: Cambridge University Press, 1997.

Cook, Noble David, and W. George Lovell, eds. *"Secret Judgments of God:" Old World Disease in Colonial Spanish America*. Norman: University of Oklahoma Press, 1992.

Cook, Sherburne F., and Woodrow Borah. *Essays in Population History: Mexico and the Caribbean*. 3 vols. Berkeley: University of California Press, 1971–79.

Corbitt, Duvon Clough. "The Administrative System in the Floridas, 1783–1821, Part 1." *Tequesta* 1 (1942): 41–62.

———. "The Administrative System in the Floridas, 1783–1821, Part 2." *Tequesta* 1 (1943): 57–67.

———. "Immigration in Cuba." *Hispanic American Historical Review* 22 (1944): 280–97.

———. "Mercedes and Realengos: A Survey of the Public Land System in Cuba." *Hispanic American Historical Review* 19 (August 1939): 269–74.

———. "Spanish Relief Policy and the East Florida Refugees of 1763." *Florida Historical Quarterly* 27 (July 1948): 67–82.

Cosner, Charlotte A. "Neither Black nor White, Slave nor Free: A Social History of the Vegueros in Cuba." Ph.D. diss., Florida International University, 2007.

Cronon, William. *Changes in the Land: Indians, Colonists, and the Ecology of New England*. New York: Hill and Wang, 1983.

———. "Modes of Prophecy and Production: Placing Nature in History." *Journal of American History* 76, no. 4 (March 1990): 1122–31.

Crosby, Alfred. The *Columbian Exchange: Biological and Cultural Consequences of 1492.* Westport, Conn.: Greenwood Press, 1976.

———. *Ecological Imperialism: The Biological Expansion of Europe, 900–1900.* 2nd ed. Cambridge: Cambridge University Press, 1993.

Cummins, Light Townsend. *Spanish Observers and the American Revolution, 1775–1783.* Baton Rouge: Louisiana State University Press, 1991.

Curtin, Philip D. *Death by Migration, Europe's Encounter with the Tropical World in the Nineteenth Century.* Cambridge: Cambridge University Press, 1989.

Cusick, James G. "Spanish East Florida in the Atlantic Economy of the Late Eighteenth Century." In *Colonial Plantations and Economy in Florida,* edited by Jane Landers, 168–88. Gainesville: University Press of Florida, 2000.

Dacy, Douglas C., and Howard Kunreuther. The *Economics of Natural Disasters: Implications for Federal Policy.* New York: Free Press, 1969.

Daniels, Christine, and Michael V. Kennedy, eds. *Negotiated Empires: Centers and Peripheries in the Americas, 1500–1820.* New York: Routledge, 2002.

Dean, Warren. *With Broadax and Firebrand: The Destruction of the Brazilian Atlantic Forest.* Berkeley: University of California Press, 1995.

Deive, Carlos Esteban. *Las emigraciones dominicanas a Cuba (1795–1808).* Santo Domingo: Fundación Cultural Dominicana, 1989.

Delgado, Jaime. "El Conde de Ricla, capitán general de Cuba." *Revista de historia de América* 55–56 (1963): 41–138.

Diaz, Henry F., and Vera Markgraf, eds. *El Niño and the Southern Oscillation: Multiscale Variability and Global and Regional Impacts.* Cambridge: Cambridge University Press, 2000.

Díaz, María Elena. The *Virgin, the King, and the Royal Slaves of El Cobre: Negotiating Freedom in Colonial Cuba, 1670–1780.* Stanford: Stanford University Press, 2000.

Dibble, Ernest F., and Earle W. Newton, eds., *Spain and Her Rivals on the Gulf Coast.* Pensacola, Fla.: Gulf Coast History and Humanities Conference, 1971.

Din, Gilbert C. *Francisco Bouligny: A Bourbon Soldier in Spanish Louisiana.* Baton Rouge: Louisiana State University Press, 1993.

Doerflinger, Thomas. *A Vigorous Spirit of Enterprise: Merchants and Economic Development in Revolutionary Philadelphia.* Chapel Hill: Published for the Institute of Early American History and Culture by the University of North Carolina Press, 1986.

Domínguez, Jorge I. *Insurrection or Loyalty: The Breakdown of the Spanish American Empire.* Cambridge, Mass.: Harvard University Press, 1980.

Drury, A. Cooper, and Richard Stuart Olson. "Disasters and Political Unrest: An Empirical Investigation." *Journal of Contingencies and Crisis Management* 6, no. 5 (September 1998): 153–61.

Duffy, John. *Sword of Pestilence: The New Orleans Yellow Fever Epidemic of 1853.* Baton Rouge: Louisiana State University Press, 1966.

Earle, Rebecca. "'A Grave for Europeans?' Disease, Death, and the Spanish-American Revolutions." In The *Wars of Independence in Spanish America*, edited by Christon I. Archer, 283–97. Wilmington, Del.: Scholarly Resources, 2000.

Eltis, David, Stephen D. Behrendt, David Richardson, and Herbert S. Klein, eds. The *Trans-Atlantic Slave Trade: A Database on CD-ROM*. Cambridge: Cambridge University Press, 2000.

Fagan, Brian M. The *Little Ice Age: How Climate Made History, 1300–1850*. New York: Basic Books, 2000.

Fenn, Elizabeth. *Pox Americana: The Great Smallpox Epidemic of 1775–82*. New York: Hill and Wang, 2001.

Ferguson, E. James. "Business, Government, and Congressional Investigation in the Revolution." *William and Mary Quarterly*, 3rd ser., 16 (July 1959): 293–318.

Fernández, José. "José Solano Bote, Marqués del Real Socorro and the Capture of Pensacola." Paper presented at the Florida Historical Society Meeting, Pensacola, May 2009.

Fernández, Juan Marchena. *Oficiales y soldados en el ejército de América*. Seville: Escuela de Estudios Hispanoamericanos, 1983.

———. "St. Augustine's Military Society." Translated by Luis Rafael Arana. *El Escribano: The St. Augustine Journal of History* 14 (1985): 43–71.

Ferrer, Ada. "El mundo cubano de azúcar frente a la revolución haitiana." In *Francisco de Arango y la invención de la Cuba azucarera*, edited by María Dolores González-Ripoll and Izaskun Álvarez Cuartero, 105–16. Salamanca: Ediciones Universidad de Salamanca, 2010.

Ferrer del Río, Antonio. *Historia del reinado de Carlos III en España*. 4 vols. Madrid: Imprenta de las Señores Matuti y Compagni, 1856. Facsimile ed., Madrid: Comunidad de Madrid Consejería de Cultura, 1988.

Fisher, John R. *Commercial Relations between Spain and Spanish America in the Era of Free Trade, 1778–1796*. Liverpool: University of Liverpool, 1985.

———. *Government and Society in Colonial Peru: The Intendant System, 1784–1814*. London: University of London, Athelone Press, 1970.

———. "Imperial 'Free Trade' and the Hispanic Economy." *Journal of Latin American Studies* 13 (May 1981): 21–56.

Fisher, John R., Allan J. Kuethe, and Anthony McFarlane, eds. *Reform and Insurrection in Bourbon New Granada and Peru*. Baton Rouge: Louisiana State University Press, 1990.

Fisher, Sarah Logan. "'A Diary of Trifling Occurrences': Philadelphia 1776–1778." Edited by Nicholas B. Wainwright. *Pennsylvania Magazine of History and Biography* 82 (October 1958): 450–65.

Fleming, James Rodger, and Roy Goodman, eds. *International Bibliography of Meteorology: From the Beginning of Printing to 1889: Four Volumes in One: Temperature, Moisture, Winds, Storms*. Upland, Pa.: Diane Publishing, 1994.

Florescano, Enrique. *Precios de maíz y crisis agrícola en México (1708–1810)*. Mexico City: El Colegio de México, 1969.

Florescano, Enrique, and Rodolfo Pastor, comps. *La crisis agrícola de 1785–1786 (Seleccion documental)*. Mexico City: Archivo General de la Nación, 1981.

Frederick, Julia Carpenter. "Luis de Unzaga and Bourbon Reform in Spanish Louisiana, 1770–1776." Ph.D. diss., Louisiana State University, 2000.

Frías Núñez, Marcelo. "El discurso médico a propósito de las fiebres y de la quina en el tratado de las calenturas (1751) de Andrés Piquer." *Asclepio* 55, no. 1 (2003): 215–33.

Funes Monzote, Reynaldo. *From Rainforest to Cane Field in Cuba: An Environmental History since 1492.* Translated by Alex Martin. Chapel Hill: University of North Carolina Press, 2008.

Galloway, Grace Growden. "Diary of Grace Growden Galloway Kept at Philadelphia from June 17th, 1778, to July 1, 1779." Introduction and notes by Raymond C. Werner. *Pennsylvania Magazine of History and Biography* 55 (1931): 35–94.

Gárate Ojanguren, Montserrat. *Comercio ultramarino e ilustración: La Real Compañía de la Habana.* Vol. 6. *Colección Ilustración Vasca.* San Sebastián: Real Sociedad Bascongada de los Amigos del País, 1993.

García Acosta, Virginia. "Introduction." In *Historia y desastres en América Latina.* Vol. 1, coordinated by Virginia García Acosta. Mexico City: Centro de Investigaciones y Estudios Superiores en Antropología Social (CIESAS), 1996.

García-Baquero González, Antonio. *Cádiz y el atlántico (1717–1778): (El comercio colonial español bajo el monopolio gaditano).* Seville: Escuela de Estudios Hispanoamericanos, 1976.

García Rodríguez, Mercedes. *Misticísmo y capitales: La compañía de Jesús en la economía habanera del siglo XVIII.* Havana: Editorial de Ciencias Sociales, 2000.

Gaspar, David Barry, and David Patrick Geggus, eds. *A Turbulent Time: The French Revolution and the Greater Caribbean.* Bloomington: Indiana University Press, 1997.

Gay-Calbó, Enrique. *Arango y Parreño: Ensayo de interpretación de la realidad económica de Cuba.* Havana: Imprenta Molina y Cía, 1938.

Geggus, David P. *Haitian Revolutionary Studies.* Bloomington: Indiana University Press, 2002.

———. "Slave Resistance in the Spanish Caribbean in the Mid-1790s." In *A Turbulent Time: The French Revolution and the Greater Caribbean,* edited by David Barry Gaspar and David Patrick Geggus, 131–55. Bloomington: Indiana University Press, 1997.

———. *Slavery, War, and Revolution: The British Occupation of St. Domingue, 1793–1798.* Oxford: Oxford University Press, 1982.

———, ed. *The Impact of the Haitian Revolution in the Atlantic World.* Columbia: University of South Carolina Press, 2001.

Gist, Richard, and Bernard Lubin. *Psychological Aspects of Disaster.* New York: Wiley, 1989.

Glantz, Michael H. *Currents of Change: El Niño's Impact on Climate and Society.* New York: Cambridge University Press, 1996.

———, ed. *Drought and Hunger in Africa: Denying Famine a Future.* New York: Cambridge University Press, 1996.

Goebel, Dorothy Burne. "British Trade to the Spanish Colonies, 1796–1823." *American Historical Review* 43 (January 1938): 288–320.

———. "The 'New England Trade' and the French West Indies, 1763–1774: A Study in Trade Policies." *William and Mary Quarterly,* 3rd ser., 20, no. 3 (July 1963): 331–72.

Gold, Robert L. "Politics and Property during the Transfer of Florida from Spanish to English Rule, 1763–1764." *Florida Historical Quarterly* 42 (July 1963): 16–34.

———. "The Settlement of East Florida Spaniards in Cuba, 1763–1766." *Florida Historical Quarterly* 42 (January 1964): 216–31.

Gomis Blanco, Alberto. "Las ciencias naturales en la expedición del Conde de Mopox a Cuba." In *La ciencia española en ultramar: Actas de las I jornadas sobre "España y las expediciones científicas en América i Filipinas,"* coordinated by Alejandro R. Díez Torre, Tomás Mallo, Daniel Pacheco, and Angeles Alonso Flecha, 308–19. Madrid: Doce Calles, 1991.

González Hernández, Pablo. "Más allá de una capitulación." Unpublished essay. Jay I. Kislak Foundation Student Essay Competition, 2003, ⟨www.kislak.com⟩.

González-Ripoll Navarro, María Dolores. "Voces de gobierno: Los bandos del capitán-general Luis de las Casas, 1790–1796." In *Cuba: La perla de las Antillas: Actas de las I jornadas sobre "Cuba y su historia,"* edited by Consuelo Naranjo Orovio and Tomás Mallo González, 149–62. Madrid: Editorial Doce Calles, 1994.

———, ed. *El rumor de Haití en Cuba: Temor, raza y rebeldía, 1789–1844.* Madrid: Consejo Superior de Investigaciones Científicas, 2004.

González-Ripoll, María Dolores, and Izaskun Álvarez Cuartero, eds. *Francisco de Arango y la invención de la Cuba azucarera.* Salamanca: Ediciones Universidad de Salamanca, 2010.

Griffin, Martin I. J. *Stephen Moylan: Muster-Master, General Secretary, and Aide-de-Camp to Washington, Quartermaster General, Colonel of Fourth Pennsylvania Light Dragoons, and Brigadier-General of the War for American Independence.* Philadelphia: N.p., 1909.

Guerra y Sánchez, Ramiro. *Azúcar y población en las Antillas.* [1944.] Havana: Editorial de Ciencias Sociales del Instituto Cubano del Libro, 1970.

———. *Manual de historia de Cuba (económica, social y política).* 2nd ed. Havana: Consejo Nacional de Cultura, 1962.

Guirao de Vierna, Angel. "La comisión real de Guantanamo en el marco de las expediciones Españolas a América." In *Cuba: La perla de las Antillas: Actas de las I jornadas sobre "Cuba y su historia,"* edited by Consuelo Naranjo Orovio and Tomás Mallo González, 85–92. Madrid: Editorial Doce Calles, 1994.

Guiteras, Pedro José. *Historia de la conquista de la Habana por los Ingleses.* [1866.] Havana: Cultural S.A., 1962.

Gurr, Ted Robert, ed. *Handbook of Political Conflict: Theory and Research.* New York: Free Press, 1980.

Gutiérrez Escudero, Antonio. *Ciencia, economía y política en hispanoamérica colonial.* Seville: Escuela de Estudios Hispanoamericanos, 2000.

Haarman, Albert W. "The Siege of Pensacola: An Order of Battle." *Florida Historical Quarterly* 44, no. 3 (January 1966): 193–99.

———. "The Spanish Conquest of British West Florida, 1779–1781." *Florida Historical Quarterly* 39, no. 2 (October 1960): 107–34.

Hall, Gwendolyn Midlo. *Africans in Colonial Louisiana: The Development of Afro-Creole Culture in the Eighteenth Century.* Baton Rouge: Louisiana State University Press, 1992.

Hart, Francis Russell. The *Siege of Havana, 1762*. Boston: Houghton Mifflin, 1931.

Hernández Palomo, José Jesús, coord. *Enfermedad y muerte en América y Andalucía (Siglos XVI-XX)*. Seville: Consejo Superior de Investigaciones Científicas, Escuela de Estudios Hispanoamericanos, 2004.

Herr, Richard, ed. Th*emes in Rural History of the Western World*. Ames: Iowa State University Press, 1993.

Hirshleifer, Jack. *Disaster and Recovery: The Black Death in Western Europe*. Santa Monica, Calif.: RAND Corporation, 1966.

Hobsbawm, Eric J. The *Age of Revolution, 1789–1848*. New York: Mentor Books, 1962.

Hoffman, Paul E. The *Spanish Crown and the Defense of the Caribbean, 1535–1568: Precedent, Patrimonialism, and Royal Parsimony*. Baton Rouge: Louisiana State University Press, 1980.

Holmes, Jack D. L. "Some Economic Problems of Spanish Governors in Louisiana." *Hispanic American Historical Review* 42 (November 1962): 521–24.

Iglesias García, Fé. *La estructura agrária en el occidente de Cuba, 1700–1750*. Manuscript in preparation. (In possession of author.)

Inglis, G. Douglas. *Constructing a Tower: The Marqués de la Torre Census of 1774*. Manuscript in preparation. (In possession of author.)

———. *New Aspects of Naval History*. Baltimore: U.S. Naval Printing Office, 1985.

Jennings, Evelyn Powell. "War as the 'Forcing House of Change': State Slavery in Late-Eighteenth-Century Cuba." *William and Mary Quarterly*, 3rd ser., 62 (July 2005): 411–40.

Jensen, Arthur L. "The Inspection of Exports in Colonial Pennsylvania." *Pennsylvania Magazine of History and Biography* 78 (July 1954): 275–97.

Jensen, Larry R. *Children of Colonial Despotism: Press, Politics, and Culture in Colonial Cuba, 1790–1840*. Tampa: University of South Florida Press, 1988.

Johnson, Sherry. "Casualties of Peace: Tracing the Historic Roots of the Florida-Cuba Diaspora, 1763–1804." *Colonial Latin American Historical Review* 10 (Winter 2002): 91–125.

———. "Climate, Community, and Commerce, among Florida, Cuba, and the Atlantic World, 1784–1800." *Florida Historical Quarterly* 80 (Spring 2002): 455–82.

———. "'La guerra contra los habitantes de los arrabales': Changing Patterns of Land Use and Land Tenancy in and around Havana, 1763–1800." *Hispanic American Historical Review* 77 (May 1997): 181–209.

———. "El Niño, Environmental Crisis, and the Emergence of Alternative Markets in the Hispanic Caribbean, 1760s-1770s." *William and Mary Quarterly*, 3rd ser., 62 (July 2005): 365–410.

———. "The Rise and Fall of Creole Participation in the Cuban Slave Trade, 1789 1796." *Cuban Studies/Estudios Cubanos* 30 (2000): 52–75.

———. "'Señoras en sus clases no ordinarias': Enemy Collaboratos or Courageous Defenders of the Family?" *Cuban Studies/Estudios Cubanos* 34 (2003): 22–23.

———. The *Social Transformation of Eighteenth-Century Cuba*. Gainesville: University Press of Florida, 2001.

———. "The St. Augustine Hurricane of 1811: Disaster and the Question of Political Unrest on the Florida Frontier." *Florida Historical Quarterly* 84 (Summer 2005): 28–56.

Jones, P. D., and M. E. Mann. "Climate over Past Millennia." *Reviews of Geophysics* 42 (2004): 1–42.

Jones, Robert F. "William Duer and the Business of Government in the Era of the American Revolution." *William and Mary Quarterly*, 3rd ser., 32 (July 1975): 393–416.

King, Lester S. The *Medical World of the Eighteenth Century*. Chicago: University of Chicago Press, 1958.

Klein, Herbert S. *African Slavery in Latin America and the Caribbean*. Oxford: Oxford University Press, 1986.

———. "The Colored Militia of Cuba, 1568–1868." *Caribbean Studies* 6 (July 1966): 17–27.

———. The *Middle Passage: Comparative Studies in the Atlantic Slave Trade*. Princeton: Princeton University Press, 1978.

———. *Slavery in the Americas: A Comparative Study of Virginia and Cuba*. [1967.] Chicago: Ivan R. Dee, 1989.

Knight, Franklin W. "Origins of Wealth and the Sugar Revolution in Cuba, 1750–1850." *Hispanic American Historical Review* 57 (May 1977): 231–53.

———. "Slave Society in Cuba during the Nineteenth Century." Madison: University of Wisconsin Press, 1970.

———. "Slavery and the Transformation of Society in Cuba, 1511–1760: From Settler Society to Slave Society." Mona, Jamaica: University of the West Indies, 1988.

Knight, Franklin W., and Peggy Liss, eds. *Atlantic Port Cities: Economy, Culture, and Society in the Atlantic World, 1650–1850*. Knoxville: University of Tennessee Press, 1991.

Kreps, Gary A., ed. *Social Structure and Disaster*. Newark: University of Delaware Press, 1989.

———. "Sociological Inquiry and Disaster Research." *Annual Review of Sociology* 10 (1984): 309–30.

Kuethe, Allan J. *Cuba, 1753–1815: Crown, Military, and Society*. Knoxville: University of Tennessee Press, 1986.

———. "The Development of the Cuban Military as a Sociopolitical Elite, 1763–83." *Hispanic American Historical Review* 61 (November 1981): 695–704.

———. "The Early Reforms of Charles III in the Viceroyalty of New Granada, 1759–1776." In *Reform and Insurrection in Bourbon New Granada and Peru*, edited by John R. Fisher, Allan J. Kuethe, and Anthony McFarlane, 21–29. Baton Rouge: Louisiana State University Press, 1991.

———. "El fin del monopolio: Los Borbones y el consulado andalúz." In *Relaciones de poder y comercio colonial*, edited by Enriqueta Vila Vilar and Allan J. Kuethe, 35–66. Sevillea and Lubbock: Escuela de Estudios Hispanoamericanos and Texas Tech University, 1999.

———. "Guns, Subsidies, and Commercial Privilege: Some Historical Factors in the Emergence of the Cuban National Character, 1763–1815." *Cuban Studies/Estudios Cubanos* 16 (1985): 123–38.

———. "Havana in the Eighteenth Century." In *Atlantic Port Cities: Economy, Culture, and Society in the Atlantic World, 1650–1850*, edited by Franklin W. Knight and Peggy Liss, 13–39. Knoxville: University of Tennessee Press, 1991.

———. "*Los llorones Cubanos*: The Socio-Military Basis of Commercial Privilege in the American Trade under Charles IV." In Th*e North American Role in the Spanish Imperial Economy, 1764–1819*, edited by Jacques A. Barbier and Allan J. Kuethe, 134–55. Manchester: Manchester University Press, 1984.

———. *Military Reform and Society in New Granada, 1773–1808*. Gainesville: University of Florida Press, 1978.

Kuethe, Allan J., and G. Douglas Inglis. "Absolutism and Enlightened Reform: Charles III, the Establishment of the Alcabala, and Commercial Reorganization in Cuba." *Past and Present* 109 (November 1985): 118–43.

Lampros, Peter J. "Merchant-Planter Cooperation and Conflict: The Havana Consulado, 1794–1832." Ph.D. diss., Tulane University, 1980.

Landers, Jane G., ed. *Colonial Plantations and Economy in Florida*. Gainesville: University Press of Florida, 2000.

———. "Rebellion and Royalism in Spanish Florida: The French Revolution on Spain's Northern Frontier." In *A Turbulent Time: The French Revolution and the Greater Caribbean*, edited by David Barry Gaspar and David Patrick Geggus, 156–77. Bloomington: Indiana University Press, 1997.

LaRosa Corzo, Gabino. *Los cimarrones de Cuba*. Havana: Editorial de Ciencias Sociales, 1988.

Larson, Harold. "Alexander Hamilton: The Fact and Fiction of His Early Years." *William and Mary Quarterly*, 3rd ser., 9 (April 1952): 139–51.

Lawson, Katherine S. "Luciano de Herrera, Spanish Spy in British St. Augustine." *Florida Historical Quarterly* 23 (1945): 170–76.

Le Riverend Brusone, Julio. *Historia económica de Cuba*. Havana: Ministerio de Educación, 1974.

Lewis, James A. "Anglo American Entrepreneurs in Havana: The Background and Significance of the Expulsion of 1784–1785." In Th*e North American Role in the Spanish Imperial Economy, 1764–1819*, edited by Jacques A. Barbier and Allan J. Kuethe, 112–26. Manchester: Manchester University Press, 1984.

———. The *Final Campaign of the American Revolution: Rise and Fall of the Spanish Bahamas*. Columbia: University of South Carolina Press, 1991.

———. "Nueva España y los esfuerzos para abastecer la Habana, 1779–1783." *Anuario de estudios americanos* 33 (1977): 501–26.

———. The *Spanish Convoy of 1750: Heaven's Hammer and International Diplomacy*. Gainesville: University Press of Florida, 2008.

Liss, Peggy K., ed. *Atlantic Empires: The Network of Trade and Revolution, 1713–1826*. Baltimore: Johns Hopkins University Press, 1983.

Lobdell, Richard. "Economic Consequences of Hurricanes in the Caribbean." *Review of Latin American Studies* 3, no. 2 (1990): 178–96.

Lockey, Joseph Byrne. *East Florida, 1783–1785: A File of Documents Assembled, and Many of Them Translated*. Edited, with a foreword, by John Walton Caughey. Berkeley: University of California Press, 1949.

López-Valdés, Rafael. "Hacía una periodización de la historia de la esclavitud en Cuba." In *La esclavitud en Cuba*, 13–29. Havana: Editorial Académica, 1986.

Lucena Salmoral, Manuel. "Las expediciones científicas en la época de Carlos III (1759–1788)." In *La ciencia española en ultramar: Actas de las I Jornadas sobre "España y las expediciones científicas en América y Filipinas,"* coordinated by Alejandro R. Díez Torre, Tomás Malló, Daniel Pacheco Fernández, and Angeles Alonzo Flecha, 49–63. Madrid: Doce Calles, 1991.

Ludlum, David M. *Early American Hurricanes, 1492–1870*. Boston: American Meteorological Society, 1963.

Lynch, John. *Bourbon Spain, 1700–1808*. Oxford: Basil Blackwell, 1989.

———. The *Spanish American Revolutions, 1808–1826*. New York: Norton, 1986.

Lyon, Eugene. Th*e Enterprise of Florida, Pedro Menéndez de Avilés and the Spanish Conquest of 1565–1568*. Gainesville: University Press of Florida, 1976.

Mann, Michael E., Zhihua Zhang, Malcom K. Hughes, Raymond S. Bradley, Sonya K. Miller, Scott Rutherford, and Fenbiao Ni. "Proxy-Based Reconstructions of Hemispheric and Global Surface Temperature Variations over the Past Two Millennia." *Proceedings of the National Academy of Sciences* 105, no. 36 (September 2008): 13252–57. Accessed on NOAA Paleoclimatology website, ⟨http://www.ncdc.noaa.gov/paleo/pubs/mann2008/mann2008.html⟩.

Marichal, Carlos, and Matilde Souto Mantecón. "Silver and Situados: New Spain and the Financing of the Spanish Empire in the Caribbean in the Eighteenth Century." *Hispanic American Historical Review* 74 (November 1994): 587–613.

Marrero y Artiles, Levi. *Cuba: Economía y sociedad*. 15 vols. Madrid: Playor S. A., 1972–88.

McAlister, Lyle N. The *"Fuero Militar" in New Spain, 1764–1800*. Gainesville: University Press of Florida, 1957.

McNeill, John R. *Atlantic Empires of France and Spain: Louisbourg and Havana, 1700–1763*. Chapel Hill: University of North Carolina Press, 1985.

———. "Yellow Jack and Geopolitics: Environment, Epidemics, and the Struggles for Empire in the American Tropics, 1640–1830." *Review: Fernand Braudel Center* 17 (2004): 355–57.

Medina Rojas, Francisco de Borja. *José de Ezpeleta: Gobernador de Mobila*. Seville: Escuela de Estudios Hispanoamericanos, 1980.

Melville, Elinor G. K. *A Plague of Sheep: Environmental Consequences of the Conquest of Mexico*. Cambridge: Cambridge University Press, 1994.

Mestre, Raul. *Arango y Parreño: El estadista sin estado*. Havana: Secretaría de Educacción, Dirección de Cultura, 1937.

Millás, José Carlos. *Hurricanes of the Caribbean and Adjacent Regions*. Miami: Academy of Arts and Sciences of the Americas, 1968.

Miller, Joseph C. *Way of Death: Merchant Capitalism and the Angola Slave Trade, 1730–1830*. Madison: University of Wisconsin Press, 1988.

Miner Solá, Edwin. *Historia de los huracanes en Puerto Rico*. 2nd ed. Puerto Rico: First Book Publishing of Puerto Rico, 1996.

Ministerio de Salud Pública. *La medicina en La Habana (cronología de los hechos médicos consignados en las Actas capitulares del Ayuntamiento de La Habana)*. 2 vols. Havana: Consejo Científico, Ministerio de Salud Pública, 1970.

Misas Jiménez, Rolando E. "La real sociedad patriótica de la Habana y las investigaciones científicas aplicadas a la agricultura (esfuerzos de institucionalización: 1793–1864." In *Cuba: La perla de las Antillas: Actas de las I jornadas sobre "Cuba y su historia*," edited by Consuelo Naranjo Orovio and Tomás Mallo González, 74–89. Madrid: Editorial Doce Calles, S. I., 1994.

Moore, John Preston. "Antonio de Ulloa: A Profile of the First Spanish Governor of Louisiana." *Louisiana History* 8 (Summer 1967): 189–218.

———. "Revolt in Louisiana: A Threat to Franco-Spanish Amistad." In *Spain and Her Rivals on the Gulf Coast*, edited by Ernest F. Dibble and Earle W. Newton, 40–55. Pensacola, Fla.: Gulf Coast History and Humanities Conference, 1971.

Morales Padrón, Francisco, ed. The *Journal of Don Francisco Saavedra de Sangronis, 1780–1783*. Translated by Aileen Moore Topping. Gainesville: University Press of Florida, 1989.

Moreno Fraginals, Manuel. *El ingenio: Complejo económico social cubano del azúcar*. 3 vols. 2nd ed. Havana: Instituto Cubano del Libro, 1978.

Mowat, Charles Loch. *East Florida as a British Province*. [1943.] Gainesville: University Press of Florida, 1964.

Mulcahy, Matthew. *Hurricanes and Society in the British Greater Caribbean, 1624–1783*. Baltimore: Johns Hopkins University Press, 2006.

Municipio de la Habana. *La dominación inglesa en la Habana: Libro de cabildos, 1762–1763*. Published under the direction of and with a preface by Emilio Roig de Leuchsenring. Havana: Municipio de la Habana, 1929.

Muñoz Pérez, J. "La publicación del reglamento de comercio libre de Indias, de 1778." *Anuario de estudios americanos* 4 (1947): 615–43.

Murphy, W. S. "The Irish Brigade of Spain at the Capture of Pensacola, 1781." *Florida Historical Quarterly* 38, no. 3 (January 1960): 216–25.

Murray, David R. *Odious Commerce: Britain, Spain, and the Abolition of the Cuban Slave Trade*. Cambridge: Cambridge University Press, 1980.

Myllyntaus, Timo. "A Natural Hazard with Fatal Consequences in Preindustrial Finland." In *Natural Disasters, Cultural Responses: Case Studies Toward a Global Environmental History*, edited by Christof Mauch and Christian Pfister, 77–102. Lanham, Md.: Lexington Books, 2009.

Nelson, George H. "Contraband Trade under the Asiento." *American Historical Review* 51 (1956): 55–67.

Nichols, Roy F. "Trade Relations and the Establishment of the United States Consulates in Spanish America." *Hispanic American Historical Review* 13 (August 1933): 289–313.

Oglesby, J. C. M. "Spain's Havana Squadron and the Balance of Power in the Caribbean, 1740–1748." *Hispanic American Historical Review* 49 (August 1969): 473–88.

Oliver Smith, Anthony. "Anthropological Research on Hazards and Disasters." *Annual Review of Anthropology* 25 (1996): 303–28.

———. "Theorizing Disasters: Nature, Power, and Culture." In *Catastrophe and Culture: The Anthropology of Disaster*, edited by Susanna M. Hoffman and Anthony Oliver Smith, 23–47. Oxford: J Currey, 2002.

Olson, Richard Stuart. "Towards a Politics of Disaster: Losses, Values, Agendas, and Blame." *International Journal of Mass Emergencies and Disasters* 18, no. 2 (August 2000): 265–87.

Olson, Richard Stuart, and A. Cooper Drury. "Un-Therapeutic Communities: A Cross-National Analysis of Post-Disaster Political Unrest." *International Journal of Mass Emergencies and Disasters* 15, no. 2 (August 1997): 221–38.

Olson, Richard Stuart, and Vincent T. Gawronski. "Disasters as Crisis Triggers for National Critical Junctures? The 1976 Guatemala Case." Paper presented at the Southeastern Council of Latin American Studies Meeting, San José, Costa Rica, 2007.

———. "Disasters as Critical Junctures? Managua, Nicaragua, 1972, and Mexico City, 1985." *International Journal of Mass Emergencies and Disasters* 21, no. 1 (March 2003): 5–35.

Ortega Pereyra, Ovidio. *La construcción naval en la Habana bajo la dominación colonial Española*. Havana: Academia de Ciencias de Cuba, 1986.

Ortiz, Fernando. *Los negros esclavos*. Havana: Editorial de Ciencias Sociales, 1975.

Ortiz de la Tabla y Ducasse, Javier. *Comercio exterior de Veracruz, 1778–1821: Crisis de dependencia*. Seville: Escuela de Estudios Hispanoamericanos, 1977.

Ortleib, L., and J. Macharé, eds. *Paleo-ENSO Records International Symposium: Extended Abstracts*. Lima: ORSTOM and CONCYTEC, 1992.

Paquette, Robert L. *Sugar Is Made with Blood: The Conspiracy of La Escalera and the Conflict between Empires over Slavery in Cuba*. Middletown, Conn.: Wesleyan University Press, 1988.

Parcero Torre, Celia María. *La pérdida de la Habana y las reformas borbónicas in Cuba (1760–1773)*. Madrid: Consejo de Castilla y León, 1998.

Pares, Richard. *War and Trade in the West Indies, 1739–1763*. Oxford: Oxford University Press, 1936.

———. *Yankees and Creoles: The Trade between North America and the West Indies before the American Revolution*. London: Longmans, Green, 1956.

Parry, John H., Philip Sherlock, and Anthony Maingot. *A Short History of the West Indies*. 4th ed. London: St. Martin's Press, 1987.

Peacock, Walter Gillis, Betty Hearn Morrow, and Hugh Gladwin, eds. *Hurricane Andrew: Ethnicity, Gender, and the Sociology of Disasters*. London: Routledge, 1997.

Pérez, Louis A., Jr. *Winds of Change: Hurricanes and the Transformation of Nineteenth-Century Cuba*. Chapel Hill: University of North Carolina Press, 2001.

Pérez Beato, Manuel. *Habana antigua: Apuntes históricos*. 2 vols. Havana: Seoane, Fernandez, y Compañía, Impresores, 1936.

Pezuela, Jacobo de la. *Diccionario geográfico estadístico, histórico de la isla de Cuba*. 4 vols. Madrid: Imprenta del Establecimiento de Mellado, 1863–66.

————. *Historia de la isla de Cuba*. 4 vols. Madrid: Bailly-Bailliere, 1878.

Pfister, Christian. "Learning from Nature-Induced Disasters: Theoretical Considerations and Case Studies from Western Europe." In *Natural Disasters, Cultural Responses: Case Studies toward a Global Environmental History*, edited by Christof Mauch and Christian Pfister, 17–40. Lanham, Md.: Lexington Books, 2009.

Pfister, Christian, Rudolph Brazdil, and Rudiger Glaser, eds. *Climatic Variability in Sixteenth-Century Europe and Its Social Dimension*. Dordrecht, Netherlands: Kluwer Academic Publishers, 1999.

Pfister, Christian, Jürg Luterbacher, Heinz Wanner, Dennis Wheeler, Rudolph Brazdil, Quansheng Ge, Zhixin Hao, Anders Moberg, Stefan Grab, and Maria Rosario del Prieto. "Documentary Evidence as Climate Proxies." White Paper written for the Proxy Uncertainty Workshop, Trieste, June 2008.

Phelan, John Leddy. *The People and the King: The Comunero Revolution in Colombia, 1781*. Madison: University of Wisconsin Press, 1978.

Pierson, William Whatley, Jr. "Francisco de Arango y Parreño." *Hispanic American Historical Review* 16 (November 1936): 451–78.

Piño-Santos, Oscar. *Historia de Cuba: Aspectos fundamentales*. Havana: Consejo Nacional de Universidades, 1964.

Piqueras, José A. "Los amigos de Arango en la corte de Carlos IV." In *Francisco de Arango y la invención de la Cuba azucarera*, edited by María Dolores González-Ripoll and Izaskun Álvarez Cuartero, 151–66. Salamanca: Ediciones Universidad de Salamanca, 2010.

Porras Muñóz, Guillermo. "El fracaso de Guarico." *Anuario de estudios americanos* 26 (1983): 569–609.

Portel Vilá, Herminio. *Historia de Cuba en sus relaciones con los Estados Unidos y España*. 4 vols. Havana: Academia de Historia de Cuba, 1935.

Portuondo Zúñiga, Olga, comp. and ed. *Nicolás José de Ribera*. Havana: Editorial de Ciencias Sociales, 1986.

Post, John D. *Food Shortage, Climatic Variability, and Epidemic Disease in Preindustrial Europe*. Ithaca, N.Y.: Cornell University Press, 1985.

————. *The Last Great Subsistence Crisis in the Western World*. Baltimore: Johns Hopkins University Press, 1977.

Priestly, Herbert Ingram. *José de Gálvez, Visitor General of New Spain (1765–1771)*. Berkeley: University of California Press, 1916.

Prieto, Maria del Rosario. "The Paraná River Floods during the Spanish Colonial Period: Impact and Reponses." In *Natural Disasters, Cultural Responses: Case Studies toward a Global Environmental History*, edited by Christof Mauch and Christian Pfister, 285–303. Lanham, Md.: Lexington Books, 2009.

Pringle, John. *Observations on the Diseases of the Army*. 3rd ed. London: A. Millar, D. Wilson, T. Durhan, and T. Payne, 1761.

Provenzo, Eugene F., Jr., and Asterie Baker Provenzo. *In the Eye of Hurricane Andrew*. Gainesville: University Press of Florida, 2002.

Puig-Samper Mulero, Miguel Angel. "Las primeras institutciones científicas en Cuba: El jardín botánico de la Habana." In *Cuba: La perla de las Antillas: Actas de las I* jornadas sobre "Cuba y su Historia," edited by Consuelo Naranjo Orovio and Tomás Mallo González, 19–34. Madrid: Editorial Doce Calles, 1994.

Puig-Samper, Miguel Angel, and Mercedes Valero. *Historia del jardín botánico de la Habana*. Madrid: Doce Calles, 2000.

Quarantelli, E. L., ed. *What Is a Disaster: Perspectives on the Question*. London: Routledge, 1998.

Quarantelli, E. L., and Russell R. Dynes. "Response to Social Crisis and Disaster." *Annual Review of Sociology* 3 (1977): 23–29.

Quinn, W. H., and V. T. Neal. "The Historical Record of El Niño Events in Climate since AD 1500." In *Climate since AD 1500 Database*, edited by R. Bradley. Boulder, Colo.: NOAA/NCDC Paleoclimatology Program, World Data Center for Paleoclimatology, ⟨www.ncdc.noaa.gov/paleo⟩.

Radding, Cynthia. *Landscapes of Power and Identity: Comparative Histories in the Sonoran Desert and the Forests of Amazonia from Colony to Republic*. Durham: Duke University Press, 2005.

———. *Wandering Peoples: Colonialism, Ethnic Spaces, and Ecological Frontiers in Northwestern Mexico, 1700–1850*. Durham: Duke University Press, 1997.

Rappaport, Edward N., and José Fernández Partagás. "The Deadliest Atlantic Tropical Cyclones, 1492–Present." National Oceanographic and Atmospheric Administration, NOAA website, ⟨http://www.nhc.noss.gov/pastdeadly1.html⟩.

Rea, Robert R. "A Distant Thunder: Anglo-Spanish Conflict in the Eighteenth Century." In *Cardenales de dos independencias* [symposium held at the Universidad Iberoamerica, November 1976], 175–87. Mexico City: Fomento Cultural Banamex, 1978.

Reparáz, Carmen de. *Yo Solo: Bernardo de Gálvez y la toma de Panzacola en 1781*. Madrid: Serba, 1986.

Richardson, Bonham C. *Economy and Environment in the Caribbean: Barbados and the Windwards in the Late 1800s*. Gainesville: University Press of Florida, 1997.

Rigau-Pérez, José G. "Smallpox Epidemics in Puerto Rico during the Prevaccine Era." *Journal of the History of Medicine and Allied Sciences* 37 (October 1982): 429–30.

Ringrose, David G. *Spain, Europe, and the "Spanish Miracle," 1700–1900*. New York: Cambridge University Press, 1996.

Risco Rodríguez, Enrique del. *Cuban Forests: Their History and Characteristics*. Translated by Fernando Nápoles Tapia. Havana: Editorial José Martí, 1999.

Rivero Muñiz, José. *Tabaco: Su historia en Cuba*. 2 vols. Havana: Instituto de Historia, Academic de Ciencias de la Republica de Cuba, 1965.

Rodríguez, Amalia A., ed. *Cinco diarios del sitio de la Habana*. Havana: Biblioteca Nacional José Martí, 1963.

Rodríguez Casado, Vicente. "El ejército y la marina en el reinado de Carlos III." *Boletín del Instituto Riva Agüero* 3 (1959): 129–56.

———. *La política marroquí de Carlos III*. Madrid: Consejo Superior de Investigaciones Científicas, 1945.

———. *La política y los politicos del reinado de Carlos III*. Madrid: Ediciones Rialp, 1962.

Rodríguez Vicente, María Encarnación. "El comercio cubano y la guerra de emancipación norteamericana." *Anuario de estudios americanos* 11 (1954): 61–106.

Salvucci, Linda K. "Anglo-American Merchants and Stratagems for Success in Spanish Imperial Markets, 1783–1807." In The *North American Role in the Spanish Imperial Economy, 1764–1819*, edited by Jacques A. Barbier and Allan J. Kuethe, 127–33. Manchester: Manchester University Press, 1984.

———. "Supply, Demand, and the Making of a Market: Philadelphia and Havana at the Beginning of the Nineteenth Century." In *Atlantic Port Cities: Economy, Culture, and Society in the Atlantic World, 1650–1850*, edited by Franklin W. Knight and Peggy Liss, 13–39. Knoxville: University of Tennessee Press, 1991.

Sánchez Ramírez, Antonio. "Notas sobre la real hacienda de Cuba." *Anuario de estudios americanos* 34 (1977): 465–83.

Santa Cruz y Mallén, Francisco Xavier de. *Historia de familias Cubanas*. 9 vols. Havana and Miami: Editorial Hércules, 1940–50.

Sauer, Carl Ortwin. The *Early Spanish Main*. Berkeley: University of California Press, 1966.

Schwartz, Stuart B. "The Hurricane of San Ciriaco: Disaster, Politics, and Society in Puerto Rico, 1899–1901." *Hispanic American Historical Review* 72, no. 3 (August 1992): 303–34.

———. "Hurricanes and the Shaping of Circum-Caribbean Cultures." Keynote address. Third Biennial Allen Morris Conference on the History of Florida and the Atlantic World. Tallahassee, Fla., 2004.

Scott, James C. *Weapons of the Weak: Everyday Forms of Peasant Resistance*. New Haven: Yale University Press, 1985.

Selkirk, Andrew. "The Last Word on Climate Change." *Current Archaeology* (May 2010): 48–49.

Sellés García, Manuel A. *Navegación astronómica en la España del siglo XVIII*. Madrid: Universidad Nacional de Educación a Distancia, 2000.

Sevilla Soler, María Rosario. *Santo Domingo: Tierra de la frontera (1750–1800)*. Seville: Escuela de Estudios Hispano-Americanos, 1980.

Shattuck, George Cheever. *Diseases of the Tropics*. New York: Appleton-Century-Crofts, 1951.

Sheridan, Richard B. "The Crisis of Slave Subsistence in the British West Indies during the American Revolution." *William and Mary Quarterly*, 3rd ser., 33, no. 4 (October 1976): 615–41.

———. "The Jamaican Slave Insurrection Scare of 1776 and the American Revolution." *Journal of Negro History* 61, no. 3 (July 1976): 290–308.

Smith, Paul H., ed. *Letters of Delegates to Congress, 1774–1789*. 11 vols. Washington, D.C.: Library of Congress, 1976–85.

Solano Pérez-Lila, Francisco de. *La pasión de reformar: Antonio de Ulloa, marino y científico, 1716–1795*. Seville: Escuela de Estudios Hispanoamericanos, 1999.

Staab, Jeffrey P., Thomas A. Grieger, Carol S. Fullerton, and Robert J. Ursano. "Acute Stress Disorder, Subsequent Posttraumatic Stress Disorder and Depression after a Series of Typhoons." *Anxiety* 2, no. 5 (1996): 219–25.

Starr, J. Barton. *Tories, Dons, and Rebels:* The *American Revolution in British West Florida.* Gainesville: University Press of Florida, 1976.

Stein, Stanley J., and Barbara H. Stein. *Apogee of Empire: Spain and New Spain in the Age of Charles III, 1759–1789.* Baltimore: Johns Hopkins University Press, 2003.

Steinberg, Ted. *Acts of God:* The *Unnatural History of Natural Disasters in America.* Oxford: Oxford University Press, 2000.

Stewart, Larry R. "The Edge of Utility: Slaves and Smallpox in the Early Eighteenth Century." *Medical History* 29 (January 1985): 54–70.

Super, John C., and Thomas C. Wright, eds. *Food, Politics, and Society in Latin America.* Lincoln: University of Nebraska Press, 1985.

Syrett, David, ed. The *Siege and Capture of Havana, 1762.* London: Navy Records Society, 1970.

TePaske, John Jay. "La política española en el Caribe durante los siglos XVII y XVIII." In *La influencia de España en el Caribe, la Florida, y la Luisiana, 1500–1800,* edited by Antonio Acosta and Juan Marchena Fernández, 61–87. Madrid: Instituto de Cooperación Iberoamericana, 1983.

Thomas, Hugh. *Cuba: Or, the Pursuit of Freedom.* New York: Harper and Row, 1971.

Tornero Tinajero, Pablo. *Crecimiento económico y transformaciones sociales: Esclavos, hacendados, y comerciantes en la Cuba colonial (1760–1840).* Madrid: Ministerio de Trabajo y Seguridad Social, 1996.

Torre, José María de la. *Lo que fuimos y lo que somos, o la Habana antigua y moderna.* [1857.] Santo Domingo: Ediciones Históricos Cubanos, 1986.

Torres Ramírez, Bibiano. "Alejandro O'Reilly en Cuba." *Anuario de estudios americanos* 24 (1967): 1357–88.

———. *Alejandro O'Reilly en las Indias.* Seville: Escuela de Estudios Americanos, 1968.

———. *La armada de barlovento.* Seville: Escuela de Estudios Americanos, 1981.

———. *La compañía gaditana de negros.* Seville: Escuela de Estudios Americanos, 1973.

Valdés, Antonio José. *Historia de la isla de Cuba y en especial de la Habana.* [1813.] Havana: Comisión Nacional Cubana de UNESCO, 1964.

Vila Vilar, Enriqueta, and Allan J. Kuethe, eds. *Relaciones de poder y comercio colonial.* Seville and Lubbock: Escuela de Estudios Hispanoamericanos and Texas Tech University, 1999.

Viñes, Benito. *Investigaciones relativas a la circulación y traslación ciclónica en los huracanes de las Antillas.* [1895.] Miami: Editorial Cubana, 1993.

Walker, Geoffrey J. *Spanish Politics and Imperial Trade, 1700–1789.* Bloomington: Indiana University Press, 1979.

Walter, John, and Roger Schofield. *Famine, Disease and the Social Order in Early Modern Society.* Cambridge: Cambridge University Press, 1989.

Watts, David. The *West Indies: Patterns of Development, Culture, and Environmental Change since 1492.* Cambridge Series in Historical Geography. Cambridge: Cambridge University Press, 1987.

Weber, David J. The *Spanish Frontier in North America.* New Haven: Yale University Press, 1992.

Weddle, Robert S. *Changing Tides: Twilight and Dawn in the Spanish Sea, 1763–1803*. College Station: Texas A&M University Press, 1995.

Woodward, Ralph Lee, Jr., ed. and trans. *Tribute to Don Bernardo de Gálvez*. New Orleans: Historic New Orleans Collection, 1979.

Worster, Donald, ed. *The Ends of the Earth: Perspectives on Modern Environmental History*. Cambridge: Cambridge University Press, 1988.

———. *The Wealth of Nature: Environmental History and the Ecological Imagination*. New York: Oxford University Press, 1993.

Wright, Irene A. *Cuba*. New York: Macmillan, 1905.

———. *The Early History of Cuba, 1492–1586*. New York: Macmillan, 1916.

Sherry Johnson, *Climate and Catastrophe in Cuba and the Atlantic World during the Age of Revolution* (2011).

Melina Pappademos, *Black Political Activism and the Cuban Republic* (2011).

Frank Andre Guridy, *Forging Diaspora: Afro-Cubans and African Americans in a World of Empire and Jim Crow* (2010).

Ann Marie Stock, *On Location in Cuba: Street Filmmaking during Times of Transition* (2009).

Alejandro de la Fuente, *Havana and the Atlantic in the Sixteenth Century* (2008).

Reinaldo Funes Monzote, *From Rainforest to Cane Field in Cuba: An Environmental History since 1492* (2008).

Matt D. Childs, The *1812 Aponte Rebellion in Cuba and the Struggle against Atlantic Slavery* (2006).

Eduardo González, *Cuba and the Tempest: Literature and Cinema in the Time of Diaspora* (2006).

John Lawrence Tone, *War and Genocide in Cuba, 1895–1898* (2006).

Samuel Farber, The *Origins of the Cuban Revolution Reconsidered* (2006).

Lillian Guerra, The *Myth of José Martí: Conflicting Nationalisms in Early Twentieth-Century Cuba* (2005).

Rodrigo Lazo, *Writing to Cuba: Filibustering and Cuban Exiles in the United States* (2005).

Alejandra Bronfman, *Measures of Equality: Social Science, Citizenship, and Race in Cuba, 1902–1940* (2004).

Edna M. Rodríguez-Mangual, *Lydia Cabrera and the Construction of an Afro-Cuban Cultural Identity* (2004).

Gabino La Rosa Corzo, *Runaway Slave Settlements in Cuba: Resistance and Repression* (2003).

Piero Gleijeses, *Conflicting Missions: Havana, Washington, and Africa, 1959–1976* (2002).

Robert Whitney, *State and Revolution in Cuba: Mass Mobilization and Political Change, 1920–1940* (2001).

Alejandro de la Fuente, *A Nation for All: Race, Inequality, and Politics in Twentieth-Century Cuba* (2001).

MIX
Paper from
responsible sources
FSC FSC® C013483
www.fsc.org